Molecular Cancer Therapeutics

MOLECULAR CANCER THERAPEUTICS

STRATEGIES FOR DRUG DISCOVERY AND DEVELOPMENT

Edited by
George C. Prendergast, Ph.D.
Lankenau Institute for Medical Research
Wynnewood, Pennsylvania
and
Department of Pathology, Anatomy, and Cell Biology
Thomas Jefferson University
Jefferson Medical College
Philadelphia, Pennsylvania

WILEY-LISS

A JOHN WILEY & SONS, INC., PUBLICATION

Published by John Wiley & Sons, Inc., Hoboken, New Jersey.
Published simultaneously in Canada.

Library of Congress Cataloging-in-Publication Data:

Molecular cancer therapeutics : strategies for drug discovery and
development / edited by George C. Prendergast.
p. cm.
Includes bibliographical references and index.
ISBN 0-471-43202-4 (Cloth)
1. Cancer—Chemotherapy. 2. Cancer—Immunotherapy. 3. Antineoplastic
agents—Design. I. Prendergast, George C.
RC 271. C5 M655 2004
616.99′4061—dc22 2003022153

Printed in the United States of America

10 9 8 7 6 5 4 3 2 1

Contents

Preface

This book draws together a diverse set of disciplines used to lay the preclinical foundation for discovering and translating new anticancer principles toward clinical testing. Cancer research has become an increasingly applied science, and it has become necessary for even basic researchers interested in general principles to monitor how their work affects broader medical issues, given major shifts in the field toward applications and emergent efforts to translate basic principles into the clinical arena.

Radical changes have occurred in both theoretical and applied concepts in cancer research in the last decade, spanning genetics, cell and animal models, drug screening, efficacy criteria, preclinical development, and clinical testing. With the completion of the human genome, and the growing sophistication of genetic concepts and technologies generally, this area in particular offers major new possibilities for cancer therapeutic discovery and development at many levels. However, during recent years market conditions have caused basic research costs to be arbitraged from many traditional pharmaceutical settings, where historically most new drugs have been discovered and developed. Furthermore, a crunch in funds for academic and biotechnology research has set in, with the completion of the doubling of the National Institutes of Health (NIH) budget and the uncertainites in financial markets after the bursting of the 1990s technology bubble. Funding issues seem likely to become more acute in coming years with the increasing political and social pressures to shift monies and resources to meet national and global health issues, including, for example, how best to distribute costly drugs and health care in both the developed and developing world. While these changes will pressure academic and industrial researchers in different ways, universal pressures will continue build to move discovery and development activity more rapidly toward practical medical applications or at least practical relevance of some kind.

Under such conditions, it is becoming increasingly important for researchers, especially younger researchers, to identify niches where they can have practical as well as scientific impact. This requires an awareness of ongoing change in the field of cancer research and also a broader awareness of how different parts of "translational" research fit together and are done in practice. It is hoped that the overview offered here, which draws together academic and industrial experts in early stage discovery and preclinical development from diverse fields, will provide individuals in all parts of the field with a broad sketch of early stages of cancer drug discovery and development.

The book focuses primarily on issues relevant to small molecule drugs, rather than biologic agents, where I believe the most significant gaps of knowledge and experience exist for most students and researchers.

Included among these areas are concepts and technologies in target discovery and validation, proof-of-concept investigations, drug "lead" screening, enzymology and medicinal chemistry, mouse model systems, preclinical pharmacokinetics and pharmacodynamics, and issues surrounding intellectual property and clinical development. A full discussion of the later stages of drug development—which would require a more comprehensive discussion of issues of clinical development, pharmacology, drug formulation, regulatory applications, patent strategies, and commercialization—deserves a separate volume of its own.

The text is directed to a broad audience of students, postdoctoral investigators, academic faculty, and scientific professionals in the biotechnology and pharmaceutical industries. Students and academic investigators typically have not had training or experience in cancer research in biotechnology/pharmacology industry. The information offered may be suited to advanced undergraduate as well as graduate courses that aim at familiarizing students with drug discovery and development issues, given the shift in career paths in recent years away from academia and towards private and commercial organizations. This book may be useful to researchers who have moved from previous training in academic settings without experience in pharmaceutical industry. Communications between workers in these industries have become important as biotechnology and biopharmacology companies increasingly provide technology, discovery, and early research for the pharmaceutical industry (which increasingly specializes in later clinical development and marketing). The text may also promote communication between preclinical investigators and clinical oncologists. Last, the principles, strategies, and pathways handled in this book are applicable more broadly to drug discovery and development, insofar as cancer research covers a broad diversity of concepts and technologies in biology. While the synthesis of such a huge and diverse area cannot help but include omissions, biases, and flaws, it is hoped that the audience reached will nevertheless benefit from seeing a broad overview of different parts of modern drug discovery, each of which contributes to bringing new ideas and discoveries in cancer research forward toward eventual, and we hope ultimately successful, clinical application.

I am grateful to the contributors to this volume, without whom the project could not have taken shape. In addition, there could have been no start or successful conclusion without Luna Han at Wiley, who helped frame the idea of a book that aimed for the first time to bring together different aspects of early phase discovery and development of cancer drugs. The best parts of the book belong to these contributors; the flaws are my own. As a cancer researcher I would never have felt remotely in the position to take on such a project, without some experience gained in pharmaceutical industry made possible by Drs. Allen Oliff and Robert Stein. Finally, I thank my wife, Kristine, and my daughter, Olivia, who continue to put up with all the excessive late night habits that derive from a career in biomedical research and the many hazards of editorial activity.

George C. Prendergast
Philadelphia, 2003

Contributors

Steven D. Averbuch, M.D., Merck Research Laboratories, Blue Bell, Pennsylvania

Pamela Benfield, Ph.D., Bristol-Myers Squibb Co., Inc., Princeton, New Jersey

W. Robert Bishop, Ph.D., Schering-Plough Research Institute, Kenilworth, New Jersey

Timothy J. Bowen, Ph.D., University of California San Diego School of Medicine, La Jolla, California

Bruce Car, Ph.D., Bristol-Myers Squibb Co., Inc., Princeton, New Jersey

Steven S. Carroll, Ph.D., Merck Research Laboratories, West Point, Pennsylvania

Paul A. Clarke, Ph.D., Institute of Cancer Research, Sutton, UK

Kim-Anh Do, Ph.D., The University of Texas MD Anderson Cancer Center, Houston, Texas

Steven F. Dowdy, M.D., Ph.D., University of California San Diego School of Medicine, La Jolla, California

Basil F. El-Rayes, M.D., Wayne State University School of Medicine, Detroit, Michigan

Sergei A. Ezhevsky, Ph.D., University of California San Diego School of Medicine, La Jolla, California

Gregory J. Hannon, Ph.D., Cold Spring Harbor Laboratory, Cold Spring Harbor, New York

James Inglese, Ph.D., Merck Research Laboratories, West Point, Pennsylvania

Eric Jonasch, M.D., The University of Texas MD Anderson Cancer Center, Houston, Texas

William G. Kaelin Jr., M.D., Ph.D., Harvard Medical School, Boston, Massachusetts

Paul Kirschmeier, Ph.D., Schering-Plough Research Institute, Kenilworth, New Jersey

Nancy E. Kohl, Ph.D., Merck Research Laboratories, West Point, Pennsylvania

Ming Liu, D.V.M., Ph.D., Schering-Plough Research Institute, Kenilworth, New Jersey

Robert B. Lobell, Ph.D., Merck Research Laboratories, West Point, Pennsylvania

Christopher Logothetis, M.D., The University of Texas MD Anderson Cancer Center, Houston, Texas

Patricia M. LoRusso, D.O., Wayne State University School of Medicine, Detroit, Michigan

Lisa Gail Malseed, J.D., Wild-Type Enterprises Worldwide, Bryn Mawr, Pennsylvania

Shi-Shan Mao, Ph.D., Merck Research Laboratories, West Point, Pennsylvania

Timothy J. McDonnell, M.D., Ph.D., The University of Texas MD Anderson Cancer Center, Houston, Texas

David B. Olson, Ph.D., Merck Research Laboratories, West Point, Pennsylvania

Patrick J. Paddison, Ph.D., Cold Spring Harbor Laboratory, Cold Spring Harbor, New York

George C. Prendergast, Ph.D., Lankenau Institute for Medical Research, and Thomas Jefferson University, Wynnewood, Pennsylvania

Laura Sepp-Lorenzino, Ph.D., Merck Research Laboratories, West Point, Pennsylvania

Beverly Teicher, Ph.D., Genzyme Corporation, Framingham, Massachusetts

Robert H. te Poele, Ph.D., Institute of Cancer Research, Sutton, UK

Paul Workman, Ph.D., Institute of Cancer Research, Sutton, UK

Yaolin Wang, Ph.D., Schering-Plough Research Institute, Kenilworth, New Jersey

Michael K. Wolf, M.D., AstraZeneca Pharmaceuticals LP, Wilmington, Delaware

Anthony Wynshaw-Boris, M.D., Ph.D., University of California San Diego School of Medicine, La Jolla, California

Guo-Jun Zhang, M.D., Ph.D., Harvard Medical School, Boston, Massachusetts

CHAPTER 1
Introduction

George C. Prendergast

The field of cancer research has evolved significantly in the past decade, essentially completing a movement started in the early 1980s that transformed the field from an largely biology-based disclipline to a molecular-based enterprise. In particular, molecular genetics had – as throughout biology – a huge impact on cancer research. The great advances made have opened a vast number of opportunities for the development of diagnostic, prognostic, and therapeutic applications. The recent goal set by the director of the U.S. National Cancer Institute to achieve effective management of cancer by 2015 reflects the wide enthusiasm for the potential of these advances to affect clinical practice at many levels.

As the field of cancer research turns increasingly toward practical applications, one issue that arises is the relative dearth of experience and training in how such applications are developed, particularly with regard to new therapeutic agents. Academic laboratories are typically in an excellent position to discover drug targets and target inhibitors, but they are often much less informed about what factors go into discovering and validating drug "leads" that would be suitable to develop (or partner with biotechnology or pharmaceutical companies to develop) for clinical testing. This situation can also prevail at small biotechnology companies, which are often seeded by academic discoveries, and at larger biotechnology and pharmaceutical companies, which must rely on (and some would say retool) young researchers, who have often trained exclusively in academic environments. In the United States, there is increasing support to drive cancer applications through greenhouse initiatives at the state level and small business grants at the federal level. Small biotechnology companies seeded by academic discoveries, benefiting from these resources, and aiming at industrial partnering or purchase may profit from the information in this book. In addition, researchers at larger biotechnology and pharmaceutical companies may benefit from the survey of strategies for target and lead drug discovery, which occur increasingly in the academic and small biotechnology sectors up to and including Phase I human clinical trials. To a growing degree, biotechnology industry provides the "R" for pharmaceutical R&D (research and development), increasing the need to promote conversation, interactions, and understanding among students and

Molecular Cancer Therapeutics: Strategies for Drug Discovery and Development, edited by George C. Prendergast
ISBN 0-471-43202-4

researchers at academic universities, medical centers, biotechnology companies, and pharmaceutical industry. This book addresses the growing interest in and need for information to develop new molecular cancer therapeutics, focusing mainly on small molecule inhibitors, where arguably the greatest gaps in information and understanding for most biologically oriented investigators occur. Some of the major issues covered in the text include

- Strategies to discover and genetically validate new drug targets.
- Drug-screening issues.
- Features of a drug lead suitable for proof of concept and further development.
- Mouse models of cancer – utility and issues of different models.
- Pharmacological validation – aligning biologic response with mechanism of action.
- Pharmacology and toxicology issues.
- Overview of clinical development and intellectual property issues.

Chapter 2 covers the changing face of cancer therapeutics research during its first 50 years as a field. Beverly Teicher introduces historical aspects of cancer drug discovery that remain relevant today, considering how classical parameters were developed to identify antitumor drugs with clinical potential. These principles were derived largely from animal-based studies. Most cytotoxic cancer drugs that are used in the clinic today were developed on the basis of these principles. In contrast, modern cancer drug discovery efforts have started with molecular targets, generally identified in cancer genetics studies, often in model systems, then moving to molecule-based screens for drug candidates, and lastly bootstrapping toward efficacy testing in cells and animals. This movement derives from the primacy that genetics has achieved in driving modern cancer research and drug discovery. Dr. Teicher discusses how the criteria for preclinical efficacy and clinical testing is shifting with the times, using illustrations from work on two classes of protein kinase inhibitors.

Most of the molecular-based therapeutics that have been clinically tested to date are cytostatic rather than cytotoxic in character. Many contributors to this book touch on the extensive preclinical and clinical experience with initial molecular therapeutics, such as the *bcr-abl* kinase inhibitor Gleevec, the epidermal growth factor (EGF) receptor antagonist Iressa, angiogenesis inhibitors, and farnesyl transferase inhibitors, many of which display mainly cytostatic properties. Because the goal is to kill cancer cells in the patient, questions about how to properly test and apply molecular cancer therapeutics in the clinic have moved to center stage. Some early progress has been made (e.g., with Gleevec), but there remain many challenges yet to be overcome.

Chapters 3 through 6 introduce concepts and technologies for the identification and validation of molecular drug targets. Chapter 3 presents a rationale behind the choice of suitable targets, based on current understanding of modern cancer genetics. The effect of intratumor and intertumor variation, multiple mutations, and tissue context on drug strategies are discussed. How the concepts of *oncogene addiction* and *synthetic lethality* may influence drug strategies are also introduced. In Chapter 4, the use of small interfering RNAs (siRNAs) for target discovery and genetic validation is presented. This technology, which was pioneered in the soil nematode *Caenorhabditis elegans,* is beginning to be widely exploited in somatic tissue culture. More recent developments marry siRNA technology to transgenic mice, as a way to achieve genetic validation of a target at the level of a whole mammalian organism. Chapter 5 presents tissue array technologies that allow one to rapidly probe hundreds of clinical tissue samples for information about the status of a molecular target in normal and malignant tissues. Tissue arrays have helped ease the bottleneck that this area has been for basic researchers interested in identifying and developing new targets. In Chapter 6, protein transduction strategies that make it possible to rapidly and directly query the function of molecular target proteins in cells are presented. Together, these strategies make it possible to efficiently probe the cancer-related functions of most any gene product in diverse model systems.

Chapters 7 and 8 introduce concepts and technologies for inhibitor screening, target and inhibitor validation, and more. Screening for small molecule inhibitors has become a field unto itself, particularly with regard to high throughput screening technology that has come to the forefront of drug discovery in recent years. Chapter 7 discusses the groundwork for designing assays that can discriminate desirable hits in an inhibitor screen. Knowing the target of a novel compound is a boon to medicinal chemists, who aim at refining the structure of a lead for improved potency, pharmacokinetic properties, and other considerations. For this reason, molecule-based screens have tended to dominate, although cell-based screens can also offer merit for medicinal chemistry development if there is a route to target identification. In addition to issues surrounding high-throughput assay development, Chapter 7 discusses common pitfalls in design and readout, as well as inhibition patterns and chemical moieties that raise red flags, signaling a problem. Chapter 8 surveys the numerous and powerful applications of gene microarrays for target discovery and validation, drug discovery and validation, drug pharmacology, and beyond. Microarray technology is perhaps the leading new technology driving cancer research forward at the current time.

Chapters 9 and 10 introduce the generation, utility, applications, and issues of mouse models of cancer for target and drug validation. Although other animals are used in cancer research, the mouse remains by far the dominant model in preclinical drug discovery and development. An overview of developments in transgenic mouse technology over the last 10 to 15 years as it pertains to cancer research is presented in Chapter 9, which focuses particularly on the generation of mice expressing oncogene and tumor-suppressor genes for cancer studies. Transgenic mouse models have significant scientific interest and potential for drug-discovery research, and their use is steadily increasing. However, some investigators have questioned whether they have lived up to expectations, including for addressing mechanistic questions, where empirical aspects of cancer related to tissue context have emerged as dominant factors. The increasing genetic sophistication being brought to engineered mice will allow their full potential, as yet unrealized, to further enhance their impact. While widely touted by academic researchers, transgenic models are used less for drug testing, particularly in industry, than the more traditional and widely established tumor xenograft models, which employ human tumor cell lines. Xenograft models have long been the major workhorse of the field. The utility of these models for predicting clinical response has been debated widely. However, some investigators with long and deep experience, such as Drs. Peter Houghton (St. Jude's Children's Hospital, Memphis) and Thomas Corbett (Wayne State University, Detroit), have made strong arguments that they indeed offer predictive utility if pharmacological and/or orthotopic principles are not violated. The advantages and disadvantages of transgenic models and xenograft models for cancer drug studies are contrasted in Chapter 10.

Chapters 11 and 12 survey pharmacodynamic and pharmacokinetic testing of novel small molecule therapeutic agents. Pharmacodynamics is described succinctly as the study "of what the drug does to the body" and pharmacokinetics as the study "of what the body does to the drug." Such work is crucial for preclinical validation and for judging the suitability of a candidate agent for clinical trials. Chapter 11 describes how pharmacodynamic studies are designed to address how the presumptive target responds to the drug in mouse models. It addresses how preclinical measurements made in mice are important to cue pharmacodynamic studies to be performed in clinical trials. Chapter 12 surveys concepts and methods used to perform preclincial pharmacokinetic and toxicology studies, which for cancer drugs are mainly performed in the mouse and rat. This chapters considers traditional areas in pharmacology – that is, absorption, dispersion, metabolism, and excretion – with discussion of the special issues related to cancer drugs. A typical scheme for pharmacokinetic analysis of a new agent is presented, and toxicities for common cancer drugs are outlined. This chapter also discusses practical considerations that derive from the combinatorial use of cancer drugs, the usual clinical situation. Together, these two chapters of the book delve into key questions that determine whether it is worthwhile to move a new therapeutic agent forward to clinical trials.

Chapters 13 and 14 survey the basic goals and issues for clinical development and the fundamental intellectual property issues that surround target and drug discovery research. As mentioned in the "Preface," this book focuses mainly on drug-discovery and -development issues at the preclinical level. These final chapters are designed to familiarize the reader with a basic understanding of clinical trials and intellectual property that are necessary for researchers at all levels, even for the investigator working at the most fundatmental levels of research. Beyond the scope of this book are further and more sophisticated discussions of clinical development, clinical pharmacology, drug formulation, regulatory applications for drug testing and approval, patent portfolio strategies, and drug launch and marketing. Large pharmaceutical companies have the most highly specialized and practical knowledge, resources, and experience in these areas. As a whole, this industry is moving to leverage these specialized areas of knowledge and expertise, providing the "D" in R&D to partner clinical development and marketing of promising novel agents that have been discovered and developed to preclinical and even early clinical stages by academic laboratories and small biotechnology/pharmacology companies. The putative economic efficiencies offered by this division of labor will prompt increasing communication among investigators working at different stages of the discovery and development process, formerly encompassed fully within a single commericial entity. Passing the baton in the relay race that makes up modern cancer drug discovery and development requires that the runners understand what their partners will be looking for.

CHAPTER 2

Molecular Cancer Therapeutics: Will the Promise Be Fulfilled?

Beverly A. Teicher

Molecular Cancer Therapeutics: Strategies for Drug Discovery and Development, edited by George C. Prendergast
ISBN 0-471-43202-4

2.1 Historical Development of Basic Concepts in Cancer Drug Development

In the modern era, drug discovery directed toward the cure of human malignant disease has completed its first half century as an organized scientific effort. Applying all of the technologies that have carried through the genomic era and into the proteomic era, what have we learned about cancer? We have learned that in certain ways malignant cells are similar to normal cells (Table 2.1). For example, there can be relatively small differences in the genes expressed in cancer cells compared to their normal counterparts (Clarke et al., 2001; Guo, 2003; Hermeking, 2003; Saha et al., 2002; Schulze and Downward, 2001; Velulescu et al., 1995). However, cancer cells frequently harbor chromosomal abnormalities and mutations not found in normal cells. Nevertheless, the most overwhelming observation remains the similarity of the wiring of the lethal malignant cell to normal cells in the host. The marked similarity in the wiring of biological response pathways used by both normal and malignant cells makes therapeutic attack of malignancy without substantial host toxicity difficult. From transcriptional analysis of many tumors, tumor cell lines and normal tissues, we have learned that although the large majority of genes expressed in malignant disease are the same as those expressed in normal tissues, small significant differences can be found. The hope of the many groups exploring molecular therapeutics for cancer treatment is that these small differences can be exploited to therapeutic advantage.

We have also learned that malignant tumors grow with understandable kinetics, as do malignant cells in culture, and we have learned that cytotoxic anticancer agents kill malignant cells with understandable kinetics and statistics. From early studies with in vivo tumor models in mice, we have learned that it is necessary to eliminate nearly every malignant cell from the host to achieve cure. Finally, from biochemical, molecular biologic, transcriptional and proteomics analyses, we have learned that cells are equipped with great plasticity and redundancy in biochemical pathways. Indeed, there seem to few critical cellular processes that are able to proceed by only a single route. From these observations and from experimental studies with inhibitors, we have learned that to have a significant effect on cell growth and, in some

Table 2.1 Cancer Therapeutics: What We Have Learned

- Malignant cells are similar to normal cells in terms of the signaling pathways they use.
- Malignant tumors have understandable growth kinetics.
- Tumor cure requires elimination of all (or nearly all) malignant cells; growth inhibition is not sufficient.
- Stopping malignant tumor growth requires $\geq 90\%$ blockade of a critical biochemical pathway; logs of cell killing are required.

Table 2.2 Cancer Therapeutics: Paradigms

- *Early paradigm:* Tumors are composed of malignant cells. All malignant cells must be killed to achieve tumor cure. The desired goal is therapeutic agents that are selectively cytotoxic toward malignant cells.
- *Current paradigm:* Tumors are composed of malignant cells and a wide variety of normal cells. These normal cells are an integral component of the malignant disease process. Therapeutic agents that selectively block important pathways in the malignant cells and/or the normal cells are desired. Antitumor activity can be produced by blockade of individual normal functions such as angiogenesis or invasion.

cases, cell survival, it is necessary to decrease the functioning of a critical pathway by $\geq 90\%$ compared to normal.

The field of anticancer therapeutics is at a critical point in its development. The traditional approach to cancer therapy has focussed on the killing of malignant cells (Table 2.2). Most of the drugs developed with this traditional goal have been cytotoxic agents with narrow therapeutic indices (disease selectivities). The skeptics have viewed many of these drugs as relatively ineffective poisons. As the field has moved away from the concept of cancer as solely malignant cells to the recognition that cancer is a disease process that is directed by the malignant cells, but that also critically requires the active involvement of a variety of "normal" cells to enable tumor growth, invasion, and metastasis, therapeutic targets have moved away from those that have as a goal killing malignant cells toward those targeted at blocking processes hypothesized to be critical to the malignant disease process (Beecken et al., 2001; Cherrington et al., 2000; Ellis et al., 2001; Gasparini, 1999; Jain, 2001; Kerbel, 2000; Kerbel et al., 2000; Miller et al., 2001; Rosen, 2000; Teicher, 1999). For example, one revolutionary concept of therapy is that directed toward the process of angiogenesis, which focuses the therapeutic attack away from the malignant cell and toward a normal cell, the endothelial cell, one of several types of stromal cells that are present in tumors and that are critical to tumor cell viability (Teicher, 2001a). Over the past ten years, many targeted therapeutic agents have been developed and entered clinical trial for testing. While these new targeted agents have, in general, proven to be better tolerated than classical cytotoxic agents, most have also proven to be less effective antitumor agents than the classical cytotoxic drugs.

The field has arrived at this dilemma, in part, because the criteria used to designate an agent active in cell culture models and in tumor models have decreased in stringency in recent years (Table 2.3). For example, many reports now describe IC_{50} (50% inhibitory concentration) rather than IC_{90} as the critical concentration for enzyme and cell culture studies and, more recently, even translating the IC_{50} levels to target plasma levels for compounds. To accommodate defining IC_{50}s as a target concentration, decreased stringency has been translated into the activity sought in in vivo tumor models, so that increase in life span (ILS) and tumor growth delay (TGD; in days), used historically, have been displaced by percent decrease in tumor volume at the

Table 2.3 Cancer Therapeutics: Criteria for Active Agents[a]

Criterion	Then	Now
Cell culture end point	IC_{90}	IC_{50}
In vivo tumor end point	ILS, TGD	T/C, none
Issues	Concentration versus dose Additivity versus synergy	

[a] *IC*, inhibitory concentration; *ILS*, increase in life span; *T/C*, treated response control; *TGD*, tumor growth delay.

maximal differential, often to no quantified end point. The strategy of using concentrations from in vitro experiments to determine target plasma levels for in vivo studies has also led to a great confusion between the applicability and definition of the terms *concentration* and *dose*. Concentration is static and useful in cell culture but varies momentarily in vivo. Dose refers to the amount of an agent administered to a host (animal or patient). Dose is dynamic with absorption distribution clearance, metabolism and excretion. Neither dose nor plasma level necessarily reflects agent levels or activity in the tumor.

The science of preclinical modeling of anticancer therapies began in the 1950s. The guidelines for experimental quality and end point rigor can be attributed in large part to the group headed by Howard Skipper at the Kettering-Meyer Laboratory affiliated with Sloan-Kettering Institute and Southern Research Institute in Birmingham, Alabama. In the mid-1960s, this group published a series of reports on the criteria of *curability*, the kinetic behavior of leukemic cells in animals, and the effects of anticancer chemotherapy. Although the fast-growing murine leukemias used in these study are now little used as primary tumor models, their value as a foundation of sound scientific in vivo methodology is undiminished. The principles put forward in these reports were derived directly from the behavior of bacterial cell populations exposed to antibacterial agents and were based on experimental findings in mice bearing intraperitoneally implanted L1210 or P388 leukemia (Himmelfarb et. al., 1967; Moore et al., 1966; Pittilo et al., 1965; Skipper, 1965, 1967, 1968, 1969, 1971a, 1971b, 1973, 1974, 1979; Skipper et al., 1965; Wilcox et al., 1965, 1966). The initial assumptions in these studies were the following. First, one living leukemic cell could be lethal to the host. Therefore, to cure experimental leukemia, it would be necessary to kill every leukemic cell in the animal, regardless of the number, anatomic distribution, or metabolic heterogeneity, with treatment that spares the host. Second, the percentage – rather than the absolute number – of in vivo leukemic cell populations of various sizes killed by a given dose of a given antileukemic drug is reasonably constant. The phenomenon of a constant percentage drug kill of a cell population, regardless of the population size, has been observed repeatedly and may be a general phenomenon. Third, the percentage of experimental leukemic cell populations killed by a single-dose drug treatment would be directly proportional to the dose level of the drug (i.e., the higher the dose, the higher the percentage of cells killed). Following these assumptions, it was obviously necessary to kill leukemic cells faster than they could be replaced by proliferation of the cells surviving the therapy, if a "cure" was to be approached (Moore et al., 1966; Pittilo et al., 1965; Wilcox et al., 1966).

The correlation between increased dose and increased cell killing or response has been questioned for newer targeted agents. For some targeted agents, it has been hypothesized that maximal dosing is not needed to produce maximal disease impact (Cristofanilli et al., 2002; Kerbel et al., 2001). Thus a discussion that is in progress in the field of cancer therapeutics is whether to back away from traditional dose escalation to maximum tolerated dose (MTD) in Phase I clinical trial and whether to back away from tumor response by decrease in volume as the most important end point in Phase II and III clinical trials (Herbst et al., 2002a; Kim and Herbst, 2002; Rosen, 2002; Scappaticci, 2002; Zhu et al., 2002).

The exponential killing of cells by drugs with time – mathematically equivalent to "a constant percentage kill of leukemic cells regardless of number" – was first observed in bacterial cell populations around 1900 (Chick, 1908) and has been investigated since that time with many antibacterial agents (Davis, 1958; Porter, 1947; Wyss, 1951). Through studies with bacterial cells exposed to anticancer agents, it was confirmed that the first-order kinetics of cell kill by anticancer agents was like that of antibacterial agents (Pittilo et al., 1965). The hypothesis that "the percentage, not the absolute number, of cells in populations of widely varying sizes killed by a given dose of a given anticancer drug is reasonably constant" was studied intensively and found, for the most part, to be valid (Pittilo et al., 1965). For antitumor drugs, this observation held true for bifunctional alkylating agents that cross-link DNA, for enzyme inhibitors, such as dihydrofolate reductase inhibitors (e.g., methotrexate), for multitargeted antifolate agents (e.g., Alimta), and for topoisomerase I inhibitors (e.g., irinotecan) (Aschele et al., 1998; Brandt and Chu, 1997; Chabot, 1997; Giovanella, 1997; McDonald et al., 1998; O'Reilly and Rowinsky, 1996; Rinaldi et al., 1995; Shih and Thornton, 1998; Takimoto, 1997; Teicher et al., 1999a).

Skipper and his group at the Kettering-Meyer Laboratory developed the murine L1210 leukemia (Law et al., 1949) as well as the murine P388 leukemia (Evans et al., 1963) into sensitive and reasonably quantitative in vivo bioassay systems, in particular to study anatomic distribution and rate of proliferation of leukemic cells and the effects of chemotherapy in tumor-bearing mice (Skipper et al., 1965). These studies were based on the notion that the drug-induced increase in host life span was achieved chiefly through leukemic cell kill, rather than through inhibition of growth of the leukemic cell population (Frei, 1964; Hananian et al., 1965; Skipper, 1964; Skipper et al., 1964). Furthermore, leukemic cells that gained access to the brain and other areas of the central nervous system (CNS) were not markedly affected by certain peripherally administered antileukemic drugs. Therefore, if there were leukemic cells in the CNS at the time when treatment was initiated, it was necessary to employ a drug that crossed the blood–brain barrier,

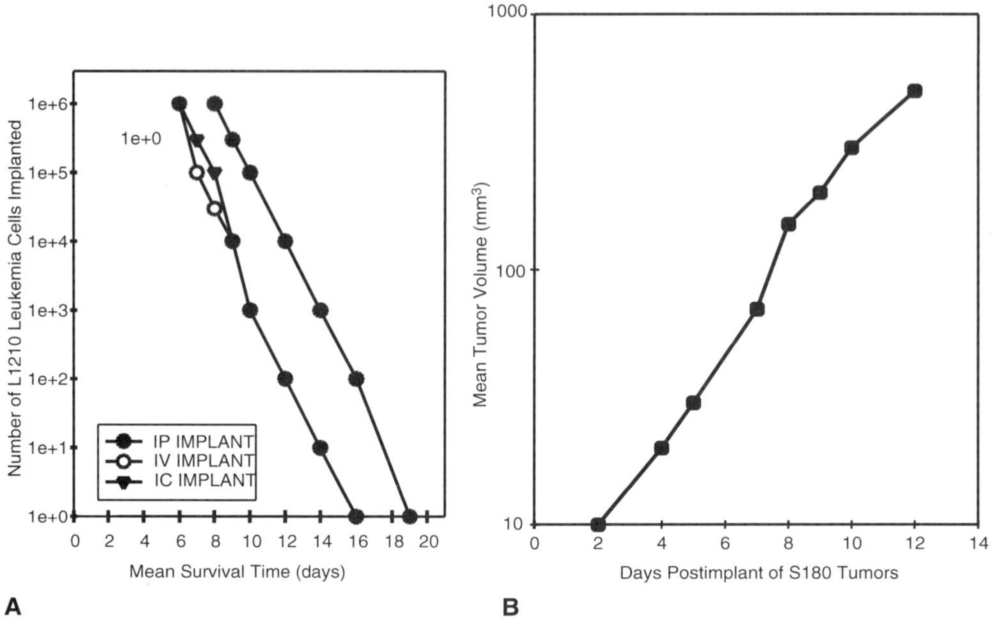

Figure 2.1 **A,** Mean survival time of mice inoculated with various numbers of murine L1210 leukemia cells injected intraperitoneally (*IP*), intravenously (*IV*), or intracranially (*IC*). These data form the basis for the in vivo bioassay method for determining the number of L1201 cells surviving after treatment of L1210 tumor-bearing mice with therapy. From these survival curves, it was determined that from IP inoculation, the L1210 cell generation time = 0.55 day and the lethal number of L1210 cells = 1.5×10^9; from IV inoculation, the L1210 cell generation time = 0.43 day; and from IC inoculation, the L1210 cell generation time = 0.46 day. (Adapted from Wilcox et al., 1965.) **B,** Exponential growth of the murine sarcoma 180 after implantation of a 2 mm^3 cube of tumor tissue by subcutaneous trocar injection. (Adapted from Wilcox et al., 1965).

if cure was to be achieved (Rall, 1965; Thomas, 1965). Antitumor activity in these early murine leukemia models was assessed on the basis of percent mean or median ILS (%ILS), net $\log_{10}$ cell kill, and long-term survivors (Bibby, 1999; Waud, 1998). The %ILS was derived from the ratio of the survival time of the treated animals (days) to the survival time of the untreated control animals (days). Calculations of net $\log_{10}$ cell kill were made from the tumor doubling time, which was determined from an internal tumor titration consisting of implants from serial 10-fold dilutions (Fig. 2.1) (Schabel et al., 1977). Long-term survivors were excluded from calculations of %ILS and net $\log_{10}$ tumor cell kill. To assess net $\log_{10}$ tumor cell kill at the end of treatment, the survival time (days) difference between treated and control groups was adjusted to account for regrowth of tumor cell populations that occurred between individual treatments (Lloyd, 1977).

Later, as syngeneic solid tumor models such as Lewis lung carcinoma and B16 melanoma were developed, the appropriate therapeutic end points devised were TGD and tumor control of a primary implanted tumor. These assays required that drugs be administered at doses producing tolerable normal tissue toxicity, so that the response of the tumor to the treatment could be observed over a relatively long period of time. Treatment with test compounds was initiated either before tumor development on the day after tumor cell implantation or after a measurable tumor nodule of a specified volume had grown. If treatment began the day after tumor cell implant, the experiment was designated a tumor growth inhibition study. If treatment began after an established tumor nodule (50–200 mm^3) had grown, the experiment was designated a TGD study. The activity of an agent in the TGD study carries more weight than the activity in the tumor growth inhibition assay, because the former assay models the situation for treating clinical disease more closely.

TGD is the difference in days for drug-treated versus control tumors to reach a specified volume, usually 500 mm^3 or 1 cm^3. Therefore, TGD is simply $T - C$ in days, where T is the mean or median time (in days) required for the treatment group tumors to reach a predetermined size and C is the mean or median time (in days) for the control group tumors to reach the same size. Tumor-free animals that are free of tumor when tumor growth delay is determined are excluded from these calculations. The TGD value coupled with the toxicity of the agent may the single most important criterion of antitumor effectiveness, because it mimics most closely the clinical end points that require observation of the host through the time of disease progression. With many of the most commonly used human tumor xenograft models, a TGD of about 20 days may be considered a probable indication of potential clinical utility.

2.2 Tyrosine Kinase Inhibitors – Initial Forays of Molecular-Targeted Cancer Therapeutics

As the understanding of cancer has increased, the breadth and complexity of the molecular events that make up malignant disease has become evident, but also daunting (Teicher, 2001a). Signaling networks that include membrane receptors, enzymes and their activators, deactivators and regulators, protein–protein interactions, protein–nucleic acid interactions, and small molecule effectors are all recognized targets for therapeutic attack. In short, antitumor agents are strategized to target specific abnormalities in the sequence or expression of genes and proteins that operate in a stepwise, combinatorial manner to permit the progression of malignant disease (Simpson and Dorow, 2001; Workman, 2001). Cell growth, motility, differentiation, and survival are regulated by signals received from the environment in either an autocrine or a paracrine manner (Heldin, 2001). Signals may come from interactions with other cells or components of the extracellular matrix or from binding of soluble signaling molecules to specific receptors at the cell membrane, thereby initiating diverse signaling pathways inside of the cell. Cancer may be visualized as a critical perturbation of signaling pathways (Arteaga et al., 2002; Bode and Dong, 2000; Elsayed and Sausville, 2001; Fodde et al., 2001; Folkman, 1971; Graff, 2002; Heymach, 2001; Hondermarck et al., 2001;

Lango et al., 2001; Lieberman et al., 2001; Reddy, 2001). Receptor tyrosine kinases (RTKs) are key mediators of many normal cellular processes but also of malignant disease processes. Several central signaling pathways controlled by tyrosine kinases – for example, those controlled by the epidermal growth factor receptor (EGFR) – have been selected as important targets for anticancer therapeutic intervention (Ciardiello and Tortora, 2001; Teicher, 1996, 1999; Zwick et al., 2001).

In the case of the EGFR, two basic strategies have been developed to block the activity of the kinase. In one strategy, monoclonal antibodies have been developed to prevent activation of the kinase by preventing binding of the EGF ligand. In a second strategy, small molecule inhibitors of the enzymatic activity of the kinase itself have been developed to inhibit autophosphorylation and the activity downstream intracellular signaling (Kari et al., 2003; Moscatello et al., 1998; Sedlacek, 2000). The inhibitors of EGFR are grouped among targeted cancer therapeutics, even though it is clear that EGFR is widely expressed in and used by normal tissues. In any case, EGFR is expressed in many tumors, for example, at fairly low levels in a variety of breast, lung, prostate, and other cancer cell lines and at higher levels in some breast (MD-MBA-468) and ovarian (OVT1) cancer cell lines.

Monoclonal antibody (MAb) 225, a mouse monoclonal antibody to EGFR, was initially shown to exhibit antitumor activity against human A431 epidermoid carcinoma and human MDA-MB-468 breast carcinoma grown as xenografts in combination with doxorubicin or cisplatin (Baselga et al., 1993; Fan et al., 1992; Mendelsohn, 1997, 2000). The humanized antibody C225 has been studied alone and in combination with gemcitabine, topotecan, paclitaxel, and radiation therapy in several human tumor xenograft models (Bruns et al., 2000; Ciardiello et al., 1999; Huang and Harari, 2000; Inoue et al., 2000). In the fast-growing genetically eugeneered organism (GEO) human colon carcinoma, C225 (10 mg/kg, intraperitoneal, 2 times/week for 5 weeks) produced a tumor growth delay of 24 days; topotecan (2 mg/kg, intraperitoneal, 2 times/week for 5 weeks), a camptothecin analog, produced a tumor growth delay of 14 days; and the combination regimen produced a tumor growth delay of 86 days (Fig. 2.2) (Ciardiello et al., 1999). It is interesting that, for reasons that are not clear, at least part of the activity of C225 could be attributed to antiangiogenic activity (Ciardiello et al., 2000a; Perrotte et al., 1999). Bruns et al. (2000) implanted L3.6pl human pancreatic carcinoma cells into the pancreas of nude mice, and beginning on day 7 posttumor cell implantation began treatment with C225 (40 mg/kg, intraperitoneal, 2 times/week for 4 weeks), gemcitabine (250 mg/kg, intraperitoneal, 2 times/week for 4 weeks), or a combination of the two. The animals were sacrificed on day 32 just after completion of the treatment regimen; therefore, no definitive end point could be assessed. Gemcitabine alone appeared to be most effective against the liver and lymph node metastases, whereas C225 alone appeared to be most effective against the primary disease. The combination regimen appeared to be the most effective of three regimens. Combination treatment regimens including C225 with radiation therapy appeared to produce at least additive tumor growth delay in two head and neck

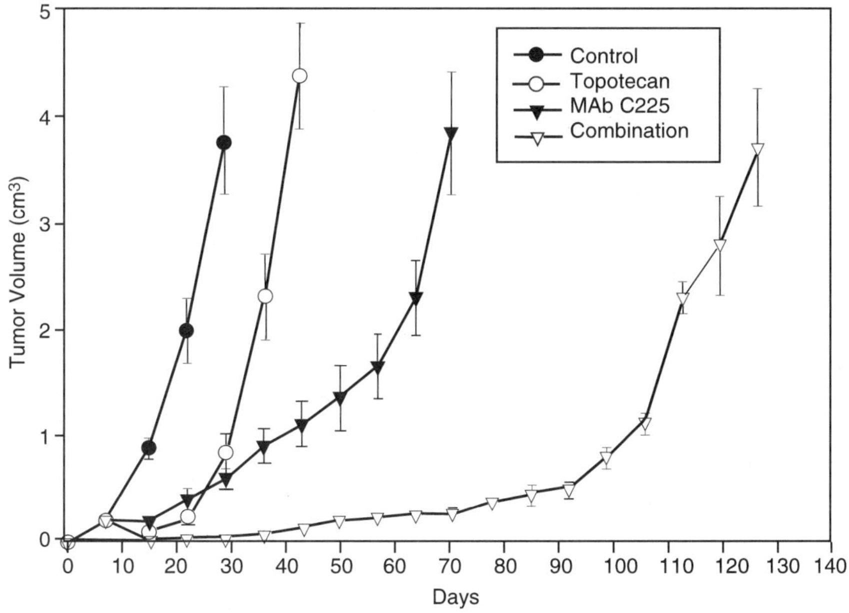

Figure 2.2 Antitumor activity of topotecan and MAb C225 on established GEO human colon carcinoma xenografts. Mice were injected subcutaneously in the dorsal flank with 10^7 human GEO colon carcinoma cells. After 7 days (average tumor size, 0.2 cm^3), mice were treated intraperitoneally with topotecan alone (2 mg/kg/dose, twice weekly on days 1 and 2 of each week for 2 weeks) or with MAb C225 alone (0.25 mg/dose, twice weekly on days 3 and 6 of each week for 5 weeks), or with both drugs on the same sequential schedule. Each group consisted of 10 mice. The experiment was repeated three times. Data represent the average of a total of 30 mice for each group. Student's t-test was used to compare tumor sizes among different treatment groups at day 29 after tumor cell implantation: MAb C225 versus control, $p < 0.001$; topotecan versus control, $p < 0.001$; topotecan followed by MAb C225 versus control, $p < 0.001$; topotecan followed by MAb C225 versus MAb C225 $p < 0.001$; topotecan followed by MAb C225 versus topotecan, $p < 0.001$. *Bars* represent SD (Ciardiello et al., 1999).

squamous carcinoma xenograft models (Huang and Harari, 2000). C225 has undergone three consecutive Phase I clinical trials, a Phase Ib clinical trial, and several single agent and combination Phase II trials. It is currently in Phase III clinical trial (Ciardiello et al., 2000a; Mendelsohn, 2000) (See Chapter 15 for more on human clinical trials.).

Several small molecule inhibitors of EGFR kinase that are competitive with ATP binding have been developed; ZD1839 (Iressa) progressed first toward clinical approval (Woodburn et al., 2000). ZD1839 has been studied in combination with cisplatin, carboplatin, oxaliplatin, paclitaxel, docetaxel, doxorubicin, etoposide, ralitrexed, and radiation therapy in human tumor xenograft models (Ciardiello et al., 2000b, 2001; Harari and Huang, 2001; Ohmori et al., 2000; Sirotnak et al., 2000; Williams et al., 2000). As observed with the EGFR Mab C225, the contribution of ZD1839 to anticancer activity of combination treatment regimens is due, at least in part, to activity as an antiangiogenic agent (Ciardiello et al., 2001; Hirata et al., 2002). When nude mice bearing the fast-growing human GEO colon carcinoma were treated with ZD1839 daily for 5 days per week for 4 weeks, at doses of 50, 100 or 200 mg/kg intraperitoneal (IP), the result was tumor growth delays of 4, 6, and 18 days, respectively (Ciardiello et al., 2000b). The 100-mg/kg dose of ZD1839 was selected for combination studies. Using the GEO colon xenograft tumor model, Ciardiello et al. (2000b) found that ZD1839 administered daily IP for 5 days per week for 4 weeks produced a 6- to 10-day tumor growth delay, whereas standard regimens for paclitaxel (20 mg/kg), topotecan (2 mg/kg), and tomudex (12.5 mg/kg) resulted in 9, 7, and 10 days of tumor growth delay, respectively. The combination treatment regimens of ZD1839 with each cytotoxic agent resulted in 33, 27, and 25 days of tumor growth delay, respectively. Sirotnak et al. (2000) administered ZD1839 (150 mg/kg) orally (PO) daily for 5 days for 2 weeks to nude mice bearing A431 human vulvar epidermoid carcinoma; A549, SK-LC-16, or LX-1 human non-small cell lung carcinomas; or PC-3 or TSU-PR1 human prostate carcinomas as a single agent or along with cisplatin, carboplatin, paclitaxel, docetaxel, doxorubicin, edatexate, gemcitabine, or vinorelbine. ZD1839 was a positive addition to all of the treatment combinations, except gemcitabine with which it did not alter the antitumor activity compared to gemcitabine alone and vinorelbine for which the combination regimen was toxic. For example, in the LX-1 non-small cell lung carcinoma xenograft, ZD1839 (150 mg/kg PO) produced a tumor growth delay of 8 days, paclitaxel (25 mg/kg IP) produced a tumor growth delay of 16 days, and the combination treatment regimens resulted in a tumor growth delay of 26 days. Working with the human GEO colon carcinoma, Ciardiello et al. (2001) found that ZD1839 (150 mg/kg IP daily for 5 days/week for 3 weeks; total dose 2250 mg/kg) was a more powerful antiangiogenic therapy than paclitaxel (20 mg/kg IP 1 day/week for 3 weeks; total dose 60 mg/kg) and that the combination treatment regimen was most effective.

Given these results, one would predict that ZD1839 would not be a highly effective single agent in the clinic, but it could be a useful component in combination treatment regimens. Expanding on these studies, Tortora et al. (2001) examined combinations of an antisense oligonucleotide targeting protein kinase A, a taxane, and ZD1839 in the fast-growing human GEO colon carcinoma xenograft. The tumor growth delays were 8 days with the taxane IDN5109 (60 mg/kg PO), 20 days with ZD1839 (150 mg/kg PO), 23 days with the antisense AS-PKAI (10 mg/kg PO), and 61 days with the three-agent combination treatment regimen. Recently, Naruse et al. (2002) found that a subline of human K562 leukemia made resistant to the phorbol ester (12-*O*-tetradecanoyl phorbol-13-acetate, TPA) and designated K562/TPA was more sensitive to ZD1839 administered intravenously(IV) or subcutaneously (SC) to nude mice bearing subcutaneensly implanted tumors than was the parental K562 line. ZD1839 has been evaluated in five Phase I clinical trials, which included 254 patients, and the response to ZD1839 apparently did not correspond to EGFR expression (Drucker et al., 2002). A Phase I study of 26 colorectal cancer patients showed that ZD1839 could be safely combined with 5-fluorouracil and leucovorin (Cho et al., 2002).

Two large multicenter Phase III clinical trials of ZD1839 (250 or 500 mg/day) in combination with carboplatin/paclitaxel or cisplatin/gemcitabine as first-line treatment in nonoperable stage III and stage IV non-small cell lung cancer patients are under way (Albanell et al., 2001; Ciardiello et al., 2001; Drucker et al., 2002). Other small molecule inhibitors of EGFR that are progressing through development are OSI-774, PD183805/CI-1033, PKI-1033, PKI166, and GW2016 (Hoekstra et al., 2002; Murren et al., 2002).

Another tyrosine kinase that has gained attention as a target for the development of molecular cancer therapeutics is the *bcr-abl* oncoprotein, a fusion protein of the ABL tyrosine kinase that is characteristic causal lesion in chronic myelogenous leukemia (CML). The BCR-ABL chimera offers an attractive protein receptor kinase target for pharmacological inhibition, because it is specifically expressed in malignant cells. STI571 (also known as Gleevec, Glivec, and CGP57148B) has been developed as a potent inhibitor of the Abl tyrosine kinase. In preclinical studies, STI571 selectively killed cells expressing retroviral *v*-Abl oncogenes or the Bcr-Abl oncogene, and it had antitumor activity as a single agent in animal models at well-tolerated doses (Gorre and Sawyer, 2002; Griffin, 2001; La Rose et al., 2002; Mauro and Druker, 2001; Mauro et al., 2002; O'Dwyer et al., 2002; Olavarria et al., 2002; Thambi and Sausville, 2002; Traxler et al., 2001). Unlike many other tyrosine kinase inhibitors that are cytostatic, STI571 is cytotoxic toward CML-derived cell lines, as demonstrated in colony formation assays using the surviving fraction end point (Liu et al., 2002). In cell culture, STI571 enhances the action of other cytotoxic agents, such as etoposide, in cells that express the *bcr-abl* oncoprotein (Liu et al., 2002; Marley et al., 2002). In cell culture studies that used the BV173 and EM-3 *bcr-abl*-positive cell lines with a growth inhibition end point, Topaly et al. (2002) found that STI571 produced greater than additive growth inhibition in combination with radiation therapy, and it produced additive to less than additive growth inhibition with busulfan and treosulfan. Mice reconstituted with *bcr-abl*-transduced bone marrow cells rapidly succumb to a fatal leukemia that is delayed significantly by treatment with STI571 (Wolff and Ilaria, 2001). Notably, in contrast to the polyclonal leukemia in control mice, STI571-treated mice develop a CML-like leukemia that is generally oligoclonal, suggesting that STI571 eliminated or severely suppressed certain leukemic clones. However, none of the STI571-treated mice was cured of the CML-like myeloproliferative disorder, and the STI571-treated CML that developed could be transplanted with high efficiency to fresh recipient animals. Thus, while it is effective, STI571 lacks the ability to efficiently control CML-like disease in all preclinical settings.

In humans, progression of CML to acute leukemia (i.e., blast crisis) has been associated with acquisition of secondary chromosomal translocations, frequently resulting in the production of a NUP98/HOXA9 fusion protein. Dash et al. (2002) developed a murine model expressing *bcr-abl* and NUP98/HOXA9 to cause blast crisis. The phenotype depends on expression of both mutant proteins, and significantly, the tumor retains sensitivity to STI571. However, despite the success of STI571 in this preclinical model of CML blast crisis, it has become clear that resistance can develop to this agent in the clinic, in many cases due to mutations in the kinase domain of *bcr-abl* that abolish STI571 binding (Krystal, 2001; Weisberg and Griffin, 2001).

STI571 is not a specific inhibitor of *bcr-abl* and is, indeed, also a potent inhibitor of other tyrosine kinases such as the receptor tyrosine kinase KIT and the platelet-derived growth factor receptor (PDGFR). This breadth of activity may be useful clinically. About 90% of malignant gastrointestinal stromal tumors (GISTs) have a mutation in the c-*kit* gene leading to KIT receptor autophosphorylation and ligand-independent activation. Notably, initial clinical studies have found that about 50% of GISTs respond to STI571 (Brahmer et al., 2002; Britten et al., 2002; Demetri, 2001; Heinrich et al., 2002; Joensuu and Dimitrijevic, 2001; Joensuu et al., 2002; Kuenen et al., 2002a; Zahalsky et al., 2002). PDGFR is expressed in several human cancers, including, for example, glioblastomas; and it is also expressed by tumor endothelial cells. These features may enable the use of STI571 for treatment of PDGFR-driven cancers, such as glioblastoma, or as a more generalized antiangiogenic agent to treat cancer.

Receptor tyrosine kinases implicated in angiogenesis are of significant interest as potential therapeutic targets in cancer, including receptors for PDGF, vascular endothelial growth factors (VEGFs), and basic fibroblast growth factor (bFGF) (Carter, 2000; Liekens et al., 2001; Mendel et al., 2000a; Rosen, 2001; Shepherd, 2001). SU5416 has been under development as a selective kinase inhibitor for Flk-1/KDR, the receptor for VEGF receptor 2 (VEGFR2). SU6668 and SU11248 are under development as broad-spectrum receptor tyrosine kinase inhibitors for VEGFR2, bFGF receptors (bFGFRs), PDGFR, and other receptor tyrosine kinases. Early in vivo work with SU5416 suffered from the use of DMSO as a vehicle for the compound administered intraperitoneally to mice once daily, beginning 1 day after tumor cell implantation (Fong et al., 1999). Using the DMSO vehicle, tumor growth delays of 0.5, 3, 6, 8, and 13 days were obtained in the human A375 melanoma xenograft with daily doses of SU5416 of 1, 3, 6, 12.5 and 25 mg/kg IP, respectively. Given these results, it appeared unlikely that SU5416 would have single agent activity in the clinic. The murine CT-26 colon carcinoma was used to assess the effect of SU5416 and SU6668 on the growth of liver metastases (Shaheen et al., 1999). CT-26 cells (10^4) were implanted beneath the capsule of the spleens of male Balb/c mice. Beginning on day 4, SU5416 (12 mg/kg) was administered in 99% PEG-300/1% Tween 80 and SU6668 (60 mg/kg) was administered in 30% PEG-300/phosphate buffered saline (pH 8.2). The compounds were injected once daily until the end of the experiment on day 22 after tumor cell implantation. The mean number of liver nodules was decreased to about 9 with SU5416 treatment, and to about 8 with SU6668 treatment, from about 19 nodules in the control animals.

SU5416 has a plasma half-life of 30 min in mice. Cell culture studies indicated that exposure to 5 μM SU5416 for 3 h inhibited the proliferation of HUVEC for 72 h (Laird et al., 2000; Mendel et al., 2000). Geng et al. (2001) found that SU5416 increased the sensitivity of murine B16 melanoma and murine GL261 glioma to radiation therapy. When the GL261 glioma was grown subcupeneously in C57BL mice, administration of SU5416 (30 mg/kg IP, twice/week for 2 weeks) produced a tumor growth delay of 4.5 days.

Fractionated radiation therapy (3 Gr for 8 days) resulted in 8.5 days of tumor growth delay. The combination regimen involving SU5416 administration along with and after completion of the radiation resulted in 16 days of tumor growth delay. SU5416 and SU6668 have been tested as single agents and in combination with fractionated radiation therapy in C3H mice bearing SCC VII squamous carcinomas (Ning et al., 2002; O'Farrell et al., 2002; Smolich et al., 2001). SU5416 (25 mg/kg, daily for 5 days) or SU6668 (75 mg/kg, daily for 5 days) was administered before or after radiation (2 Gr daily for 5 days). The tumor growth delay with SU5416 was 2 days, which increased to 6.5 days when combined with radiation therapy. The tumor growth delay with SU6668 was 3.3 days, which increased to 11.9 days when combined with radiation therapy. Administration of the compounds before or after radiation delivery did not affect the tumor response. SU6668 and SU11248, compounds with relatively broad selectivity, are undergoing clinical trials (Abrams et al., 2002a, 2002b; Brahmer et al. 2002; Britten et al., 2002; Krystal et al., 2001; Kuenen et al., 2002b; Mendel et al., 2003; Potapova et al., 2002; Raymond et al., 2002; Zahalsky et al., 2002).

Like STI571, the SU5416, SU6668, and SU11248 compounds have been found to inhibit the receptor tyrosine kinase encoded by c-*kit* (KIT) (Abrams et al., 2002a, 2002b; Fiedler et al., 2001; Heinrich et al., 2002; Hoekman, 2001; Mendel et al., 2003; Potapova et al., 2002; Raymond et al., 2002). KIT is essential for the development of normal hematopoietic cells and has been proposed to play a functional role in acute myeloid leukemia (AML). Mesters et al. (2001) reported a 4-month response in a patient with acute myeloid leukemia after treatment with SU5416. SU5416 and similar agents may also be useful for the treatment of von Hippel-Lindau syndrome patients (Harris, 2000). While SU5416 and similar agents appear to be quite tolerable as single agents, SU5416 was difficult to administer in combination with cisplatin and gemcitabine, due to the incidence of thromboembolic events (Aklilu et al., 2002; Hoekman et al., 2002; Kuenen et al., 2002a; Rosen, 2002). Other small molecule tyrosine kinase inhibitors showing promise in early clinical trial include OSI774 (Tarceva), PTK787/ZK222584, and ZD6474. PTK787/ZK 222584 has shown activity in several solid tumor models (Desai et al., 2002; Drevs et al., 2000, 2002a, 2002b; Hurwitz et al., 2002; Mita et al., 2002; Morgan et al., 2002; Patnaik et al., 2002; Thomas et al., 2002; Townsley et al., 2002; Wood et al., 2000; Yung et al., 2002). When the RENCA murine renal cell carcinoma was grown in the subrenal capsule of Balb/c mice, the animals developed a primary tumor as well as metastases to the lung and to the abdominal lymph nodes. Daily oral treatment with PTK787/ZK222584 (50 mg/kg) resulted in a decrease of 61% and 67% in primary tumors after 14 and 21 days, respectively. The occurrence of lung metastases was reduced 98% and 78% on days 14 and 21, respectively; and lymph node metastases appeared only on day 21 (Fig. 2.3) (Drevs et al., 2000). The major alternative therapeutic methodology being developed to inhibit the VEGF signaling pathway is anti-VEGF neutralizing monoclonal antibodies (Borgstroem et al., 1999; Schlaeppi and Wood, 1999; Townsley et al., 2002; Yang et al., 2002).

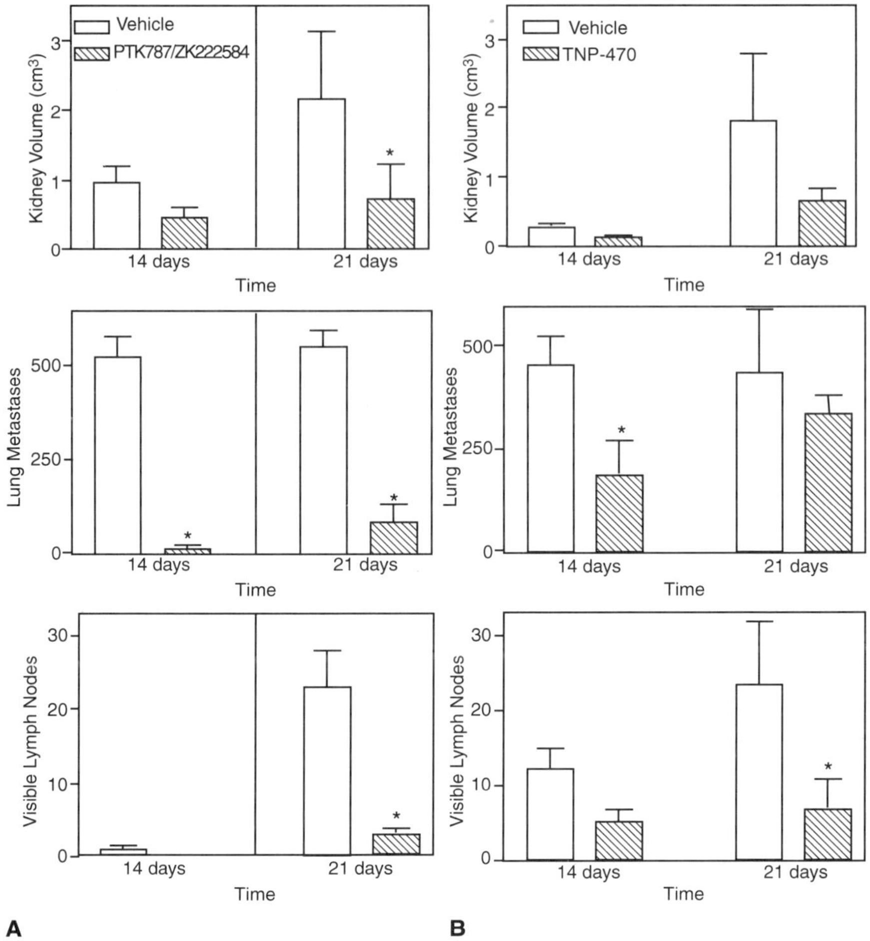

Figure 2.3 **A,** Effect of PTK787/ZK 222584 on tumor volume and number of metastases in murine renal cell carcinoma. PTK787/ZK 222584 was administered daily at 50 mg/kg PO. Therapy was initiated 1 day after inoculation of RENCA cells into the subcapsular space of the left kidney of syngeneic BALB/c mice. Animals were sacrificed after either 14 ($n = 12$) or 21 ($n = 20$) days. Primary tumor volume, number of lung metastases, and number of visible lymph nodes were assessed. **B,** Effects of TNP-470 on tumor volume and number of metastases. BALB/c mice were sacrificed 14 ($n = 10$) or 21 ($n = 10$) days after inoculation of RENCA with TNP-470 (30 mg/kg SC, administered every other day) was initiated 1 day after inoculation of RENCA cells. The control group received vehicle only. In the group that was sacrificed after 21 days, TNP-470 treatment had to be discontinued in all animals on day 13 because of strong side effects, such as weight loss > 20% and ataxia. Values are means, and the bars are SEM. Significance (*) calculated by comparing means of the treated group and means of the control group using the Mann Whitney *t*-test. (Drevs et al., 2000).

2.3 Serine-Threonine Kinase Inhibitors: Focus on Protein Kinase C as a Paradigm

Progress in the development of tyrosine kinase inhibitors reinforces interest in the potential of serine-threonine kinases as targets for molecular cancer therapeutics. One example to illustrate the exploration of this theme can be drawn from studies of protein kinase C (PKC), several isoforms of which are centrally involved in signaling transduction pathways that control cell cycle, apoptosis, angiogenesis, differentiation, invasiveness, senescence, and drug efflux (Blumberg et al., 2000; Goekjian and Jirousek, 2001; Nishizuka, 1992; O'Brian et al., 2001; Shen et al., 1999; Swannie and Kaye, 2002; Way et al., 2000). The interface of PKC signaling with angiogenesis is an area of particular interest. For example, activation of PKC pathways in human glioblastoma U973 cells by phorbol 12-myristate 13-acetate (PMA) leads to upregulation of VEGF expression, via an mRNA stabilization mechanism (Shih et al., 1999). Other recent results suggest the involvement of PKC in the invasiveness of breast cancer cells through regulation of urokinase plasminogen activator (Bhat-Nakshatri et al., 2002; Kim et al., 2001; Silva et al., 2002). Several studies have associated specific isoforms of PKC with important metabolic pathways in prostate cancer cells (Flescher and Rotem, 2002; Ghosh et al., 2002; Lin et al., 2001; Sumitomo et al., 2002) as well as malignant gliomas (Andratschke et al., 2001; Da Rocha et al., 2002). In regard to angiogenesis, the factor most closely associated in cancer patients is VEGF (Andratschke et al., 2001; Carter, 2000). The signal transduction pathways of the KDR/Flk-1 and Flt-1 receptors include tyrosine phosphorylation but also downstream activation of PKC and the MAP kinase pathway (Buchner, 2000; Ellis et al., 2000; Guo et al., 1995; Martelli et al. 1999; McMahon, 2000; Sawano et al., 1997; Xia et al., 1996).

To assess the contribution of PKC activation to VEGF signal transduction, studies were made of the effects of LY333531, an inhibitor that blocks the kinase activity of conventional and novel PKC isoforms, particularly the PKC-β isoform (Aiello et al., 1997; Danis et al., 1998; Ishii et al., 1996; Jirousek et al., 1996; Yoshiji et al., 1999). At concentrations predicted to selectively and completely inhibit PKC-β, the compound abrogated the growth of bovine aortic endothelial cells stimulated by VEGF (Jirousek et al., 1996). Oral administration of the inhibitor also decreased neovascularization in an ischemia-dependent model of in vivo retinal angiogenesis; furthermore, blocking increases in retinal vascular permeability stimulated by the intravitreal instillation of VEGF (Aiello et al., 1997; Danis et al., 1998; Ishii et al., 1996). Similarly, administration of LY333531 to animals bearing BNL-HCC hepatocellular carcinoma xenografts transfected with the VEGF gene under tetracycline control, markedly decreased the growth of subcutaneous or orthotopic tumors in a manner that was associated with decreased VEGF expression in the tumors (Yoshiji et al., 1999). LY333531 has demonstrated antitumor activity alone and in combination with standard cancer therapies in the murine Lewis lung carcinoma and in several human tumor xenografts (Teicher et al., 1999b). In related studies of a different agent, the National Cancer Institute 60-cell line panel was used to identify UCN-01, or 7-hydroxystaurosporine, a compound that inhibits PKC and other kinases. UCN-01, which has undergone a Phase I clinical trial (Dees et al., 2000; Grosios, 2001; Sausville et al., 2001), has been shown to inhibit the in vitro and in vivo growth of many types of tumor cells, including breast, lung, and colon cancers (Abe et al., 2001; Akinaga et al., 1991, 1997; Busby et al., 2000; Chen et al., 1999; Graves et al., 2000; Kruger et al., 1999; Sarkaria et al., 1999; Senderowicz and Sausville, 2000; Sugiyama et al., 1999).

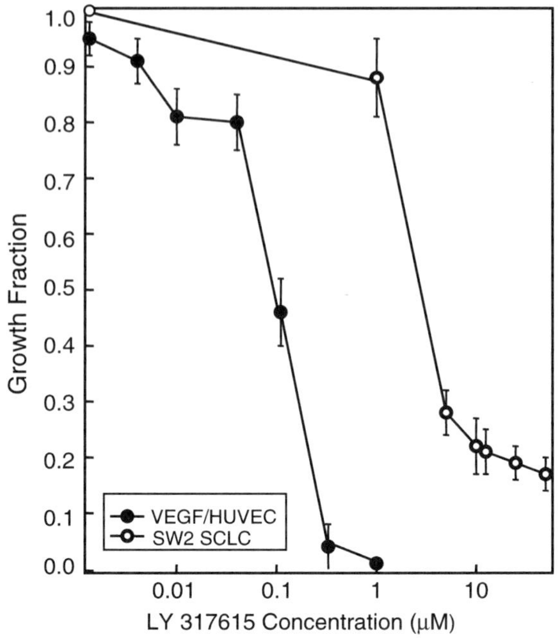

Figure 2.4 Concentration-dependent growth inhibition of human umbilical vein endothelial cells and human SW2 small cell lung carcinoma cells after 72 h. exposure to various concentrations of LY317615 as determined by WST-1 assay. Points are the means of three determinations, and bars are SEM. (Teicher et al., 2002b).

The compound LY317615 is another potent and selective inhibitor of PKC-β (Teicher et al., 2002b). When various concentrations of LY317615 were added to cultures of VEGF-stimulated human umbilical vascular endothelial cells (HUVECs), cell proliferation was profoundly inhibited (Fig. 2.4). In a control experiment, the exposure of human SW2 small cell lung carcinoma cells to LY317615 did not have a similarly potent growth inhibitory effect. In vivo tests that delivered LY317615 orally twice per day for 10 days after surgical implant of VEGF-impregnated filters resulted in markedly decreased vascular growth in the corneas of Fisher 344 female rats. Similarly, LY317615 decreased vascular growth in a dose-dependent manner to a level as low as that displayed by the unstimulated surgical control (Fig. 2.5) (Teicher et al., 2002b). In the same assay, LY317615 also decreased vascular growth 74% relative to control, under conditions in which bFGF was used to drive the assay (Fig. 2.5).

Tumor xenograft experiments confirmed the expectation that LY317615 could impede or reverse tumor angiogenesis. Nude mice bearing human tumor xenografts were treated with LY317615 orally twice daily on days 4–14 or 14–30 after tumor cell implantation. Using CD105 or CD31 as markers of endothelial cells, the number of intratumoral vessels in the samples was quantified by counting immunohistochemically stained regions in 10 microscope fields. In this assay, LY317615 delivered at 30 mg/kg decreased the number of intratumoral vessels by 50–75% of the control group (Table 2.4) (Teicher et al., 2001a, 2001b, 2001c, 2001d, 2002b). Although LY317615

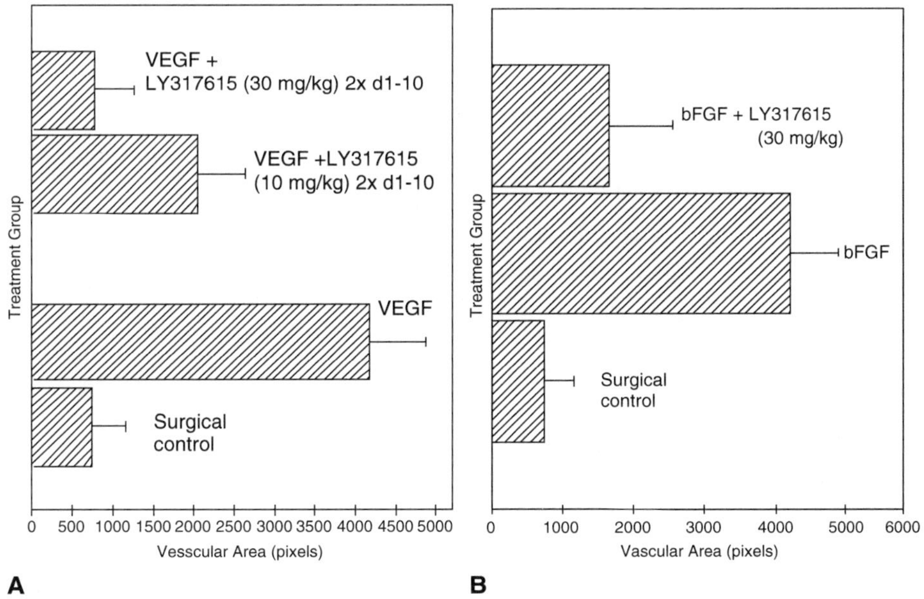

Figure 2.5 Vascular area determined by image analysis and described in pixel number for Fisher 344 female rats implanted with a small filter disc (inside diameter of a 20-g needle) impregnated with VEGF or bFGF (except the surgical control). Animals were untreated or treated with LY317615 (10 or 30 mg/kg) administered orally twice per on days 1–10. Data are the means of four to six determinations from photographs on day 14, and the bars are SEM. (Teicher et al., 2002b).

responses clearly included an antiangiogenic component, in no case was angiogenesis completely blocked as in the cornel micropocket neoangiogenesis model. Moreover, the tumor growth delay in the tested tumors did not correlate with the decrease in the number of intratumoral vessel (Table 2.4). The plasma levels of VEGF in mice bearing the human SW2 SCLC and Caki-1 renal cell carcinomas treated or untreated with LY317615 were measured by the Luminex assay (Keyes et al., 2002; Thornton et al., 2002). Plasma VEGF

Table 2.4 PKC Inhibitor LY317615

	Intratumoral Vessels					
	Control		LY317615			
Tumor	CD31	CD105	CD31	CD105	Mean (% normal)	Tumor Growth Delay (days)
SW2	80	50	24	28	43	9.7
MX-1	26	7	17	4	61	21
HS746T	19	11	15	7	71	15
Calu-6	17	20	8	10	48	9
T98G	12	7.5	4.5	4	45	8.7
CaKi1	10.5	11	1.5	2	16	15
HT29	9.5	11	3	4.5	36	14
Hep3B	7	4	3	1.5	40	20
SKOV-3	5	4	2	1	33	—

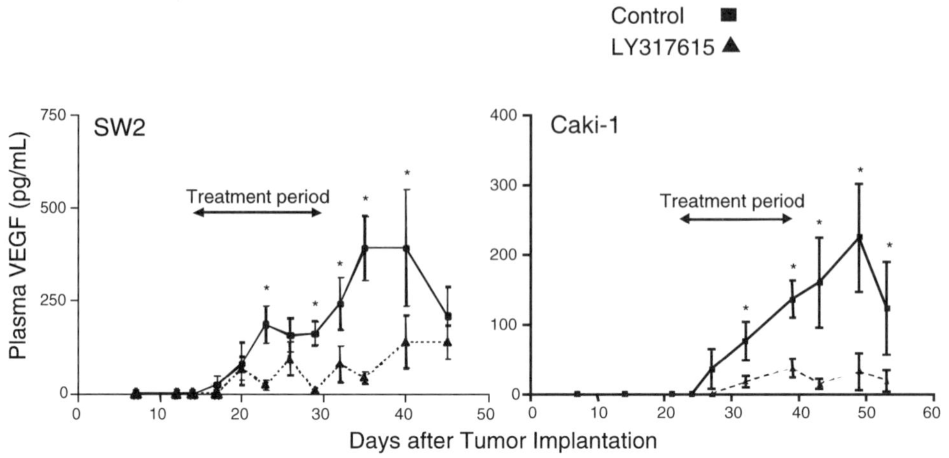

Figure 2.6 Plasma VEGF levels in nude mice bearing human SW2 SCLC, Caki-1 renal cell carcinoma or HCT116 colon carcinoma xenograft tumors, either untreated controls or treated with LY317615 orally twice daily on days 14–30 days (21–39 for Caki-1 bearing mice). The data represent the average results for three trials. Each point is the average of nine individual tumors, bars represent SEM, and Asterisk (*) indicates statically significant differences ($p < 0.05$).

levels were undetectable until tumor volumes were 500–600 mm^3 (Fig. 2.6). Using the Luminex assay, plasma VEGF levels were found to be similar between the treated and untreated groups through day 20 (at 75 pg/mL), after which the SW2 or Caki-1 control groups continued to increase throughout the study, reaching values of 400 pg/mL or 225 pg/mL, at day 40 postimplantation, respectively, whereas plasma VEGF levels in the treatment group remained suppressed throughout the treatment regimen. The plasma VEGF levels, reaching a maximum of 37 pg/mL, remained suppressed out to day 53, which was 14 days after terminating treatment (Keyes et al., 2002; Thornton et al., 2002). These observations supported the idea that PKC targeting could offer a viable antiangiogenesis strategy as an antitumor therapy.

Combination regimens of kinase inhibitors are increasingly being explored as a way to potentiate responses and enhance antitumor efficacy. In the present case, a sequential treatment regimen was used to examine the efficacy of the PKC inhibitor LY317615 in the xenograft model for SW2 small cell lung cancer. Administration of LY317615 alone on days 14–30 after tumor implantation over a dosage range from 3 to 30 mg/kg produced tumor growth delays between 7.4 and 9.7 days in the SW2 small cell lung cancer. The SW2 tumor responds to paclitaxel and treatment with that drug alone produced a 25-day tumor growth delay. Sequential treatment of paclitaxel followed by LY317615 (30 mg/kg) resulted in > 60 days of tumor growth delay, a 2.5-fold increase in the duration of tumor response. Using carboplatin, to which SW2 cancer cells are less responsive, produced a tumor growth delay of only 4.5 days in that tumor; however, sequential treatment with LY317615 also enhanced the response, resulting in 13.1 days of tumor growth delay (Teicher et al., 2001d). The antitumor activity of LY317615 alone and in combination with cytotoxic antitumor agents has been explored in several human tumor xenografts (Keyes et al., 2002; Teicher et al., 2001d; Thornton et al., 2002).

While in most cases the tumor growth delay produced by LY317615 as a single agent was insufficient to predict single agent activity in the clinic, combination regimens incorporating LY317615 proved to be a useful addition to the therapeutic regimen. LY317615 is currently ongoing Phase I clinical trials (Herbst et al., 2002a).

2.4 New Target Discovery Methods

Methods to identify new drug targets are changing rapidly. For example, many molecules that are currently being explored as potential drug targets were discovered by searching for transcripts that are more highly expressed in cancer cells than in normal cells. However, recent studies have shown that cancer cells and normal cells in culture do not provide a good representation of gene expression in vivo (Armstrong et al., 2002; Bhattacharjee et al., 2001; Golub, 2001; Golub et al., 1999; LaTulippe et al., 2002; Pomeroy et al., 2002; Ramaswamy and Golub, 2002; Ramaswamy et al., 2001; Singh et al., 2002; Van de Vijver et al., 2002). Target discovery is moving closer to clinical disease by examining gene expression in clinical samples that have not been cultured. Genomics arrays such as those from Affymetrix Inc. or Agilent Technologies Inc. have been useful tools in this effort (Hermeking, 2003; Saha et al., 2002; Velulescu et al., 1995). A growing number of retrospective clinical studies have examined the gene expression signature for various tumor types, with the following studies as examples. Van de Vijver et al. (2002) found gene expression profiles to be a powerful predictor of disease outcome in young patients with breast cancer in a study of 295 patients. LaTuippe et al. (2002) found > 3,000 tumor-intrinsic genes that differ among nonrecurrent primary prostate cancers and metastatic prostate cancers. Pomeroy et al. (2002) found that the clinical outcome of children with medulloblastomas was highly predictable on the basis of the gene expression profiles of their tumors at diagnosis. Among the questions raised by these findings are, What is the minimum number of genes whose expression can be used to determine the diagnosis for the patient? and How can these findings be applied in the standard clinical setting? These questions are being addressed by activity in the field.

In addition to genomic arrays, which are limited by the number of genes that can be included on the array, serial analysis of gene expression (SAGE) provides an alternate and unbiased method to identifygenes that are differentially expressed in tumor cells (Clarke et al., 2001; Guo, 2003; Schulze and Downward, 2001). In terms of angiogenic targets, St. Croix et al. (2000) have reported the isolation of endothelial cells from a sample of colon carcinoma and a sample of normal colon mucosa. SAGE analysis of RNA isolated from the samples encompassed the expression of about 20,000 genes. Data analysis identified 800 genes that were differentially expressed at significant levels between the tumor and normal colon endothelium, 500 of which were higher in the tumor endothelium and 300 of which were higher in the normal endothelium. Although genes related to the VEGF pathway were present, they were not among the most differentially expressed. Similar SAGE

Table 2.5 Current EC SAGE Portfolio of Endothelial Cells

Clinical Sample	Description	Tags Generated
Colon N	Normal colon mucosa ECs	96,000
Colon T	Primary colon carcinoma ECs	96,000
Brain N1	Normal temporal lobectomy ECs	43,000
Brain N2	Normal temporal lobectomy ECs	49,000
Brain T1	Grade IV glioma ECs	46,000
Brain T2	Grade III glioma ECs	50,000
Brain T3	Grade IV glioma ECs	58,000
Breast N1	Normal mammary reduction ECs	50,000
Breast T1	Primary breast cancer ECs	50,000
Breast T2	Primary breast cancer ECs	50,000
Breast T	Breast cancer bone metastases ECs	in progress
NSCLC	Non-small cell lung carcinoma ECs	in progress
Colon T	Colon cancer liver metastases ECs	in progress

analyses have been performed on endothelial cells from brain cancers and normal brain and from primary breast cancers and normal breast tissue (Table 2.5). Several interesting observations emerge from these data. First, tumor endothelial cells have more abnormal gene expression than has been generally hypothesized – that is, tumor angiogenesis is rather abnormal with regard to gene expression. Second, endothelial cells from different normal tissues have different patterns of gene expression. Third, tumor endothelial cells from tumors of different tissues or organs have unique gene expression profiles. When endothelial gene expression profiles for the most differentially expressed genes ($\geq$ 98% confidence) were compared, there was about a 20% overlap in the genes expressed at highest levels between breast and brain cancers or between colon and brain cancers (Fig. 2.7). There was a slightly greater overlap between the genes expressed between the endothelial cells from the primary breast cancers and primary colon cancer (Fig. 2.8). However, there was only about a 10% overlap in the genes expressed at the highest

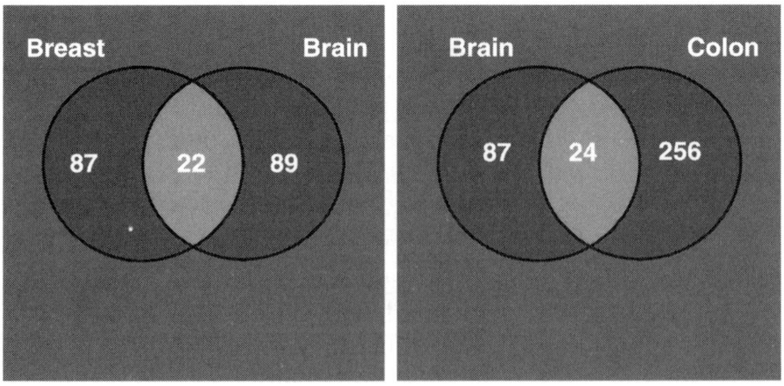

Figure 2.7 Venn diagram showing overlap of genes expressed at higher levels in endothelial cells isolated from surgical samples of primary human breast cancers and human brain cancers and human colon cancer and human brain cancers. Data were obtained from SAGE analysis of the transcriptomes of the endothelial cells from surgical samples of the human tumors and corresponding normal tissues. The SAGE data were subjected to statistical analyses. Shown are numbers of genes that are expressed at higher levels in the tumor endothelial cells with $\geq$ 98% confidence by χ^2 analysis.

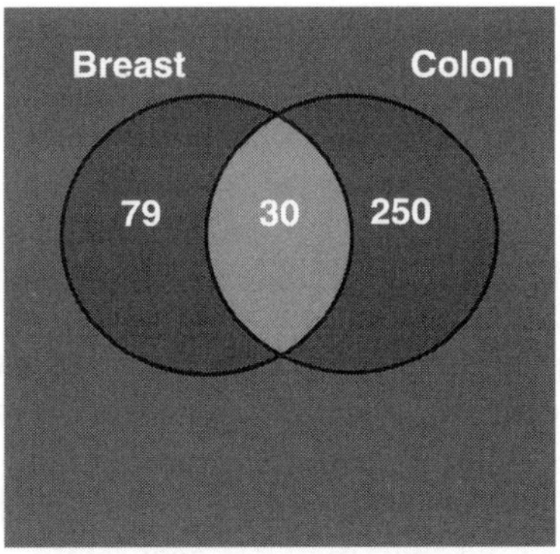

Figure 2.8 Venn diagram showing overlap of genes expressed at higher levels in endothelial cells isolated from surgical samples of primary human breast cancers and human colon cancer. Data were obtained from SAGE analysis of the transcriptomes of the endothelial cells from surgical samples of the human tumors and corresponding normal tissues. The SAGE data were subjected to statistical analyses. Shown are numbers of genes that are expressed at higher levels in the tumor endothelial cells with $\geq$ 98% confidence by χ^2 analysis.

levels when the three tumor endothelial cell SAGE libraries were compared (Fig. 2.9). Together, these results suggest broad variation in endothelial gene expression patterns, although some overlap can be identified between different normal and malignant settings.

2.5 New Tumor Models

There are four general types of in vivo models that are available for the assessment of efficacy of experimental therapeutics in cancer: syngeneic graft models, transgenic and knockout gene mutant models (genetically engineered

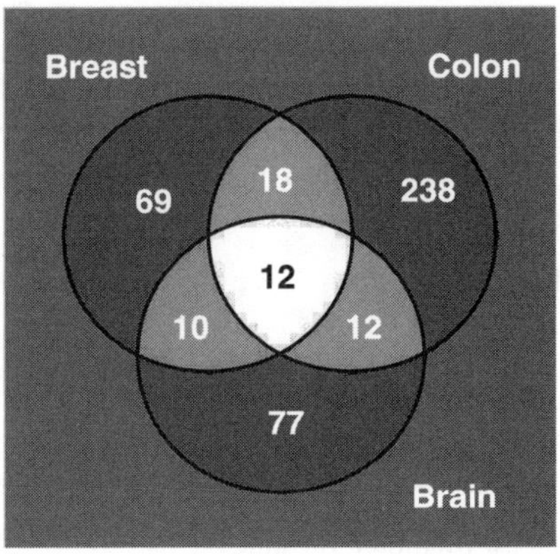

Figure 2.9 Venn diagram showing overlap of genes expressed at higher levels in endothelial cells isolated from surgical samples of primary human breast cancers, human brain cancers and human colon cancer. Data were obtained from SAGE analysis of the transcriptomes of the endothelial cells from surgical samples of the human tumors and corresponding normal tissues. The SAGE data were subjected to statistical analyses. Shown are numbers of genes that are expressed at higher levels in the tumor endothelial cells with $\geq$ 98% confidence by χ^2 analysis.

mice, sometimes referred to as "autocthonous" models), human xenograft models, and carcinogen-induced models. Each of these models has inherent advantages and disadvantages, discussed briefly here and in more depth in Chapter 13.

Syngeneic graft models that are available in rats and mice offer the advantages of an immunocompetent settting: a wide variety of tumor types that can be studied, timely assay, reliability, and low cost. The response of many of these models to current anticancer therapies is often well established so that one can compare the efficacy of new experimental therapies. One can also readily obtain sufficient numbers of test animals for valid statistical analysis. The disadvantage is that these models are rodent based and the tumors are usually fast growing, so they may not accurately model human disease. Johnson et al. (2001) reviewed compounds that entered into clinical trial on the basis of activity in syngeneic models. The correlation between compound activity in a particular tumor in mice and activity in the homologous tumor type in humans was low, with only $\sim$ 50% of compounds displaying activity in $>$ 33% of the mouse models tested displaying activity in at least two disease types in humans.

Genetically engineered mice that develop tumors are, generally, immunocompetent and develop tumors that can be described as syngeneic and orthotopic (Table 2.6). The term *autocthonous* has been suggested for genetically engineered mice that develop spontaneous, orthotopic tumors (e.g., transgenic oncomouse models). The disadvantages of these models are their relative expense, related to the requirements for breeding and housing the animals (and, frequently, the need to obtain a license to use them). Tumors usually develop late in the animal's life span, so these models are also relatively slow. In addition, there may be few histologies available and obtaining sufficient animals to establish valid statistics may be an issue. It is important that few of these models have been validated as representative of the human disease through molecular markers and response to current anticancer therapies (Bergers et al., 1999; Van Dyke and Jacks, 2002). There

Table 2.6 Transgenic and Knockout Mutant Mouse Tumor Models

Advantages
- Immunocompetent
- Syngeneic
- Orthotopic (autochthonous)

Disadvantages
- Require breeding (and frequently licensing) – high cost
- Usually develop tumors late in life span – slow
- Relatively few histologies available
- Difficult to obtain many animals – affects statistical considerations
- Validated as models of human disease?
- Validated as models for treatment response?

Examples
- TRAMP model – chemoprevention studies
- RIP/Tag model – antiangiogeneic studies
- K14-HPV16 model – molecular studies

are a more limited number of reports using genetically engineered mice for testing potential cancer therapeutic agents, as compared to syngeneic graft models or human xenograft models. However, two models that have received significant attention are the TRAMP model for prostate cancer, which has been used in a variety of chemoprevention studies (Gupta et al., 2000, 2001; Huss et al., 2001; Mentor-Marcel et al., 2001; Raghow et al., 2000, 2002), and the RIP-Tag model for pancreatic cancer, which has been used to study angiogenesis (Qian et al., 2001).

Human tumor xenograft models have been used very widely to study potential cancer therapeutic agents. One advantage offered is that human malignant cells of a wide variety of tissues have been studied and, in many cases, documented in terms of tumor growth, making tumor growth or tumor delay assays reliable to run. The response of many of these tumor models to current anticancer therapies is often also well established. The disadvantages of this type of model are that the hosts are immunodeficient (usually nude or SCID mice), the tumors are generally slow growing, the stromal component is murine, and the animals are costly to obtain and require special housing. The cost of the studies often leads to fewer animals being used, sometimes negatively affecting the statistical analysis of the results. Subcutaneous human solid tumor xenografts are also often resistant to currently used standard agents in the clinic, at least when assessed in terms of the induction of partial or complete responses (Dykes et al., 2001; Plowman et al., 1997).

Carcinogen-induced tumor models are used less widely to assess potential therapeutic agents. These models include induction of oral cancer in hamsters by 7,12-dimethylbenz(a)anthracene (1 mg/mL to cheek pouch 3 day/week for 16 weeks), induction of mammary carcinoma in rats by *N*-nitroso-*N*-methylurea (50 mg/kg IP, then wait 35 days), and induction of colon carcinoma in rats by azoxymethane (15 mg/kg SC, once/week for 2 weeks, then wait 40–50 weeks). Mixed variations of these models also exist, such as, the treatment of *min* mouse, which is prone to intestinal lesions, with azoxymethane to promote lesions. In many cases, carcinogen-induced tumor models are more widely used for studies of chemoprevention, as they have been relatively difficult to apply to therapeutic research.

Models to analyze the treatment of metastases can be obtained in two ways. First, syngeneic models that are naturally metastatic can be used. One variation being explored recently by many investigators is to generate metastases with genetically engineered tumors. In this approach, metastases are generated after resection of a primary syngeneic tumor, itself derived by subcutaneous injection of cells cultured from a spontaneous-arising tumor in the genetically engineered model. Another method to generate metastases is to inject syngeneic tumor cells intravenously for lung metastases, intrasplenically for liver metastases, intracardially or intratibially for bone metastases, or into the internal carotid artery for brain metastases. One benefit to these studies has been the development of bioluminescence methods to image metastases in animals. To facilitate analysis, there are now a variety of human cancer cell lines available that express green fluorescent protein, including glioblastoma, pancreatic cancer, prostate cancer, and colon cancer cell lines; there are also human cancer cell lines that express luciferase, including lung cancer, prostate cancer, and other lines, all of which can be used with the necessary instrumentation to detect metastases by fluorescence (Teicher, 2001b). Genetically engineered variants that offer tagged cells are also available but at greater cost. The increasing use of metastatic and orthotopically implanted tumor models have reinvigorated the evidence that tumor response to therapy can vary markedly, depending on the anatomical location of the disease.

2.6 Summary

The debate in the field of anticancer therapeutics over whether to alter the criteria for success extends from the earliest cell-based assays to the results of clinical trials. Decreasing the criteria for success in cell-based assays from IC_{90} to IC_{50} and the criteria for success in preclinical tumor models from tumor growth delays to T/C% at maximal points has led to clinical testing of agents that have shown only limited anticancer activity. There has been a drive to modify the criteria for clinical success from tumor shrinkage to surrogate end points, such as blood flow imaging or change in a plasma marker (Cristofanilli et al., 2002; Kerbel et al., 2001). In addition, some investigators have argued against using traditional dose escalation end points to dose-limiting toxicity (DLT) in early clinical trials with new targeted therapeutics, instead suggesting that plasma IC_{50} levels and mouse plasma levels should be sufficient to determine efficacious doses. Several of the molecular targets of these new agents, such as EGFR and VEGFR, are widely found in normal tissues. In light of the above arguments, one implication is that these targeted therapeutics might have very small therapeutic indices if dose escalation to antitumor activity is attempted. The discussion of how to best apply targeted therapeutics will no doubt be influenced by empirical clinical experience, as different viewpoints are assessed by experiment. Molecular understanding of malignancy is improving rapidly; however, improvement in cancer therapy seems likely to remain a step-by-step process, as long as distinguishing normal cells from malignant cells in a therapeutically meaningful way continues to be a challenge.

References

Abe, S., Kubota, T., Otani, Y., et al. UCN-01 (7-hydroxystaurosporine) inhibits in vivo growth of human cancer cells through selective perturbation of G1 phase checkpoint machinery. *Jpn. J. Cancer Res.* **92,** 537–545 (2001).

Abrams, T. J., Lee, L. B., Murray, L. J., et al. Su11248 and STI-571, small molecule inhibitors of Kit and PDGFR inhibit growth of SCLC preclinical models. *Eur. J Cancer* **38 (suppl 7),** 80 (2002b).

Abrams, T. J., Murray, L. J., Pryer, N. K., et al. Preclinical evaluation of the tyrosine kinase inhibitor SU11248 for the treatment of breast cancer. *Eur. J Cancer* **38 (suppl 7),** 75 (2002a).

Aiello, L. P., Bursell, S. E., Clermont, A, et al. Vascular endothelial growth factor-induced retinal permeability is mediated by protein kinase C in vivo and suppressed by an orally effective beta-isoform-selective inhibitor. *Diabetes* **46,** 1473–1480 (1997).

Akinaga, S., Gomi, K., Morimoto, M., et al. Antitumor activity of UCN-01, a selective inhibitor of protein kinase C, in murine and human models. *Cancer Res.* **51,** 4888–4892 (1991).

Akiyama, T., Yoshida, T., Tsujita, T., et al. G1 phase accumulation induced by UCN-01 is associated with dephosphorylation of Rb and CDK2 proteins as well as induction of CDK inhibitor p21/Cip1/WAF1/Sdi1 in p53-mutated human epidermoid carcinoma A431 cells. *Cancer Res.* **57,** 1495–1501 (1997).

Aklilu, M., Kindler, H. L., Gajewski, T. F., et al. Toxicities of the antiangiogenic agent SU5416 in phase II studies [Abstract]. *Proc. Am. Soc. Clin .Oncol.* 1921 (2002).

Albanell, J., Rojo, F., and Baselga, J. Pharmacodynamic studies with the epidermal growth factor receptor tyrosine kinase inhibitor ZD1839. *Semin. Oncol.* **28(5 Suppl 16),** 56–66 (2001).

Andratschke, N., Grosu, A. L., Molls, M., and Nieder, C. Perspectives in the treatment of malignant gliomas in adults. *Anticancer Res.* **21,** 3541–3550 (2001).

Armstrong, S. A., Staunton, J. E., Silverman, L. B., et al. MLL translocations specify a distinct gene expression profile that distinguishes a unique leukemia. *Nature Genet.* **30,** 41–47 (2002).

Arteaga, C. L., Khuri, F., Krystal, G., and Sebti, S. Overview of rationale and clinical trials with signal transduction inhibitors in lung cancer. *Semin. Oncol.* **29(1 Suppl 4),** 15–26 (2002).

Aschele, C., Baldo, C., Sobrero, A. F., et al. Schedule-dependent synergism between ZD1694 (ralititrexed) and CPT-11 (irinotecan) in human colon cancer in vitro. *Clin. Cancer Res.* **4,** 1323–1330 (1998).

Baselga, J., Norton, L., Masui, H., et al. Antitumor effects of doxorubicin in combination with anti-epidermal growth factor receptor monoclonal antibodies. *J. Natl. Cancer Inst.* **85,** 1327–1333 (1993).

Beecken, W.-D. C., Fernandez, A., Joussen, A. M., et al. Effect of antiangiogenic therapy on slowly growing, poorly vascularized tumors in mice. *J. Natl. Cancer Inst.* **93,** 382–387 (2001).

Bergers, G., Javaherian, K., Lo, K. M., et al. Effects of angiogenesis inhibitors on multistage carcinogenesis in mice. *Science* **284,** 808–812 (1999).

Bhat-Nakshatri, P., Sweeney, C. J., and Nakshatri, H. Identification of signal transduction pathways involved in constitutive NF-kappaB activation in breast cancer cells. *Oncogene* **21,** 2066–2078 (2002).

Bhattacharjee, A., Richards, W. G., Staunton, J., et al. Classification of human lung carcinomas by mRNA expression profiling reveals distinct adenocarcinoma subclasses. *Proc. Natl. Acad. Sci. USA* **98,** 13790–13795 (2001).

Bibby, M. C. Making the most of rodent tumor systems in cancer. *Br. J. Cancer* **79,** 1633–1640 (1999).

Blumberg, P. M., Acs, P., Bhattacharyya, D. K., and Lorenzo, P. S. Inhibitors of protein kinase C and related receptors for the lipophilic second-messenger sn-1,2-diacylglycerol. In: J. S. Gutkind, ed., *Cell Cycle Control: The Molecular Basis of Cancer and Other Diseases.* Totowa, NJ, Humana Press (2000).

Bode, A. M., and Dong, Z. Signal transduction pathways: Targets for chemoprevention of skin cancer. *Lancet Oncol.* **1,** 181–188 (2000).

Borgstroem, P., Gold, D. P., Hillan, K. J., and Ferrara, N. Importance of VEGF for breast cancer angiogenesis in vivo: Implications from intravital microscopy of combination treatments with an anti-VEGF neutralizing monoclonal antibody and doxirubicin. *Anticancer Res.* **19,** 4203–4214 (1999).

Brahmer, J. R., Kelsey, S., Scigalla, P., et al. A phase I study of SU6668 in patients with refractory solid tumors [Abstract]. *Proc. Am. Soc. Clin. Oncol.* **38,** 335 (2002).

Brandt, D. S., and Chu, E. Future challenges in the clinical development of thymidylate synthase inhibitor compounds. *Oncol. Res.* **9,** 403–410 (1997).

Britten, C. D., Rosen, L. S., Kabbinavar, F., et al. Phase I trial of SU6668, a small molecule receptor tyrosine kinase inhibitor, given twice daily in patients with advanced cancer [Abstract]. *Proc. Am. Soc. Clin. Oncol.* **38,** 1922 (2002).

Bruns, C. J., Harbison, M. T., Davis, D. W., et al. Epidermal growth factor receptor blockade with C225 plus gemcitabine results in regression of human pancreatic carcinoma growing orthotopically in nude mice by antiangiogenic mechanisms. *Clin. Cancer Res.* **6,** 1936–1948 (2000).

Buchner, K. The role of protein kinase C in the regulation of cell growth and in signaling to the cell nucleus. *J. Cancer Res. Clin. Oncol.* **126,** 1–11 (2000).

Busby, E. C., Leistritz, D. F., Abraham, R. T., et al. The radiosensitizing agent 7-hydroxystaurosporine (UCN-01) inhibits the DNA damage checkpoint kinase hChk1. *Cancer Res.* **60,** 2108–2112 (2000).

Carter, S. K. Clinical strategy for the development of angiogenesis inhibitors. *Oncologist* **5,** 51–54 (2000).

Chabot, G. C. Clinical pharmacokinetics of irinotecan. *Clin. Pharmacokinet.* **33,** 245–259 (1997).

Chen, X., Lowe, M., and Keyomarsi, K. UCN-01-mediated G1 arrest in normal but not tumor breast cells is Rb-dependent and p53-independent. *Oncogene* **18,** 5691–5702 (1999).

Cherrington, J. M., Strawn, L. M., and Shawver, L. K. New paradigms for the treatment of cancer: The role of anti-angiogenesis agents. *Adv. Cancer Res.* **56,** 1–37 (2000).

Chick, H. An investigation of the laws of disinfection. *J. Hyg. (London)* **8,** 92–158 (1908).

Cho, C. D., Fisher, G. A., Halsey, J. Z., et al. A phase I study ZD1839 (Iressa) in combination with oxaliplatin, 5-fluororuacil (%-FU) and leucovorin (LV) in advanced solid malignancies [Abstract]. *Proc. Am. Soc. Clin. Oncol.* **38,** 38 (2002).

Ciardiello, F., Bianco, R., Damiano, V., et al. Antiangiogenic and antitumor activity of anti-epidermal growth factor receptor C225 monoclonal antibody in combination with vascular endothelial growth factor antisense oligonucleotide in human GEO colon cancer cells. *Clin. Cancer Res.* **6,** 3739–3747 (2000a).

Ciardiello, F., Bianco, R., Damiano, V., et al. Antitumor activity of sequential treatment with topotecan and anti-epidermal growth factor receptor monoclonal antibody C225. *Clin. Cancer Res.* **5,** 909–916 (1999).

Ciardiello, F., Caputo, R., Bianco, R., et al. Antitumor effect and potentiation of cytotoxic drug activity in human cancer cells by ZD-1839 (Iressa), an epidermal growth factor receptor-sensitive tyrosine kinase inhibitor. *Clin. Cancer Res.* **6,** 2053–2063 (2000b).

Ciardiello, F., Caputo, R., Bianco, R., et al. Inhibition of growth factor production and angiogenesis in human cancer cells by ZD1839 (Iressa), a selective epidermal growth factor receptor tyrosine kinase inhibitor. *Clin. Cancer Res.* **7,** 1459–1465 (2001).

Ciardiello, F., and Tortora, G. A novel approach in the treatment of cancer: Targeting the epidermal growth factor receptor. *Clin. Cancer Res.* **7,** 2958–2970 (2001).

Clarke, P. A., te Poele, R., Wooster, R., et al. Gene expression microarray analysis in cancer biology, pharmacology and drug development: progress and potential. *Biochem. Pharmacol.* **62,** 1311–1336 (2001).

Cristofanilli, M., Charnsangavej, C., and Hortobagyi, G. N. Angiogenesis modulation in cancer research: Novel clinical approaches. *Nature Rev. Drug Discovery* **1,** 415–426 (2002).

Danis, R. P., Bingaman, D. P., Jirousek, M., and Yang, Y. Inhibition of intraocular neovascularization caused by retinal ischemia in pigs by PKCbeta inhibition with LY333531. *Invest. Ophthalmol. Visual Sci.* **39,** 171–177 (1998).

Da Rocha, A. B., Mans, D. R., Regner, A., and Schwartsmann, G. Targeting protein kinase C: A new therapeutic opportunity against high-grade malignant glioma? *Oncologist* **7,** 17–33 (2002).

Dash, A. B., Williams, I. R., Kutok, J. L., et al. A murine model of CML blast crisis induced by cooperation between BCR/ABL and NUP98/HOXA9. *Proc. Natl. Acad. Sci. USA* **99,** 7622–7627 (2002).

Davis, B. D. In: *Bacterial and Mycotic Infections in Man*, 3rd ed. Ed. R. J. Dubos. Philadelphia, Lippincott (1958).

Dees, E., O'Reilly, S., Figg, W., et al. A phase I and pharmacologic study of UCN-01, a protein kinase C inhibitor. *Proc. Am. Soc. Clin. Oncol.*, **36,** 355 (2000).

Demetri, G. D. Targeting *c*-kit mutations in solid tumors: Scientific rationale and novel therapeutic options. *Semin. Oncol.* **28 (5 suppl 17),** 19–26 (2001).

Desai, B., Higgins, B., and Smith, M. Antitumor activity of the EGFR/TK inhibitor Tarceva (erlotinib, OSI-774) in tumors models. *Eur. J. Cancer* **38 (suppl 7),** 63 (2002).

Drevs, J., Hofmann, I., Hugenschmidt, H., et al. Effects of PTK787/ZK 222584, a specific inhibitor of vascular endothelial growth factor receptor tyrosine kinases, on primary tumor, metastasis, vessel density and blood flow in a murine renal cell carcinoma model. *Cancer Res.* **60,** 4819–4824 (2000).

Drevs, J., Schmidt-Gersbach, C. I., Mross, K., et al. Surrogate markers for the assessment of biological activity of the VEGF-receptor inhibitor PTK787/ZK222584 (PTK/ZK) in two clinical phase I trials [Abstract]. *Proc. Am. Soc. Clin. Oncol.* **38,** 337 (2002a).

Drevs, J., Steward, W. P., Dugan, M., et al. Biomarkers (VEGF, bFGF) for assessing the biological activity of PTK787/zk222584 (PTK/ZK), a vascular endothelial growth factor (VEGF) receptor5 inhibitor, in tumors known to overexpress VEGF. *Eur. J. Cancer* **38 (suppl 7),** 78 (2002b).

Drucker, B. J., Schwartz, L., Marion, S., and Motzer, R. Phase II trial of ZD1839 (Iressa), an EGF receptor inhibitor, in patients with advanced renal cell carcinoma [Abstract]. *Proc. Am. Soc. Clin. Oncol.* **38,** 720 (2002).

Dykes, D. J., Abbott, B. J., Mayo, J. G., et al. Development of human tumor xenograft models for in vivo evaluation of new antitumor drugs. In: H. H. Fiebig and D. P. Berger, eds. *Immunodeficient Mice in Oncology.* Basel, Karger (2001).

Ellis, L. M., Liu, W., Ahmad, S. A., et al. Overview of angiogenesis: Biologic implications for antiangiogenic therapy. *Semin. Oncol.* **28 (5 Suppl 16),**94–104 (2001).
Ellis, L. M., Takahashi, Y., Liu, W., and Shaheen, R. M. Vascular endothelial growth factor in human colon cancer: Biology and therapeutic implications. *Oncologist* **5 (suppl 1),** 11–15 (2000).
Elsayed, Y. A., and Sausville, E. A. Selected novel anticancer treatments targeting cell signaling proteins. *Oncologist* **6,** 517–537 (2001).
Evans, V. J., LaRock, J. F., Yosida, T. H., and Potter, M. A new tissue culture isolation and explanation of the P388 lymphocytic neoplasm in a chemically characterized medium. *Exp. Cell Res.* **32,** 212–217 (1963).
Fan, Z., Baselga, J., Masui, H., and Mendelsohn, J. Antitumor effect of anti-epidermal growth factor receptor monoclonal antibodies plus cis-diamminedichloroplatinum on well established A431 cell xenografts. *Cancer Res.* **53,** 4637–4642 (1992).
Fiedler, W., Staib, P., Kuse, R., et al. Role of angiogenesis inhibitors in acute myeloid leukemia. *Cancer J.* **7 (suppl 3),** S129–S133 (2001).
Flescher, E., and Rotem, R. Protein kinase C epsilon mediates the induction of P-glycoprotein in LNCaP prostate carcinoma cells. *Cell. Signal.* **14,** 37–43 (2002).
Fodde, R., Smits, R., and Clevers, H. APC, signal transduction and genetic instability in colorectal cancer. *Nature Rev. Cancer* **1,** 55–67 (2001).
Folkman, J. Tumor angiogenesis: Therapeutic implications. *N. Engl. J. Med.* **285,** 1182–1186 (1971).
Fong, T. A., Shawver, L. K., Sun, L., et al. SU5416 is a potent and selective inhibitor of the vascular endothelial growth factor (Flk-1/KDR) that inhibitors tyrosine kinase catalysis, tumor vascularization and growth of multiple tumor types. *Cancer Res.* **59,** 99–106 (1999).
Frei, E. III. Potential for eliminating leukemic cells in childhood acute leukemia [Abstract]. *Proc. Am. Assoc. Cancer Res.* **5:** 20 (1964).
Gasparini, G. The rationale and future potential of angiogenesis inhibitors in neoplasia. *Drugs* **58,** 17–38 (1999).
Geng, L., Donnelly, E., McMahon, G., et al. Inhibition of vascular endothelial growth factor receptor signaling leads to reversal of tumor resistance to radiotherapy. *Cancer Res.* **61,** 2413–2419 (2001).
Ghosh, P. M., Bedolla, R., Mikhailova, M., and Kreisberg, J. I. RhoA-dependent murine prostate cancer cell proliferation and apoptosis: role of protein kinase C zeta. *Cancer Res.* **62,** 2630–2636 (2002).
Giovanella, B. C. Topoisomerase I inhibitors. In: *Cancer Therapeutics: Experimental and Clinical Agents.* Ed. B. A. Teicher. Totowa, NJ, Humana Press (1997).
Goekjian, P. G., and Jirousek, M. R. Protein kinase C inhibitors as novel anticancer drugs. *Exp. Opin. Invest. Drugs* **10,** 2117–2140 (2001).
Golub, T. R. Genomic approaches to the pathogenesis of hematologic malignancy. *Curr. Opin. Hematol.* **8,** 252–261 (2001).
Golub, T. R., Slonim, D. K., Tamayo, P., et al. Molecular classification of cancer: Class discovery and class prediction by gene expression monitoring. *Science* **286,** 531–537 (1999).
Gorre, M. E., and Sawyer, C. L. Molecular mechanisms of resistance to STI571 in chronic myeloid leukemia. *Curr. Opin. Hematol.* **9,** 303–307 (2002).
Graff, J. R. Emerging targets in the AKT pathway for treatment of androgen-independent prostatic adenocarcinoma. *Exp. Opin. Therap. Targets* **6,** 103–113 (2002).
Graves, P. R., Yu, L., Schwarz, J. K., et al. The Chk1 protein kinase and the Cdc25C regulatory pathways are targets of the anticancer agent UCN-01. *J. Biol. Chem.* **275,** 5600–5605 (2000).
Griffin, J. The biology of signal transduction inhibition: Basic science to novel therapies. *Semin. Oncol.* **28 (5 suppl 17),** 3–8 (2001).
Grosios K. UCN-01. *Curr. Opin. Invest. Drugs* **2,** 287–297 (2001).
Guo, D., Jia, Q., Song, H. Y., et al. Vascular endothelial cell growth factor promotes tyrosine phosphorylation of mediators of signal transduction that contain SH2 domains. Association with endothelial cell proliferation. *J. Biol. Chem.* **270,** 6729–6733 (1995).
Guo, Q. M. DNA microarray and cancer. *Curr. Opin. Oncol.* **15,** 36–43 (2003).
Gupta, S., Ahmad, N., Marengo, S. R., et al. Chemoprevention of prostate carcinogenesis by alpha-difluoromethylornithine in TRAMP mice. *Cancer Res.* **60,** 5125–5133 (2000).
Gupta, S., Hastak, K., Ahmad, N., et al. Inhibition of prostate carcinogenesis in TRAP mice by oral infusion of green tea polyphenols. *Proc. Natl. Acad. Sci. USA* **98,** 10350–10355 (2001).
Hananian, J., Holland, J. F., and Sheehe, P. Intensive chemotherapy of acute lymphocytic leukemia in children [Abstract]. *Proc. Am. Assoc. Cancer Res.* **6,** 26 (1965).
Harari, P. M., and Huang, S. M. Radiation response modification following molecular inhibition of epidermal growth factor receptor signaling. *Semin. Radiol. Oncol.* **11,** 281–289 (2001).
Harris, A. L. Von Hippel-Landau syndrome: Target for anti-vascular endothelial growth factor (VEGF) receptor therapy. *Oncologist* **5 (suppl 1),** 32–36 (2000).

Heinrich, M. C., Blanke, C. D., Druker, B. J., and Corlees, C. L. Inhibition of KIT tyrosine kinase activity: A novel molecular approach to the treatment of KIT-positive malignancies. *J. Clin. Oncol.* **20,** 1692–1703 (2002).

Heldin, C. H. Signal transduction: multiple pathways, multiple options for therapy. *Stem Cells* **19,** 295–303 (2001).

Herbst, R. S., Hidalgo, M., Pierson, A. S., et al. Angiogenesis inhibitors in clinical development for lung cancer. *Semin. Oncol.* **29 (Suppl 4),** 66–77 (2002a).

Herbst, R. S., Thornton, D. E., Kies, M. S., et al. Phase 1 study of LY317615, a protein kinase C inhibitor. *Proc. Am. Soc. Clin. Oncol.* **38,** 326 (2002b).

Hermeking, H. Serial analysis of gene expression and cancer. *Curr. Opin. Oncol.* **15**, 44–49 (2003).

Heymach, J. V. Angiogenesis and antiangiogenic approaches to sarcomas. *Curr. Opin. Oncol.* **13,** 261–269 (2001).

Himmelfarb, P., Thayer, P. S., and Martin, H. Growth of colonies of murine leukemia L1210 in vitro. *Cancer Chemother. Rep.* **51,** 451–453 (1967).

Hirata, A., Ogawa, S.-I., Kometani, T., et al. ZD1839 (Iressa) induces antiangiogenic effects through inhibition of epidermal growth factor receptor tyrosine kinase. *Cancer Res.* **62,** 2554–2560 (2002).

Hoekman, K. SU6668, a multitargeted angiogenesis inhibitor. *Cancer* **7 (suppl 3),** 134–138 (2001).

Hoekstra, R., Dumez, H., van Oosterom, A. T., et al. A phase I and pharmacological study of PKI166, an epidermal growth factor receptor (EGFR) tyrosine kinase inhibitor, administered orally in a two weeks on, two weeks off scheme to patients with advanced cancer [Abstract]. *Proc. Am. Soc. Clin. Oncol.* Abstr 340 (2002).

Hoekman, K., Kuenen, B., Levi, M., et al. Activation of the coagulation cascade and endothelial cell perturbation during treatment with cisplatin, gemcitabine, and the angiogenesis inhibitor SU5416. *Proc. Am. Soc. Clin. Oncol.* **38,** 21 (2002).

Hondermarck, H., Vercoutter-Edouart, A. S., Revillion, F., et al. Proteomics of breast cancer for marker discovery and signal pathway profiling. *Proteomics* **1,** 1216–1232 (2001).

Huang, S.-M., and Harari, P. Modulation of radiation response after epidermal growth factor receptor blockade in squamous cell carcinomas: Inhibition of damage repair, cell cycle kinetics and tumor angiogenesis. *Clin. Cancer Res.* **6,** 2166–2174 (2000).

Hurwitz, H., Holden, S. N., Eckhardt, S. G., et al. Clinical evaluation of ZD6474, an orally active inhibitor of VEGF signaling, in patients with solid tumors [Abstract]. *Proc. Am. Soc. Clin. Oncol.* **38,** 325 (2002).

Huss, W. J., Hanrahan, C. F., Barrios, R. J., et al. Angiogenensis and prostate cancer: Identification of a molecular progression switch. *Cancer Res.* **61,** 2736–2743 (2001).

Inoue, K., Slaton, J. W., Perrotte, P., et al. Paclitaxel enhances the effects of the anti-epidermal growth factor receptor monoclonal antibody ImClone C225 in mice with metastatic human bladder transitional cell carcinoma. *Clin. Cancer Res.* **6,** 4874–4884 (2000).

Ishii, H., Jirousek, M. R., Koya, D., et al. Amelioration of vascular dysfunctions in diabetic rats by an oral PKC beta inhibitor. *Science* **272,** 728–731 (1996).

Jain, R. K. Normalizing tumor vasculature with anti-angiogenic therapy: A new paradigm for combination therapy. *Nature Med.* **7,** 987–989 (2001).

Jirousek, M. R., Gillig, J. R., Gonzalez, C. M., et al. (S)-13-[(dimethylamino)methyl]-10,11,14, 15-tetrahydro-4,9:16, 21-dimetheno-1H, 13H-dibenzo[e,k]pyrrolo[3,4-h][1,4,13]oxadiazacyclohexadecene-1,3(2H)-dione (LY333531) and related analogues: Isozyme selective inhibitors of protein kinase C beta. *J. Med. Chem.* **39,** 2664–2671 (1996).

Joensuu, H., and Dimitrijevic, S. Tyrosine kinase inhibitor imatinib (STI571) as an anticancer agent for solid tumors. *Ann. Med.* **33,** 451–455 (2001).

Joensuu, H., Krause, A., Demetri, G. D., et al. Gastrointestinal stromal tumor (GIST) patients who respond to imatinib (STI571, Gleevec) show marked decline of circulating levels of VRGF, KIT, and bFGF in serum, but not stem cell factor (SCF) levels [Abstract]. *Proc. Am. Soc. Clin. Oncol.* **38,** 552 (2002).

Johnson, J. I., Decker, S., Zaharevitz, D., et al. Relationships between drug activity in NCI preclinical in vitro and in vivo models and early clinical trials. *Br. J. Cancer* **84,** 1424–1431 (2001).

Kari, C., Chan, T. O., de Quadros, M. R., and Rodick, U. Targeting the epidermal growth factor receptor in cancer: Apoptosis takes center stage. *Cancer Res.* **63,** 1–5 (2003).

Kerbel, R. S. Tumor angiogenesis: Past, present and the near future. *Carcinogenesis* **21,** 505–515 (2000).

Kerbel, R. S., Viloria-Petit, A., Klement, G., and Rak, J. Accidental antiangiogenic drugs, anti-oncogene directed signal transduction inhibitors and conventional chemotherapeutic agents as examples. *Eur. J. Cancer* **36,** 1248–1257 (2000).

Kerbel, R. S., Yu, J., Tran, J., et al. Possible mechanisms of acquired resistance to anti-angiogenic drugs: implications for the use of combination therapy approaches. *Cancer Metastasis Rev.* **20,** 79–86 (2001).

Keyes, K., Cox, K., Treadway, P., et al. An In vitro tumor model: Analysis of angiogenic factor expression after chemotherapy. *Cancer Res. 62,* 5597–5602 (2002).

Keyes, K., Mann, L., Cox, K., et al. Circulating angiogenic growth factor levels in mice bearing human tumors using Luminex multiplex technology. *Cancer Chemo. Pharmacol,* **51,** 321–327 (2003).

Kim, E. S., and Herbst, R. S. Angiogenesis inhibitors in lung cancer. *Curr. Oncol. Rep.* **4,** 325–333 (2002).

Kim, Y. W., Hur, S.Y., Kim, T. E., et al. Protein kinase C modulates telomerase activity in human cervical cancer cells. *Exp. Molec. Med.* **33,** 156–163 (2001).

Kruger, E. A., Blagosklonny, M. V., Dixon, S. C., and Figg, W. D. UCN-01, a protein kinase C inhibitor, inhibits endothelial cell proliferation and angiogenic hypoxic response. *Invasion Metastasis* **18,** 209–218 (1999).

Krystal, G. W. Mechanisms of resistance to imatinib (STI571) and prospects for combination with conventional chemotherapeutic agents. *Drug Resistance Updates* **4,** 16–21 (2001).

Krystal, G. W., Honsawek, S., Kiewlich, D., et al. Indoline tyrosine kinase inhibitors block kit activation and growth of small cell lung cancer cells. *Cancer Res.* **61,** 3660–3668 (2001).

Kuenen, B. C., Rosen, L., Smit, E. F., et al. Dose-finding and pharmacokinetics study of cisplatin, gemcitabine and SU5416 in patients with solid tumors. *J. Clin. Oncol.* **20,** 1657–1667 (2002a).

Kuenen, B., Ruijter, R., Hoekman, K., et al. Dose finding study of SU6668 given thrice daily by oral route under fed conditions in patients with advanced malignancies [Abstract]. *Proc. Am. Soc. Clin. Oncol.* **38,** 437 (2002b).

Laird, A. D., Vajkoczy, P., Shawver, L. K., et al. SU6668 is a potent antiangiogenic and antitumor agent that induces regression of established tumors. *Cancer Res.* **60,** 4152–4160 (2000).

Lango, M. N., Shin, D. M., and Grandis, J. R. Targeting growth factor receptors: Integration of novel therapeutics in the management of head and neck cancer. *Curr. Opin. Oncol.* **13,** 168–175 (2001).

La Rose, P., O'Dwyer, M. E., and Druker, B. J. Insights from pre-clinical studies for new combination treatment regimens with the BCR-ABL kinase inhibitor imatinib mesylate (Gleevec/Glivec) in chronic myelogenous leukemia: A translational perspective. *Leukemia* **16,** 1213–1219 (2002).

LaTulippe, E., Satagopan, J., Smith, A., et al. Comprehensive gene expression analysis of prostate cancer reveals distinct transcriptional programs associated with metastatic disease. *Cancer Res.* **62,** 4499–4506 (2002).

Law, L. W., Dunn, T. B., Boyle, P. J., and Miller, J. H. Observations on the effects of a folic-acid antagonists on transplantable lymphoid leukemias in mice. *J. Natl. Cancer Inst.* **10,** 179–195 (1949).

Lieberman, S. M., Horig, H., and Kaufman, H. L. Innovative treatments for pancreatic cancer. *Surg. Clin. North Am.* **81,** 715–739 (2001).

Liekens, S., De Clercq, E., and Neyts, J. Angiogenesis: Regulators and clinical applications. *Biochem. Pharmacol.* **61,** 253–270 (2001).

Lin, M. F., Zhang, X. Q., Dean, J., and Lin, F. F. Protein kinase C pathway is involved in regulating the secretion of prostatic phosphatase in human prostate cancer cells. *Cell Biol. Int.* **25,** 1139–1148 (2001).

Liu, W. M., Stimson, L. A., and Joel, S. P. The in vitro activity of the tyrosine kinase inhibitor STI571 in BCR-ABL positive chronic myeloid leukemia cells: Synergistic interactions with anti-leukemic agents. *Br. J. Cancer* **86,** 1472–1478 (2002).

Lloyd, H. H. Application of tumor models toward the design of treatment schedules for cancer chemotherapy. In: *Growth Kinetics and Biochemical Regulation of Normal and Malignant Cells.* Ed. B. Drewinko and R. M. Humphrey. Baltimore, Williams & Wilkins (1977).

Marley, S. B., Davidson, R. J., Goldman, J. M., and Gordon, M. Y. Effects of combinations of therapeutic agents on the proliferation of progenitor cells in chronic myeloid leukemia. *Br. J. Hematol.* **116,** 162–165 (2002).

Martelli, A. M., Sang, N., Borgatti, P., et al. Multiple biological responses activated by nuclear protein kinase C. *J. Cellular Biochem.* **74,** 499–521 (1999).

Mauro, M. J., and Druker, B. J. STI571: Targeting BCR-ABL as therapy for CML. *Oncologist* **6,** 233–238 (2001).

McDonald, A. C., Vasey, P. A., Adams, L., et al. A phase I and pharmacokinetic study of LY231514, the multitargeted antifolate. *Clin. Cancer Res.* **4,** 605–610 (1998).

Mendel, D. B., Laird, A. D., Smolich, B. D., et al. Development of SU5416, a selective small molecule inhibitor of VEGF receptor tyrosine kinase activity, as an anti-angiogenic agent. *Anti-Cancer Drug Design* **15,** 29–41 (2000a).

Mendel, D. B., Laird, A. D., Xin, X., et al. In vivo antitumor activity of SU11248, a novel tyrosine kinase inhibitor targeting vascular endothelial growth factor and platelet-derived growth factor receptors: Determination of a pharmacokinetic/pharmacodynamic relationship. *Clin. Cancer Res.* **9,** 327–337 (2003).

Mendel, D. B., Schreck, R. E., West, D. C., et al. The angiogenesis inhibitor SU5416 has long-lasting effects on vascular endothelial growth factor receptor phosphorylation and function. *Clin. Cancer Res.* **6,** 4848-4858 (2000b).

Mendelsohn, J. Epidermal growth factor receptor inhibition by a monoclonal antibody as anticancer therapy. *Clin. Cancer Res.* **3,** 2703–2707 (1997).

Mendelsohn, J. Blockade of receptors for growth factors: An anticancer therapy. *Clin. Cancer Res.* **6,** 747–753 (2000).

Mentor-Marcel, R., Lamartiniere, C. A., Eltoum, I. E., et al. Genistein in the diet reduces the incidence of poorly differentiated prostatic adenocarcinoma in transgenic mice (TRAMP). *Cancer Res.* **61,** 6777–6782 (2001).

Mesters, R. M., Padro, T., Bieker, R., et al. Stable remission after administration of the receptor tyrosine kinase inhibitor SU5416 in a patient with refractory acute myeloid leukemia. *Blood* **98,** 241–243 (2001).

Miller, K. D., Sweeney, C. J., and Sledge, G. W. Redefining the target: Chemotherapeutics as antiangiogenics. *J. Clin. Oncol.* **19,** 1195–1206 (2001).

Mita, A., Forouzesh, B., Hidalgo, M., et al. Phase I, pharmacokinetic (PK), and biological studies of the epidermal growth factor receptor-tyrosine kinase (EGFR-TK) inhibitor OSI-774 (Erlotinib or Tarceva) in combination with docetaxel. *Eur. J. Cancer* **38 (suppl 7),** 53 (2002).

Moore, G. E., Sandberg, A. A., and Ulrich, K. Suspension cell culture and in vivo and in vitro chromosome constitution of mouse leukemia 1210. *J. Natl. Cancer Inst.* **36,** 405–421 (1966).

Morgan, B., Drevs, J., Steward, W. P., et al. Defining the biologically active dose for PTK787/ZK222584 (ptk/zk), a vascular endothelial growth factor (VEGF) receptor inhibitor, based on the assessment of two biomarkers [dynamic contrast enhanced magnetic resonance imaging (DCE-MRI), plasma VEGF] in two phase I studies. *Eur. J. Cancer* **38 (suppl 7),** 75 (2002).

Moscatello, D. K., Holgado-Madruga, M., Emlet, D. R., et al. Constitutive activation of phosphatidylinositol 3-kinase by a naturally occurring mutant epidermal growth factor receptor. *J. Biol. Chem.* **273,** 200–206 (1998).

Naruse, I., Ohmori, T., Ao, Y., et al. Antitumor activity of the selective epidermal growth factor receptor-tyrosine kinase inhibitor (EGFR-TKI) Iressa (ZD1839) in an EGFR-expressing multidrug-resistant cell line in vitro and in vivo. *Int. J. Cancer* **98,** 310–315 (2002).

Ning, S., Laird, D., Cherrington, J. M., and Knox, S. J. The antiangiogenic agents SU5416 and SU6668 increase the antitumor effects of fractionated irradiation. *Radiat. Res.* **157,** 45–51 (2002).

Nishizuka, Y. Intracellular signaling by hydrolysis of phospholipids and activation of protein kinase C. *Science* **8,** 607–614 (1992).

O'Brian, C. A., Ward, N. E., Stewart, J. R., and Chu, F. Prospects for targeting protein kinase C isozymes in the therapy of drug-resistant cancer – An evolving story. *Cancer Metastasis Rev.* **20,** 95–100 (2001).

O'Dwyer, M. E., Mauro, M. J., and Druker, B. J. Recent advances in the treatment of chronic myelogenous leukemia. *Ann. Rev. Med.* **53,** 369–381 (2002).

O'Farrell, A.-M., Yuen, H. A., Louie, S. G., et al. Analysis of mechanism of action and biomarkers for kinase inhibitor SU5416 in AML patients [Abstract]. *Proc. Am. Soc. Clin. Oncol.* **38,** 1077 (2002).

Ohmori, T., Ao, Y., Nishio, K., et al. Low dose cisplatin can modulate the sensitivity of human non-small cell lung carcinoma cells to EGFR tyrosine kinase inhibitor (ZD1839; Iressa) in vivo. *Proc. Am. Assoc. Cancer Res.* **41,** 482 (2000).

Olavarria, E., Craddock, C., Dazzi, F., et al. Imatinib mesylate (STI571) in the treatment of relapse of chronic myeloid leukemia after allogenic stem cell transplantation. *Blood* **99,** 3861–3862 (2002).

O'Reilly, S., and Rowinsky, E. C. The clinical status of irinotecan (CPT-11), a novel water soluble camptothecin analogue. *Crit. Rev. Oncol. Hematol.* **24,** 47–70 (1996).

Patnaik, A., Goetz, A., Hammond, L. A., et al. Phase I, pharmacokinetic (PK) and biologic study of OSI-774, a selective epidermal growth factor receptor (EGFR) tyrosine kinase (TK) inhibitor in combination with paclitaxel and carboplatin in patients with advanced solid malignancies. *Eur. J. Cancer* **38 (suppl 7),** 54 (2002).

Perrotte, P., Matsumoto, T., Inoue, K., et al. Anti-epidermal growth factor receptor antibody C225 inhibits angiogenesis in human transitional cell carcinoma growing orthotopically in nude mice. *Clin. Cancer Res.* **5,** 257–265 (1999).

Pittilo, R. F., Schabel, F. M., Wilcox, W. S., and Skipper, H. E. Experimental evaluation of potential anticancer agents. XVI. Basic study of effects of certain anticancer agents on kinetic behavior of model bacterial cell populations. *Cancer Chemother. Rep.* **47,** 1–26 (1965).

Plowman, J., Dykes, D., Hollingshead, M., et al. Human tumor xenografts in NCI drug development. In: B. A. Teicher, ed., *Anticancer Drug Development Guide to Preclinical Screening, Clinical Trials, and Approval.* Totowa, NJ, Humana Press (1997).

Pomeroy, S. L., Tamayo, P., Gaasenbeek, M., et al. Prediction of central nervous system embryonal tumor outcome based on gene expression. *Nature* **415,** 436–442 (2002).

Porter, J. R. *Bacterial Chemistry and Physiology*. New York, Wiley (1947).

Potapova, O., Laird, A., Nannini, M., et al. Mechanism of action and biomarker studies of SU11248, a selective inhibitor of split kinase domain receptor tyrosine kinases (including VEGF receptors, PDGF receptors, c-Kit and Flt3). *Eur. J. Cancer* **38 (suppl 7),** 61 (2002).

Qian, F., Hanahan, D., and Weissman, I. L. L-Selectin can facilitate metastasis to lymph nodes in a transgenic mouse model of carcinogenesis. *Proc. Natl. Acad. Sci. USA* **98,** 3976–3681 (2001).

Raghow, S., Hooshdaran, M. Z., Katiyar, S., and Steiner, M. S. Toremifene prevents prostate cancer in the transgenic adenocarcinoma of mouse prostate model. *Cancer Res.* **62,** 1370–1376 (2002).

Raghow, S., Kuliyev, E., Steakley, M., et al. Efficacious chemoprevention of primary prostate cancer by flutamide in an autochthonous transgenic model. *Cancer Res.* **60,** 4093–4097 (2000).

Rall, D. P. Experimental studies of the blood-brain barrier. *Cancer Res.* **25,** 1572–1577 (1965).

Ramaswamy, S., and Golub, T. DNA microarrays in clinical oncology. *J. Clin. Oncol.* **20,** 1932–1941 (2002).

Ramaswamy, S., Tamayo, P., Rifkin, R., et al. Mutliclass cancer diagnosis using tumor gene expression signatures. *Proc. Natl. Acad. Sci. USA* **98,** 15149–15154 (2001).

Raymond, E., Faivre, S., Vera, K., et al. First results of a phase I and pharmacokinetic study of SU011248, a novel oral antiangiogenic agent, in patients with advanced solid tumors. *Eur. J. Cancer* **38 (suppl 7),** 17 (2002).

Reddy, S. A. Signaling pathways in pancreatic cancer. *Cancer J.* **7,** 274–286 (2001).

Rinaldi, D. A., Burris, H. A., Dorr, F. A., et al. Initial Phase I evaluation of the novel thymidylate synthase inhibitor, LY231514, using the modified continual reassessment method for dose escalation. *J. Clin. Oncol.* **13,** 2842–2850 (1995).

Rosen, L. Antiangiogenic strategies and agents in clinical trials. *Oncologist* **5,** 20–27 (2000).

Rosen, L. S. Angiogenesis inhibition in solid tumors. *Cancer J.* **7 (suppl 3),** S120–S128 (2001).

Rosen, L. S. Clinical experience with angiogenesis signaling inhibitors: Focus on vascular endothelial growth factor (VEGF) blockers. *Cancer Control* **9,** 36–44 (2002).

Saha, S., Sparks, A. B., Rago, C., et al. Using the transcriptome to annote the genome. *Nature Biotechnol.* **20,** 508–512 (2002).

Sarkaria, J. N., Busby, E. C., Tibbetts, R. S., et al. Inhibition of ATM and ATR kinase activities by the radiosensitizing agent, caffeine. *Cancer Res.* **59,** 4375–4382 (1999).

Sausville, E. A., Arbuck, S. G., Messmann, R., et al. Phase I trial of 72-hour continuous infusion UCN-01 in patients with refractory neoplasms. *J. Clin. Oncol.* **19,** 2319–2333 (2001).

Sawano, A., Takahashi, T., Yamaguchi, S., and Shibuya, M. The phosphorylated 1169-tyrosine containing region of flt-1 kinase (VEGFR-1) is a major binding site for PLCgamma. *Biochem. Biophys. Res. Commun.* **238,** 487–491 (1997).

Scappaticci, F. A. Mechanisms and future directions for angiogenesis-based cancer therapies. *J. Clin. Oncol.* **20,** 3906–3927 (2002).

Schabel, F. M. Jr., Griswold, D. P. Jr., Laster, W. R. Jr., et al. Quantitative evaluation of anticancer agent activity in experimental animals. *Pharmacol. Ther. (A)* **1,** 411–435 (1977).

Schlaeppi, J. M., and Wood, J. M. Targeting vascular endothelial growth factor (VEGF) for anti-tumor therapy, by anti-VEGF neutralizing monoclonal antibodies or by VEGF receptor tyrosine-kinase inhibitors. *Cancer Metastasis Rev.* **18,** 473–481 (1999).

Schulze, A., and Downward, J. Navigating gene expression using microarrays: A technology review. *Nature Cell Biol.* **3,** E190–E195 (2001).

Sedlacek, H. H. Kinase inhibitors in cancer therapy: A look ahead. *Drugs* **59,** 435–476 (2000).

Senderowicz, A. M., and Sausville, E. A. Preclinical and clinical development of cyclin-dependent kinase modulators. *J. Natl. Cancer Inst.* **92,** 376–387 (2000).

Shaheen, R. M., Davis, D. W., Liu, W., et al. Antiangiogenic therapy targeting the tyrosine kinase receptor for vascular endothelial growth factor receptor inhibits the growth of colon cancer liver metastasis and induces tumor and endothelial cell apoptosis. *Cancer Res.* **59,** 5412–5416 (1999).

Shen, B.-Q., Lee, D. Y., and Zioncheck, T. F. Vascular endothelial growth factor governs endothelial nitric-oxide synthase expression via a KDR/flk-1 receptor and protein kinase C signaling pathway. *J. Biol. Chem.* **274,** 33057–33063 (1999).

Shepherd, F. A. Angiogenesis inhibitors in the treatment of lung cancer. *Lung Cancer* **34 (suppl 3),** S81–S89 (2001).

Shih, C., and Thornton, D. E. Preclinical pharmacology studies and the clinical development of a novel multitargeted antifolate, MTA (LY231514). In: *Anticancer Drug Development Guide: Antifolate Drugs in Cancer Therapy*. Ed. A. L. Jackman. Totowa, NJ, Humana Press (1998).

Shih, S.-C., Mullen, A., Abrams, K., Mukhopadhyay, M., and Claffey, K. P. Role of protein kinase C isoforms in phorbol ester-induced vascular endothelial growth factor expression in human glioblastoma cells. *J. Biol. Chem.* **274,** 15407–15414 (1999).
Silva, D., English, D., Lyons, D., and Lloyd, F. P. Protein kinase C induces moltility of breast cancers by upregulating secretion of urokinase-type plsaminogen activator through activation of AP-1 and NF-kappaB. *Biochem. Biophys. Res. Commun.* **290,** 552–557 (2002).
Singh, D., Febbo, P. G., Ross, K., et al. Gene expression correlates of clinical prostate cancer behavior. *Cancer Cell* **1,** 203–209 (2002).
Simpson, R. J., and Dorow, D. S. Cancer proteomics from signaling networks to tumor markers. *Trends Biotechnol.* 19, S40–S48 (2001).
Sirotnak, F. M., Zakowsky, M. F., Miller, V. A., et al. Efficacy of cytotoxic agents against human tumor xenografts is markedly enhanced by coadministration of ZD 1839 (Iressa) an inhibitor of EGFR tyrosine kinase. *Clin. Cancer Res.* **6,** 4885–4892 (2000).
Skipper, H. E. Biochemical, biological, pharmacologic, toxicologic, kinetics and clinical (subhuman and human) relationships. *Cancer* **21,** 600–610 (1968).
Skipper, H. E. Cancer chemotherapy is many things [G. H. A. Clowes Memorial Lecture]. *Cancer Res.* **31,** 1173–1180 (1971b).
Skipper, H. E. Criteria associated with destruction of leukemia and solid tumor cells in animals. *Cancer Res.* **27,** 2636–2645 (1967).
Skipper, H. E. The effects of chemotherapy on the kinetics of leukemic cell behavior. *Cancer Res.* **25,** 1544–1550 (1965).
Skipper, H. E. Historic milestones in cancer biology: A few that are important in cancer treatment (revisited). *Semin Oncol.* **6,** 506–514 (1979).
Skipper, H. E. Improvement of the model systems. *Cancer Res.* **29,** 2329–2333 (1969).
Skipper, H. E. Kinetics of mammary tumor cell growth and implications for therapy. *Cancer* **28,** 1479–1499 (1971a).
Skipper, H. E. Perspectives in cancer chemotherapy: therapeutic design. *Cancer Res.* **24,** 1295–1302 (1964).
Skipper, H. E. Successes and failures at the preclinical level; where now? *Proceedings of the Seventh National Cancer Conference.* **7,** 109–121 (1972).
Skipper, H. E. Thoughts on cancer chemotherapy and combination modality therapy. *JAMA* **230,** 1033–1035 (1974).
Skipper, H. E., Schabel, F. M., Wilcox, W. S., et al. Experimental evaluation of potential anticancer agents. XVIII. Effects of therapy on viability and rate of proliferation of leukemia cells in various anatomic sites. *Cancer Chemother. Rep.* **47,** 41–64 (1965).
Skipper, H. E., Schabel, F. M., and Wilocox, W. S. Experimental evaluation of potential anticancer agents. XIII. On the criteria and kinetics associated with "curability" of experimental leukemia. *Cancer Chemother. Rep.* **35,** 1–111 (1964).
Smolich, B. D., Yuen, H. A., West, K. A., et al. The antiangiogenic protein kinase inhibitors SU5416 and SU6668 inhibit the SCF receptor (c-kit) in a human myeloid leukemia cell line and in acute myeloid leukemia blasts. *Blood* **97,** 1413–1421 (2001).
St. Croix, B., Rago, C., Velculescu, V., et al. Genes expressed in human tumor endothelium. *Science* **289,** 1197–1202 (2000).
Sugiyama, K., Akiyama, T., Shimizu, M., et al. Decrease in susceptibility toward induction of apoptosis and alteration in G1 checkpoint function as determinants of resistance of human lung cancer cells against the antisignaling drug UCN-01 (7-hydroxystaurosporine). *Cancer Res.* **59,** 4406–4412 (1999).
Sumitomo, M., Ohba, M., Asakuma, J., et al. Protein kinase C delta amplifies ceramide formation via mitochondrial signaling in prostate cancer cells. *J. Clin. Invest.* **109,** 827–836 (2002).
Swannie, H. C., and Kaye, S. B. Protein kinase C inhibitors. *Curr. Oncol. Rep.* **4,** 37–46 (2002).
Takimoto, C. H. Antifolates in clinical development. *Semin. Oncol.* **24 (Suppl 18),** 40–51 (1997).
Teicher, B. A., ed. *Antiangiogenic Agents in Cancer Therapy.* Totowa, NJ, Humana Press (1999).
Teicher, B. A. Malignant cells, directors of the malignant process: Role of transforming growth factor beta. *Cancer Metastasis Rev.* **20,** 133–143 (2001a).
Teicher, B. A. Systems approach to cancer therapy (antiangiogenics + standard cytotoxics' mechanism(s) of interaction). *Cancer Metastasis Rev.* **15,** 247–272 (1996).
Teicher, B. A., ed. *Tumor Models in Cancer Research.* Totowa, NJ, Humana Press (2001b).
Teicher, B. A., Alvarez, E., Liu, P., et al. MTA (LY231514) in combination treatment regimens using human tumor xenografts and the EMT6 murine mammary carcinoma. *Semin. Oncol.* **26, (Suppl 6),** 55–62 (1999a).
Teicher, B. A., Alvarez, E., Mendelsohn, L. G., et al. Enzymatic rationale and preclinical support for a potent protein kinase Cb inhibitor in cancer therapy. *Adv. Enzyme. Reg.* **39,** 313–327 (1999b).

Teicher, B. A., Alvarez, E., Menon, K., et al. Antiangiogenic effects of a protein kinase C beta-selective small molecule. *Cancer Chemother. Pharmacol.* **49,** 69–77 (2002b).

Teicher, B. A., Menon, K., Alvarez, E., et al. Antiangiogenic and antitumor effects of a protein kinase C beta inhibitor in human hepatocellular and gastric cancer xenografts. *In Vivo* **15,** 185–193 (2001c).

Teicher, B. A., Menon, K., Alvarez, E., et al. Antiangiogenic and antitumor effects of a protein kinase C beta inhibitor in human HT-29 colon carcinoma and human Caki-1 renal cell carcinoma xenografts. *Anticancer Res.* **21,** 3175–3184 (2001a).

Teicher, B. A., Menon, K., Alvarez, E., et al. Antiangiogenic and antitumor effects of a protein kinase C beta inhibitor in human T98G glioblastoma multiforme xenografts. *Clin. Cancer Res.* **7,** 634–640 (2001d).

Teicher, B. A., Menon, K., Alvarrez, E., et al. Antiangiogenic and antitumor effects of a protein kinase C beta inhibitor in murine Lewis lung carcinoma and human Calu-6 non-small cell lung carcinoma xenografts. *Cancer Chemother. Pharmacol.* **48,** 473–480 (2001b).

Thambi. P., and Sausville, E. A. STI571 (imatinib mesylate): The tale of a targeted therapy. *Anti-Cancer Drugs* **13,** 111–114 (2002).

Thomas, L. B. Pathology of leukemia in the brain and meninges: postmortem studies of patients with acute leukemia and of mice inoculated with L1210 leukemia. *Cancer Res.* **25,** 1555–1571 (1965).

Thomas, A., Morgan, B., Decatris, A., et al. A phase I study of the oral vascular endothelial growth factor (VEGF) receptor tyrosine kinase inhibitor PTK787/222584 on atwice daily schedule in patients with advanced cancer. *Eur. J. Cancer* **38 (suppl 7),** 77 (2002).

Thornton, D. E., Keyes, K., Mann, L., et al. Determination of cancer growth factor biomarkers in plasma from mice bearing subcutaneous human tumor xenografts using Luminex Multiplex technology. *Proc. Am. Assoc. Cancer Res.* **43,** 897 (2002).

Topaly, J., Fruehauf, S., Ho, A. D., and Zeller, W. J. Rationale for combination therapy of chronic myelogenous leukemia with imatinib and irradiation or alkylating agents: Implications for pre-transplant conditioning. *Br. J. Cancer* **86,** 1487–1493 (2002).

Tortora, G., Caputo, R., Damiano, V., et al. Oral administration of a novel taxane, an antisense oligonucelotide targeting protein kinase A, and the epidermal growth factor receptor inhibitor Iressa causes cooperative antitumor and antiangiogenic activity. *Clin. Cancer Res.* **7,** 4156–4163 (2001).

Townsley, C., Major, P., Siu, L. L., et al. Phase II study of OSI-774 in patients with metastatic colorectal cancer. *Eur. J. Cancer* **38 (suppl 7),** 57 (2002).

Traxler, P., Bold, G., Buchdunger, E., et al. Tyrosine kinase inhibitors: From rational design to clinical trials. *Med. Res. Rev.* **21,** 499–512 (2001).

Van de Vijver, M. J., He, Y. D., van't Veer, L. J., et al. A gene-expression signature as a predictor of survival in breast cancer. *N. Engl. J. Med.* **347,** 1999–2009 (2002).

Van Dyke, T., and Jacks, T. Cancer modeling in the modern era: Progress and challenges. *Cell* **108,** 135–144 (2002).

Velulescu, V. E., Zhang, L., Vogelstein, B., et al. Serial analysis of gene expression. *Science* **270,** 484–487 (1995).

Waud, W. R. Murine L1210 and P388 leukemias. In: *Anticancer Drug Development Guide: Pre-clinical Screening, Clinical Trials and Approval.* Ed. B. Teicher. Totowa, NJ, Humana Press (1998).

Way, K. J., Chou, E., and King, G. L. Identification of PKC-isoform-specific biological actions using pharmacological approaches. *Trends Pharmaceut. Sci.* **21,** 181–187 (2000).

Weisberg, E., and Griffin, J. D. Mechanisms of resistance imatinib (STI571) in preclinical models and in leukemia patients. *Drug Resistance Updates* **4,** 22–28 (2001).

Wilcox, W. S., Griswold, D. P., Laster, W. R., et al. Experimental evaluation of potential anticancer agents. XVII. Kinetics of growth and regression after treatment of certain solid tumors. *Cancer Chemother. Rep.* **47,** 27–39 (1965).

Wilcox, W. S., Schabel, F. M., and Skipper, H. E. Experimental evaluation of potential anticancer agents XV: On the relative rates of growth and hoast kill of "single" leukemia cells that survive in vivo cytoxan therapy. *Cancer Res.* **26,** 1009–1014 (1966).

Williams, K., Telfer, B. A., Stratford, I. J., and Wedge, S. R. An evaluation of the EGFR tyrosine kinase inhibitor ZD1839 (Iressa) in combination with ionizing radiation. Paper presented at the 11th NCI-EORTC-AACR Symposium on New Drugs in Cancer Therapy. Amsterdam (Nov. 7–10, 2000).

Wolff, N. C., and Ilaria, R. L. Establishment of a murine model for therapy-treated chronic myelogenous leukemia using the tyrosine kinase inhibitor STI571. *Blood* **98,** 2808–2816 (2001).

Wood, J. M., Bold, G., Buchdunger, E., et al. PTK787/ZK 222584, a novel and potent inhibitor of vascular endothelial growth factor receptor tyrosine kinases, impairs vascular endothelial growth factor-induced responses and tumor growth after oral administration. *Cancer Res.* **60,** 2178–2189 (2000).

Woodburn, J. R., Kendrew, J., Fennell, M., end Wakeling, A. E. ZD1839 ('Iressa') a selective epidermal growth factor receptor tyrosine kinase inhibitor (EGFR-TKI): Inhibition of c-*fos* mRNA, an intermediate marker of EGFR activation, correlates with tumor growth inhibition. *Proc. Am. Assoc. Cancer Res.* **41,** 402 (2000).

Workman, P. Scoring a bull's-eye against cancer genome targets. *Curr. Opin. Pharmacol.* 1, 342–352 (2001).

Wyss, O. Chemical factors affecting growth and death. In: *Bacterial Physiology*. Ed. C. H. Werkman and P. W. Wilson. New York, Academic Press (1951).

Xia, P., Aiello, L. P., Ishii, H., et al. Characterization of vascular endothelial growth factor's effect on the activation of protein kinase C, its isoforms, and endothelial cell growth. *J. Clin. Invest.* **98,** 2018–2026 (1996).

Yang, J. C., Haworth, L., Steinberg, S. M., et al. A randomized double-blind placebo-controlled trial of bevacizumab (anti-VEGF antibody) demonstrating a prolongation in time to progression in patients with matastatic renal cancer [Abstract]. *Proc. Am. Soc. Clin. Oncol.* **38,** 15 (2002).

Yoshiji, H., Kuriyama, S., Ways, D. K., et al. Protein kinase C lies on the signaling pathway for vascular endothelial growth factor-mediated tumor development and angiogenesis. *Cancer Res.* **59,** 4413–4418 (1999).

Zahalsky, A. J., Wong, R. J., Lis, E., et al. Phase II trial of SU5416 in patients with advanced incurable head and neck cancer [Abstract]. *Proc. Am. Soc. Clin. Oncol.* **38,** 902 (2002).

Zhu, Z., Bohlen, P., and Witte, L. Clinical development of angiogenesis inhibitors to vascular endothelial growth factor and its receptors as cancer therapeutics. *Curr. Cancer Drug Targets* **2,** 135–156 (2002).

Zwick, E., Bange, J., and Ullrich, A. Receptor tyrosine signaling as a target for cancer intervention strategies. *Endocrine-Related Cancer* **8,** 161–173 (2001).

CHAPTER 3

Cancer Genetics and Drug Target Selection

Guo-Jun Zhang and William G. Kaelin Jr.

Molecular Cancer Therapeutics: Strategies for Drug Discovery and Development, edited by George C. Prendergast
ISBN 0-471-43202-4

One challenge in treating cancer may stem not so much from a failure to identify small molecules that can kill cancer cells but, instead, from a failure to identify small molecules that can kill cancer cells while sparing normal cells. Indeed, perhaps as many as 0.1–1% of the molecules in a typical pharmaceutical company chemical compound library will kill cancer cells when tested in standard high-throughput cytotoxicity assays at low micromolar concentrations. Thus >1000 potential anticancer drugs might be discovered in a chemical library containing >100,000 compounds (most major pharmaceutical companies have access to between 100,000 and 1,000,000 compounds). Unfortunately, the vast majority of such compounds will also kill normal cells and, accordingly, most of the anticancer drugs discovered using this paradigm (which includes the majority of anticancer drugs in use today) have remarkably low therapeutic indices (defined as toxic dose/therapeutic dose). Moreover, the sheer number of compounds scoring positively in such high-throughput cytotoxicity assays has dictated that decisions be made with respect to which compounds to analyze and develop further. In past years, these decisions were influenced by factors such as potency, ease of synthesis, novelty, and drug-like properties based on established empirical criteria. While important, none of these considerations necessarily speaks to selectivity. As a result, it is possible that compounds capable of selectively killing cancer cells relative to normal cells were overlooked in the course of the countless high-throughput screens for anticancer drugs performed in both the public and private sectors over the past several decades.

Our growing understanding of the genetic alterations that cause cancer is beginning to lay the foundation for new paradigms of anticancer drug discovery. Some of the first generation of anticancer drugs that have been based on cancer genetics include Herceptin, which is a humanized monoclonal antibody directed against Her2/Neu (McKeage and Perry, 2002), and small molecule kinase inhibitors such as Gleevec and Iressa (Fabbro et al., 2002). Gleevec inhibits c-Abl, c-Kit, and the platelet-derived growth factor (PDGF) receptor, whereas Iressa inhibits the epithelial growth factor (EGF) receptor. Indeed, the success of Gleevec against chronic myelogenous leukemia, which is characterized by activation of c-Abl by virtue of the Bcr-Abl translocation associated with this disorder, validates this general approach (Druker and Lydon, 2000; Druker et al., 2001a, 2001b; Kantarjian et al., 2002; Talpaz et al., 2002). Nonetheless, there are concerns as to whether the startling success of Gleevec in chronic myelogenous leukemia (CML) will be repeated with other such targeted therapies in the future, especially in light of the modest activity of Herceptin (as a single agent) in breast cancer and the rather disappointing results obtained so far with Iressa in lung cancer. This review focuses on emerging genetic paradigms in cancer relevant to development of anticancer drugs.

3.1 Cancer as a Genetic Disease

It has become clear that cancer is caused by the accumulation of mutations in a cell that is permissive for expression of the transformed phenotype. In rare cases, certain mutations are inherited in the germline, where they become manifested as a hereditary predisposition to cancer. Analysis of families bearing such mutations, as well as genetically engineered mice, suggests that the genetic alterations required for transformation are profoundly influenced by the cell of origin. For example, *APC* inactivation plays an intimate role in the development of colonic neoplasms in both mice and humans, but not in other common epithelial neoplasms. Heterozygosity for the *RB-1* tumor-suppressor gene predisposes individuals to retinoblastomas and sarcomas in humans and pituitary tumors in mice. This later observation, along with similar comparisons for other cancer genes (such as the *p*53 and *VHL* tumor-suppressor genes) indicates that the cancer phenotype is also influenced by the species context. Thus it is possible for the same genotype to cause two different cancer phenotypes in mice and humans and for the same phenotype in mice and humans to be caused by different genotypes. This nuance is important in considering the development of mouse models for testing molecularly targeted therapies, because the working assumption for such therapies is that their action is predicated on genotype.

Cancer-causing mutations include gain-of-function mutations that convert proto-oncogenes into oncogenes and loss-of-function mutations that inactivate tumor-suppressor genes. Epigenetic changes, such as changes in DNA methylation that alter gene expression patterns, also contribute to malignant transformation. Collectively, these genetic and epigenetic changes are responsible for the hallmarks of cancer, which include growth factor independence, diminished susceptibility to programmed cell death (apoptosis), the ability to invade and metastasize, induction of angiogenesis, and escape from immunity and senescence (Hanahan and Weinberg, 2000). Mathematical models based on age-specific cancer incidence and assumptions with respect to spontaneous mutation rates have led to the conjecture that most adult solid tumors require 5–10 rate-limiting, causal, mutations (Renan, 1993). In keeping with this idea, the most intensively studied common epithelial tumors are associated with multiple recurrent genetic abnormalities. For example, inactivation of the *APC* and *p*53 tumor-suppressor genes, as well as one or more tumor suppressor genes on 18q, in conjunction with activation of K-*ras,* are recurrent causal mutations in colorectal cancer (Kinzler and Vogelstein, 1996). Although the precise number of rate-limiting, causal, mutations in solid tumors can be debated due to the assumptions made in the models above, it is nonetheless useful to distinguish between the mutations that cause the cancer phenotype (or, put another way, the mutations that have been selected for in vivo) and the many other mutations present in a cancer cell that might be considered epiphenomenal, or bystander, mutations (Fig. 3.1). Examples of such mutations might include amplification of genes that are contiguous to an oncogene as a result of chromosomal gains or deletion of genes that are contiguous to a tumor suppressor gene as a result of chromosomal loss. From a therapeutic point of view, it is also useful to subdivide causal mutations that are required to initiate malignancy but not maintain it from those mutations that are required continually to maintain the malignant phenotype. An example of the former might include a mutation that increases the probability of sustaining a mutation (i.e., a mutation in a so-called caretaker gene), whereas the latter might include certain

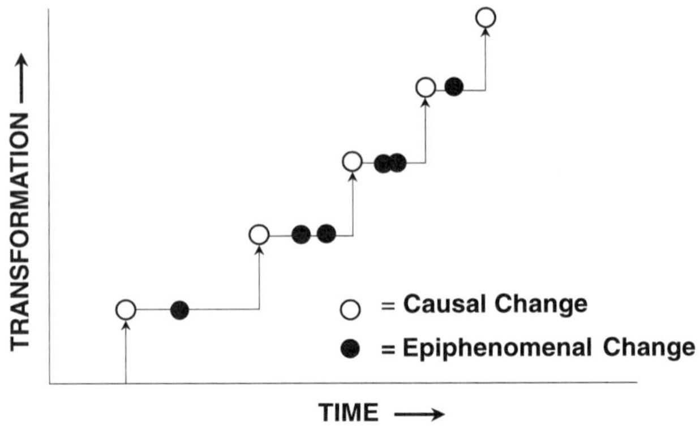

Figure 3.1 Conversion of a normal cell to a malignant cell is due to accumulation of genetic damage in a susceptible cell. Some changes contribute in a causal manner (*open circles*) to the transformed phenotype, whereas others can be viewed as epiphenomenal (*closed circles*). During tumor progression, which can take years, the rate at which mutations are acquired often accelerates owing to genomic instability. Genomic instability is common in cancer and is caused by mutations affecting the surveillance and repair of DNA damage.

antiproliferative and proapoptotic gatekeeper genes (Kinzler and Vogelstein, 1997).

3.2 Intratumor and Intertumor Heterogeneity

Clinicians and pathologists have appreciated for decades that no two tumors are alike (intertumor heterogeneity) and that significant heterogeneity exists within the same tumor (intratumor heterogeneity). Techniques such as gene-expression profiling, comparative genomic hybridization, and spectral karyotyping have provided striking confirmation of intertumor heterogeneity at the molecular level. At first blush, this heterogeneity presents significant challenges to the development of rational therapeutics (Table 3.1). For example, intertumor heterogeneity might imply the need for sophisticated genotyping of every tumor with an eye toward tailoring individual therapeutic cocktails for individual patients. Subdividing tumor types might also lead to some tumors and targets being abandoned due to concerns regarding market size. With respect to intratumor heterogeneity, there is concern for rapid selection of resistant clones that do not share the targeted genotype.

Table 3.1 Potential Impediments to Anticancer Drug Discovery

Intratumor heterogeneity
Intertumor heterogeneity
Small market size for some cancers
Phenotype due to multiple mutations in the same cell
Many cancer-causing mutations induce loss of function
Many cancer-causing mutations affect genes that are also important for normal cells

Fortunately, there are several approaches to intertumor heterogeneity. The first is to focus on causal genetic abnormalities and to ignore the epiphenomenal changes. The second approach, at the risk of oversimplification, is to focus on molecular pathways rather than on individual genes, because it is likely that the selection pressure for mutations in human cancer acts on pathways rather than individual proteins. Thus, for example, the vast majority of tumors have mutations that directly or indirectly compromise the pRB pathway, which is a negative regulator of cell proliferation, and the p53 pathway, which induces apoptosis in response to oncogenic signals and DNA damage. Components of the pRB pathway include the p16/INK4A tumor-suppressor protein, the oncoproteins cyclin D1 and cyclin-dependent kinase 4 (CDK4), and the pRB protein itself, whereas components of the p53 pathway include the p19ARF tumor-suppressor protein, the MDM2 oncoprotein, and p53. Thus a tumor with amplified cyclin D1 can be seen to be similar to a tumor loss of p16/INK4A (at least with respect to control of the cell-cycle by pRB) and a tumor with amplified *MDM2* can be seen to be similar to a tumor that has loss p53. In reality, such pathways are part of even more complex molecular networks. Fortunately, algorithms for analyzing data from genomewide assays (such as gene-expression profiles) can be used to group or cluster similar tumors based on shared higher-order features, even if many dissimilarities exist at the level of individual genes.

It seems likely, a priori, that causal genetic changes would be less heterogeneous within a given tumor than epiphemonal changes, since the latter, by definition, are not under selection pressure. Accordingly, therapies based on epiphenomenal changes that play no role in maintaining the malignant phenotype are more likely to be associated with the emergence of resistant subclones, compared to therapies based on causal changes. With respect to causal changes that are essential for maintenance, one might anticipate that the earliest changes would mark every malignant cell, even if different subclones with different constellations of late changes emerged over time. Accordingly, an effort to target early, causal, mutations would arguably minimize concerns related to intratumor heterogeneity. We would argue that some late mutations (whether causal or epiphenomenal) are merely advantageous, or indeed even tolerated, because of the mutations that preceded them. For example, it is thought that activation of certain oncogenes is tolerated only in cells that lack p53 function. Thus targeting early mutations would decrease the likelihood for selecting for resistant subclones (based on tumor heterogeneity), as well as increase the likelihood of unmasking deleterious consequences (from the perspective of the malignant clone) of late mutations.

3.3 Do Multiple Mutations Imply the Need for Combination Therapy?

It is widely believed that the existence of multiple mutations within a cancer cell necessarily implies the need for combination drug therapy. Even focusing on the causal mutations might leave one with the daunting task of pharmacologically attacking 5–10 molecular pathways. Indeed, there is lingering concern that the striking success of Gleevec in CML is because CML, in contrast to most adult solid tumors, is a genetically simple neoplasm. Indeed, in animal models, the Bcr-Abl fusion protein is sufficient to induce a CML-like condition, suggesting that CML might be a one-hit neoplasm (Van Etten, 2001).

Fortunately, the argument that multiple mutations logically necessitates combination therapy can be disputed both on theoretical grounds and empirically. With respect to theory, it seems likely that the need for multiple mutations to transform a normal cell into a fully malignant cell reflects, at least in part, redundancy in tumor-suppression pathways. In short, cancer cells have multiple causal mutations because they need multiple causal mutations to escape the multiple constraints that would otherwise prevent their successful expansion. An analogy might be that of a combination lock that has five tumblers. The fact that all five tumblers need to fall in place for the lock to open does not logically necessitate that all five tumblers be out of place to prevent it from opening. Instead, the premise implies that the lock will not open if any of the five tumblers is out of place.

In keeping with the lock analogy, there are now numerous examples in which correction of a single genetic defect in a genetically complex cancer cell has led to impaired tumor cell growth in vitro or in vivo. For example, restoring the function of a single tumor-suppressor gene, such as *p53*, *RB-1*, *VHL*, or *PTEN*, is sufficient to inhibit tumorigenesis in relevant cancer models as is conditional inactivation of certain oncogenes such as c-*myc* and H-*ras* (Baker et al., 1990; Chin et al., 1999; Felsher and Bishop, 1999; Furnari et al., 1997; Huang et al., 1988; Iliopoulos et al., 1995; Jain et al., 2002). Even the experience with Gleevec in the clinic supports the notion that multiple mutations does not inherently dictate the need for combination therapy. First, Gleevec has activity (albeit with somewhat reduced response rates) in accelerated phase and blast phase CML (Druker et al., 2001b; Talpaz et al., 2002). These conditions, in contrast to stable phase CML, are clearly characterized by the existence of multiple mutations in addition to the canonical *bcr-abl* translocation associated with this disease. It is important that the emergence of resistance in accelerated or blast phase CML is often due to subtle mutations within the Bcr-Abl ATP-binding site that render the kinase insensitive to Gleevec (Gorre et al., 2001). This both genetically validates Bcr-Abl as the therapeutically relevant target of Gleevec in CML and also establishes that there is an ongoing requirement for Bcr-Abl activity in acclerated and blast phase CML, despite the presence of mutations at other genetic loci. Finally, and more important still, Gleevec is very active against a solid tumor called gastrointestinal stromal cell tumor (GIST), which is characterized by activating mutations in c-*kit* (Joensuu et al., 2001; van Oosterom et al., 2001). Cytogenetic analyses of GISTs has identified multiple recurrent abnormalities, including deletion of 14q, deletion of chromosome 22, and deletion of 1p (Andersson et al., 2002). Moreover, germline activating mutations in c-*kit* give rise to hereditary GISTs in humans (Nishida et al., 1998). Tumor development in this setting follows a predictable course, beginning with hyperplasia of the interstitial cells of Cajal followed by emergence of benign GISTs, which can then give rise to malignant GISTs. These tumors develop over decades and are associated with the cytogenetic abnormalities described above. These findings, collectively, suggest that c-*kit* activation is an early, causal, mutation in GIST and that conversion from hyperplasia to frank malignancy is associated with the acquisition of multiple mutations over time, much as is thought to occur in other adult solid neoplasms.

In summary, the existence of multiple causal genetic changes in a cancer does not imply, prima facie, the need for combination therapy. For example, a small molecule that either mimicked a critical activity of p53 or inhibited an enzyme required for survival of p53-defective cells (see below) should, in theory, be effective in the ~50% of human tumors that lack p53 (barring extragenic suppressor mutations). On the other hand, use of combination therapy will probably be required to minimize the emergence of pharmacological resistance, much as is done in the treatment of HIV and tuberculosis.

3.4 Oncogene Addiction

A special consideration with respect to the sensitivity of cancer cells to agents (genes, drugs) directed against single genetic alterations relates to the emerging concept of "oncogene addiction" (Mills et al., 2001; Reddy and Kaelin, 2002; Weinstein, 2002). It was noted early on that hematopoetic cells engineered to produce Bcr-Abl were killed by Gleevec even under conditions where the parental (Bcr-Abl negative) cells were not (Druker et al., 1996). Similar findings have been made with activated versions of phosphatidylinositol 3-kinase (PI3K) and drugs that inhibit PI3K signaling (Mills et al., 2001). This has led to the notion that constitutive, high-level signaling through an oncogenic pathway might render a cell dependent on that pathway for survival. This notion might also help explain the involution of tumors observed following conditional inactivation of oncogenes in vivo, although these later experiments are somewhat confounded because the corresponding normal cells, by definition, would not be expected to survive at ectopic sites. Several non-mutually exclusive models have been invoked to account for the apparent oncogene addiction. One model suggests that constitutive, high-level signaling through a particular pathway leads to silencing of collateral pathways that would otherwise promote the survival of such cells in the face of an inhibitor. A second model suggests that constitutive, high-level signaling by oncogenic pathways leads to both antiapoptotic and proapoptotic signals. Cell death will ensue following oncogene withdrawal if the later decay with a longer half-life than the former. Finally, dependence on an oncogene might be due to synthetic interactions between the oncogene and secondary mutations at other genetic loci. Whatever the precise mechanism, oncogene addiction would, if generalizable, obviously simplify the choice of anticancer drug targets. Thus, for example, the finding that B-*raf* is mutationally activated in ~70% of melanomas would suggest, *prima facie*, that B-*raf* is a potential drug target in melanoma (Brose et al., 2002; Davies et al., 2002).

3.5 The Loss-of-Function Problem

Many cancer-associated mutations are loss-of-function mutations affecting tumor-suppressor genes. This presents a pharmacological problem, because most successful drugs, including drugs such as Gleevec, inactivate, rather than reactivate, their protein targets. More problematic still are situations in which the loss-of-function mutation leads to complete absence of its normal protein target (e.g., following homozygous deletion of a tumor-suppressor locus). One approach to this problem would be to look for drugable downstream proteins that deliver oncogenic signals following tumor-suppressor protein inactivation. For example, it has been argued that the enzyme CDK4 would be a rational target for cells lacking the CDK inhibitor p16/INK4A and that the vascular endothelial growth factor (VEGF) receptor KDR would be a rationale target for cells lacking pVHL (Kaelin, 1999). This general approach, however, presumes that a single (or at least tractable number of) critical downstream target(s) can be identified for the tumor-suppressor protein of interest. A second approach to the loss-of-function problem would be to exploit synthetic lethal interactions, as described below.

3.6 Synthetic Lethality

A theoretical approach to obtaining selective anticancer drugs would be to identify a target that is not essential for survival in normal cells but is essential for survival in cells that harbor mutations in a specific cancer gene or set of genes. In theory, an inhibitor of such a target should kill cancer cells with the relevant mutation(s) while not killing normal cells. This idea is motivated by studies of genes that are said to be synthetic lethal. Two genes are synthetic lethal if mutation of either gene alone is compatible with viability but mutation of both genes lead to cell death. Hartwell's group proposed that if one of these genes were a cancer gene, then the product of the other gene would be a potential drug target (Hartman et al., 2001; Hartwell et al., 1997).

Although this approach is conceptually attractive, it presumes that one can systematically identify the gene (or genes) that are synthetic lethal to a given human cancer gene. Until recently this was not possible. Now, however, at least two approaches can be envisioned for identifying such synthetic lethal interactions in mammalian cells. The first would be to carry out high-throughput cytotoxicity assays using matched (isogenic) cell lines that do or do not carry a cancer-relevant mutation with the goal of identifying compounds that selectively kill the former. In this way, it might be possible to identify a compound that behaves as though it were inhibiting a target that is synthetic lethal to the cancer gene in question. The existence of such compounds has already been established by prior studies. For example, it appears that p53(-/-) cells are more sensitive than their wild-type counterparts to the ATR/ATM inhibitor caffeine with respect to induction of lethal premature chromosomal condensation as well as induction of radiosensitivity and *PTEN-/-* cells are more sensitive than their normal counterparts to the mTOR inhibitor rapamycin (Neshat et al., 2001; Nghiem et al., 2001; Podsypanina et al., 2001; Powell et al., 1995). Torrance et al. (2001) used a high-throughput screen with isogenic cell line pairs to identify small molecules that selectively inhibit the growth of cancer cells carrying mutant K-*ras*. The main challenge with such high-throughput screens is to identify the relevant protein target(s) of the compounds that score positively.

An alternative strategy, at least in principal, would be to systematically disrupt the functions of known genes (or at least those genes predicted to encode drugable targets) in such isogenic cell line screens with genetic tools rather than relying on chemical compound libraries. In this regard, RNA interference (RNAi) is a powerful technique for disrupting the function of a given gene of interest in *Drosophila* and *Caenorhabditis elegans*. This technique has lately been optimized in *C. elegans* by feeding the worms bacteria that express dsRNA corresponding to the targeted mRNA. This feeding method was used successfully in a functional genomic RNAi analysis of *C. elegans* (Kamath et al., 2003). Recently, the use of RNAi has been successfully applied to human cells through the use of synthetic short, interfering RNAs (siRNAs) (Elbashir et al., 2001) as well as vectors encoding short hairpin RNAs (shRNAs), which mimic siRNA in cells (Abbas-Terki et al., 2002; Brummelkamp et al., 2002; Lee et al., 2002; Miyagishi and Taira, 2002; Paddison et al., 2002; Yu et al., 2002). An extensive discussion of RNAi technology is presented in Chapter 2.

3.7 Context and Selectivity

As mentioned at the onset, the challenge in cancer medicine is to identify and develop drugs that will kill cancer cells while sparing normal tissues. In general, there are two ways to obtain selectivity with a systemically administered drug (Fig. 3.2) (Kaelin, 1999). The first, and simplest, would be to exploit targets that are present in the diseased cells or tissues but not in their normal counterparts. For example, this paradigm is frequently exploited in the treatment of infectious diseases. There is a common misconception that this paradigm applies to Gleevec, since its intended target, the Bcr-Abl fusion oncoprotein, is present in CML cells but not normal cells. However, as described above, Gleevec also inhibits the c-Abl protein (in addition to c-Kit and PDGF receptor). Thus the target-driven model can not readily account for the selectivity of Gleevec. On the other hand, selectivity can also be observed when the requirement for a particular target is quantitatively or qualitatively altered in the context of the diseased cells/tissues compared to their normal counterparts. The differential requirement might be due to cell intrinsic changes (e.g., genetic and epigenetic changes within a cancer cell) or extrinsic changes (e.g., when a cancer cell is growing in an abnormal microenvironment that would be incompatible with the growth of a normal cell). A synthetic lethal interaction between a drug target and a cancer-causing

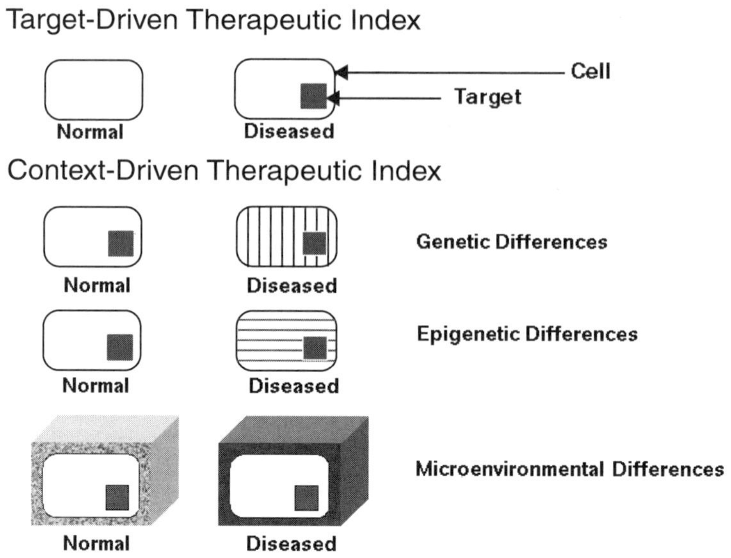

Figure 3.2 Enhanced therapeutic index (toxic dose/therapeutic dose) can be achieved if the target (*black square*) is unique to the disease cells (target-driven therapeutic index) or if contextual differences lead to an increased requirement for the target in the disease cells relative to normal cells (context-driven therapeutic index). The contextual differences can include cell-intrinsic differences, such as genetic (vertical bars) and epigenetic (horizontal bars) differences, or cell-extrinsic differences, such as microenvironmental differences.

mutation is a specific example of a cell-intrinsic, context-dependent, basis for selectivity. The synthetic lethal paradigm can be extrapolated to any situation in which the requirement for a target in a cancer cell has been altered by virtue of one or more of the mutations that mark that cell, whether causal or epiphenomenal. Oncogene addiction is another example of cell-intrinsic changes leading to selectivity. The importance of microenvironmental cues is illustrated well in the process of anoikis, whereby normal epithelial cells undergo apoptosis after detachment from their underlying stroma and matrix (Frisch and Screaton, 2001). As cancer cells must compensate for the loss of mitogenic and survival signals that accompany ectopic growth, their dependence on particular molecular pathways may differ from that of their normal, orthotopically growing, counterparts.

In considering the potentially central importance of context, it is also important to bear in mind that germline disruption in mice (i.e., conventional knockouts) has often been used to predict the toxicity of small molecule antagonists directed against their protein products. Such studies, however, reveal the requirement for specific genes/proteins during development and do not necessarily speak to their requirement in an intact adult. Accordingly, such studies can grossly overestimate the potential toxicity of a pharmacological agent. For example, Gleevec is remarkably well tolerated in adult humans despite the importance of c-Abl during murine development (Tybulewicz et al., 1991).

3.8 Summary

Recent advances in molecular oncology are laying the foundation for the development of rational anticancer drugs. The success of Gleevec against CML and, perhaps more important, GIST supports the idea that cancer genetics can be exploited to select drug targets. Proof of concept experiments in the laboratory indicate that correction of a single genetic abnormality can translate into a therapeutic outcome in cancer, although it is likely that combination therapy will be needed in the clinic to minimize the emergence of drug-resistant clones. The genes that are recurrently mutated in a particular cancer earmark the molecular pathways that were essential for the evolution of that cancer in vivo. In some cases, cancer cells become addicted to activation of such pathways. Knowledge that cancer genes define critical pathways and networks can be used to expand the search for drugable cancer targets beyond their protein products. The mutations that have occurred in cancer cells, whether causal or epiphenomenal, place targets in a different context than their normal counterparts and hence may serve as the basis for selective killing of cancer cells relative to normal cells. Targeting early, causal, genetic alterations in cancer cells is likely to minimize the emergence of resistance as well as maximize the likelihood of unmasking synthetic interactions with subsequent genetic and epigenetic alterations that arose during tumor progression. Tools such as siRNA and chemical genetics should aid in the prospective identification of novel targets that, when inhibited, selectively kill cancer cells due to such contextual differences. The next 10–20 years should witness a fundamental shift in which the treatment of cancer is no longer largely empiric but instead informed by the genetic abnormalities that underlie the transformed phenotype.

References

Abbas-Terki, T., Blanco-Bose, W., Deglon, N., et al. Lentiviral-mediated RNA interference. *Hum. Gene Ther.* **13,** 2197–2201 (2002).

Andersson, J., Sjogren, H., Meis-Kindblom, J., et al. The complexity of KIT gene mutations and chromosome rearrangements and their clinical correlation in gastrointestinal stromal (pacemaker cell) tumors. *Am. J. Pathol.* **160,** 15–22 (2002).

Baker, S. J., Markowitz, S., Fearon, E., Willson, B., and Vogelstein B. Suppression of human colorectal carcinoma cell growth by wild-type p53. *Science* **249,** 912–915 (1990).

Brose, M. S., Volpe, P., Feldman, M., et al. BRAF and RAS mutations in human lung cancer and melanoma. *Cancer Res.* **62,** 6997–7000 (2002).

Brummelkamp, T. R., Bernards, R., and Agami, R. A system for stable expression of short interfering RNAs in mammalian cells. *Science* **296,** 550–553 (2002).

Chin, L., Tam, A., Pomerantz, J., et al. Essential role for oncogenic Ras in tumour maintenance. *Nature* **400,** 468–472 (1999).

Davies, H., Bignell, G. R., Cox, C., et al. Mutations of the BRAF gene in human cancer. *Nature* **417,** 949–954 (2002).

Druker, B., and Lydon, N. Lessons learned from the development of an abl tyrosine kinase inhibitor for chronic myelogenous leukemia. *J. Clin. Invest.* **105,** 3–7 (2000).

Druker, B., Sawyers, C., Kantarjian, H., et al. Activity of a specific inhibitor of the BCR-ABL tyrosine kinase in the blast crisis of chronic myeloid leukemia and acute lymphoblastic leukemia with the Philadelphia chromosome. *N. Engl. J. Med.* **344,** 1038–1042 (2001b).

Druker, B., Talpaz, M., Resta, D., et al. Efficacy and safety of a specific inhibitor of the BCR-ABL tyrosine kinase in chronic myeloid leukemia. *N. Engl. J. Med.* **344,** 1031–1037 (2001a).

Druker, B., Tamura, S., Buchdunger, E., et al. Effects of a selective inhibitor of the Abl tyrosine kinase on the growth of Bcr-Abl positive cells. *Nature Med.* **2,** 561–566 (1996).

Elbashir, S., Harborth, J., Lendeckel, W., et al. Duplexes of 21-nucleotide RNAs mediate RNA interference in cultured mammalian cells. *Nature* **411,** 494–498 (2001).

Fabbro, D., Parkinson, D., and Matter, A. Protein tyrosine kinase inhibitors: New treatment modalities? *Curr. Opin. Pharmacol.* **2,** 374–381 (2002).

Felsher, D., and Bishop, J. Reversible tumorigenesis by MYC in hematopoietic lineages. *Mol. Cell* **4,** 199–207 (1999).

Frisch, S., and Screaton, R. Anoikis mechanisms. *Curr. Opin. Cell Biol.* **13,** 555–562 (2001).

Furnari, F., Lin, H., Huang, H., and Cavenee, W. Growth suppression of glioma cells by PTEN requires a functional phosphatase catalytic domain. *Proc. Natl. Acad. Sci. USA* **94,** 12479–12484 (1997).

Gorre, M., Mohammed, M., Ellwood, K., et al. Clinical resistance to STI-571 cancer therapy caused by BCR-ABL gene mutation or amplification. *Science* **293,** 876–880 (2001).

Hanahan, D., and Weinberg, R. The hallmarks of cancer. *Cell* **100,** 57–70 (2000).

Hartman, J. T., Garvik, B., and Hartwell, L. Principles for the buffering of genetic variation. *Science* **291,** 1001–1004 (2001).

Hartwell, L., Szankasi, P., Roberts, C., et al. Integrating genetic approaches into the discovery of anticancer drugs. *Science* **278,** 1064–1068 (1997).

Huang, H.-J. S., Yee, J.-K., Shew, J.-Y., et al. Suppression of the neoplastic phenotype by replacement of the RB gene in human cancer cells. *Science* **242,** 1563–1566 (1988).

Iliopoulos, O., Kibel, A., Gray, S., and Kaelin, W. G. Tumor suppression by the human von Hippel-Lindau gene product. *Nature Medicine* **1,** 822–826 (1995).

Jain, M., Arvanitis, C., Chu, K., et al. Sustained loss of a neoplastic phenotype by brief inactivation of MYC. *Science* **297,** 102–104 (2002).

Joensuu, H., Roberts, P., Sarlomo-Rikala, M., et al. Effect of the tyrosine kinase inhibitor STI571 in a patient with a metastatic gastrointestinal stromal tumor. *N. Engl. J. Med.* **344,** 1052–1056 (2001).

Kaelin, W. J. Choosing anticancer drug targets in the postgenomic era. *J. Clin. Invest.* **104,** 1503–1506 (1999).

Kamath, R. S., Fraser, A. G., Dong, Y., et al. Systematic functional analysis of the *Caenorhabditis elegans* genome using RNAi. *Nature* **421,** 231–237 (2003).

Kantarjian, H., Sawyers, C., Hochhaus, A., et al. Hematologic and cytogenetic responses to imatinib mesylate in chronic myelogenous leukemia. *N. Engl. J. Med.* **346,** 645–652 (2002).

Kinzler, K., and Vogelstein, B. Cancer-susceptibility genes: Gatekeepers and caretakers. *Nature* **386,** 761–763 (1997).

Kinzler, K., and Vogelstein, B. Lessons from hereditary colorectal cancer. *Cell* **87,** 159–170 (1996).

Lee, N. S., Dohjima, T., Bauer, G., et al. Expression of small interfering RNAs targeted against HIV-1 rev transcripts in human cells. *Nature Biotechnol.* **20,** 500–505 (2002).

McKeage, K., and Perry, C. Trastuzumab: A review of its use in the treatment of metastatic breast cancer overexpressing HER2. *Drugs* **62,** 209–243 (2002).

Mills, G., Lu, Y., and Kohn, E. Linking molecular therapeutics to molecular diagnostics: Inhibition of the FRAP/RAFT/TOR component of the PI3K pathway preferentially blocks PTEN mutant cells in vitro and in vivo. *Proc. Natl. Acad. Sci. USA* **98,** 10031–10033 (2001).

Miyagishi, M., and Taira, K. U6 promoter-driven siRNAs with four uridine 3′ overhangs efficiently suppress targeted gene expression in mammalian cells. *Nature Biotechnol.* **20,** 497–500 (2002).

Neshat, M., Mellinghoff, I., Tran, C., et al. Enhanced sensitivity of PTEN-deficient tumors to inhibition of FRAP/mTOR. *Proc. Natl. Acad. Sci. USA* **98,** 10314–10319 (2001).

Nghiem, P., Park, P., Kim, Y., et al. ATR inhibition selectively sensitizes G1 checkpoint-deficient cells to lethal premature chromatin condensation. *Proc. Natl. Acad. Sci. USA* **98,** 9092–9097 (2001).

Nishida, T., Hirota, S., Taniguchi, M., et al. Familial gastrointestinal stromal tumours with germline mutation of the KIT gene. *Nature Genet.* **19,** 323–324 (1998).

Paddison, P. J., Caudy, A. A., Bernstein, E., et al. Short hairpin RNAs (shRNAs) induce sequence-specific silencing in mammalian cells. *Genes Dev.* **16,** 948–958 (2002).

Podsypanina, K., Lee, R., Politis, C., et al. An inhibitor of mTOR reduces neoplasia and normalizes p70/S6 kinase activity in Pten+/− mice. *Proc. Natl. Acad. Sci. USA* **98,** 10320–10325 (2001).

Powell, S., DeFrank, J., Connell, P., et al. Differential sensitivity of p53(−) and p53(+) cells to caffeine-induced radiosensitization and override of G2 delay. *Cancer Res.* **55,** 1643–1648 (1995).

Reddy, A., and Kaelin, W. G. Using cancer genetics to guide the selection of anticancer drug targets. *Curr. Opin. Pharmacol.* **2,** 366–373 (2002).

Renan, M. How many mutations are required for tumorigenesis? *Mol. Carcinog.* **7,** 139–146 (1993).

Talpaz, M., Silver, R., Druker, B., et al. Imatinib induces durable hematologic and cytogenetic responses in patients with accelerated phase chronic myeloid leukemia: Results of a phase 2 study. *Blood* **99,** 1928–1937 (2002).

Torrance, C. J., Agrawal, V., Vogelstein, B., and Kinzler, K. W. Use of isogenic human cancer cells for high-throughput screening and drug discovery. *Nature Biotechnol.* **19,** 940–945 (2001).

Tybulewicz, V. L., Crawford, C. E., Jackson, P. K., et al. Neonatal lethality and lymphopenia in mice with a homozygous disruption of the c-*abl* proto-oncogene. *Cell* **65,** 1153–1163 (1991).

Van Etten, R. Pathogenesis and treatment of Ph+ leukemia: Recent insights from mouse models. *Curr. Opin. Hematol.* **8,** 224–230 (2001).

van Oosterom, A., Judson, I., Verweij, J., et al. Safety and efficacy of imatinib (STI571) in metastatic gastrointestinal stromal tumours: A phase I study. *Lancet* **358,** 1421–1423 (2001).

Weinstein, I. B. Addiction to oncogenes – The Achilles' heel of cancer. *Science* **297,** 63–64 (2002).

Yu, J. Y., DeRuiter, S. L., and Turner, D. L. RNA interference by expression of short-interfering RNAs and hairpin RNAs in mammalian cells. *Proc. Natl. Acad. Sci. USA* **99,** 6047–6052 (2002).

CHAPTER 4

RNA Interference in Mammals: Journey to the Center of Human Disease

Patrick J. Paddison and Gregory J. Hannon

Molecular Cancer Therapeutics: Strategies for Drug Discovery and Development, edited by George C. Prendergast
ISBN 0-471-43202-4

Historians of Western civilization often cite the arch in building design as a good indicator of technological advancement, since the construction of even small arches requires rudimentary knowledge of engineering and mathematics. In the biological sciences a similar measure of technological advancement is the ability to remove gene products from cells and model organisms; for in many cases it is only when a gene is removed, or at least has its activity affected, that its function can be properly assigned. While genetic manipulations for selected gene removal have become routine in any number of invertebrate systems, mammalian-based systems have lagged behind, being almost impervious to similar techniques. As a result, many basic questions regarding the function of molecular pathways have gone unanswered in mammals. For drug discovery, the deletion or modification of a gene can, in many cases, represent the ideal activity of a small molecule inhibitor. Thereby, our ability to model idealized drug targets and rational therapies through removal of cellular gene products has, to date, been limited in scope. This may be about to change, however; a new gene knock-down technology has appeared, timed perfectly with the completion of the human and mouse genomes, which should readily allow the functional exploration of mammalian genomes. This technology is based on a conserved biological response known as dsRNA-dependent, sequence-specific gene silencing or RNA interference (RNAi).

RNAi emerged out of the pioneering work of Fire (1998) in the nematode *Caenorhaditis elegans*. Attempting to use antisense RNA to knock down gene expression, they found synergistic effects on gene silencing when antisense and sense RNA strands where delivered together. Although at first RNAi seemed a peculiarity of nematodes, dsRNA-dependent gene silencing has since become one of the biggest surprises in the past decade of research in eukaryotic cells. The core machinery that underlies RNAi is conserved in virtually every experimental eukaryotic system (with the notable exception of the yeast *Saccharomyces cerevisiae*) and has been co-opted in most of them to trigger gene silencing.

With the advent of RNAi in mammals and the refinement of techniques to trigger gene silencing, we have reached a point at which any gene in the human or mouse genome can conceivably be targeted using small dsRNA gene silencing triggers – synthetic small interfering RNAs (siRNAs) or expressed short hairpin RNAs (shRNAs). In the next few years in the biomedical sciences, siRNAs and shRNAs will be employed to validate disease models in vitro in cell-based systems and in vivo in rodent and primate systems; to validate drug activities through the remove of suspected targets; to identify new drug candidates in genomewide, functional genomic screens; and to combat disease directly as therapeutic molecules in the clinic. In this review we provide an overview of the RNAi pathway and the extent to which the RNAi pathway can be co-opted in mammals to evoke gene silencing. While the field of RNAi in mammals in still in its infancy in regard to genomewide applications, we also discuss various possible screening strategies in mammals, with a special emphasis on drug discovery.

4.1 Mechanics of RNA Interference

The phenomenology of RNAi from early experiments in *C. elegans,* plants, and *Drosophila* suggested a homology-based silencing mechanism that somehow used dsRNA to seek and, in most cases, destroy cognate targets. In the past 4 years uncovering and characterizing many of the underlying components and biochemical determinants of RNAi in invertebrate systems has helped translate RNAi into a genetic tool in mammals (Hannon, 2002). At least two core components of RNAi pathway appear to be generally required for dsRNA dependent silencing phenomena: dicer and argonaute (Ago) gene family members. Dicer and dicer-related proteins sit atop the RNAi pathway in the first catalytic step that converts various forms of dsRNA into smaller, guide dsRNAs of 21–25 nt. Dicer-related genes harbor four conserved sequence motifs: a DExH/DEAH ATPase/RNA helicase domain, a PAZ domain (unique to RNAi genes), an RNAseIII dsRNA nuclease domain, and dsRNA binding domains. Argonaute proteins, which are components of the RNA-induced silencing complex (RISC), contain a PAZ domain and a carboxyl-terminal PIWI domain (Hannon, 2002).

The current model for RNAi begins with the conversion of the dsRNA silencing "trigger" into small RNAs (siRNAs) by dicer (Bernstein et al., 2001a). These small RNAs (~ 22–25 nt in size) become incorporated into a RISC, which uses the sequence of the siRNAs as a guide either to identify homologous mRNAs (Hammond et al., 2000; Nykanen et al., 2001; Tuschl et al., 1999; Zamore et al., 2000) or in some invertebrate systems, to identify similar regions in euchromatin (Fig. 4.1). Depending on the organism and the cellular context, different Ago-associated "effector" complexes trigger mRNA destruction (i.e., RISC), translational inhibition (Grishok et al.,

RNAi Trigger	dsRNA	RNA Species	Synthesis	Delivery
siRNA	19 nt		Chemical synthesis	Transfection
Diced dsRNA	19-27 nt		In vitro enzymatic cleavage	Transfection
Expressed siRNAs	19-23 nt	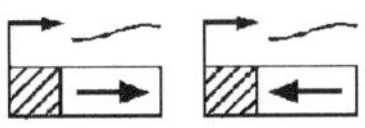	In vivo expression; RNA pol III promoter	Transfection/ viral
shRNA	19-29 nt		Chemical synthesis; in vitro transcript	Transfection
Expressed shRNAs	19-29 nt	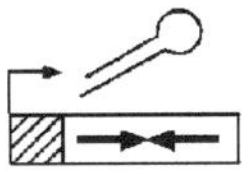	In vivo expression; RNA pol II and III promoter	Transfection/ viral

Figure 4.1 The basic mechanism of RNAi-mediated gene silencing in mammals. Dicer processes shRNAs and miRNA into ~21 nt guide RNAs, which are taken up by one or more of the RNAi effector complexes to target cognate RNA transcripts. siRNAs presumably bypass the requirement for dicer.

2001), or transcriptional gene silencing (Hall et al., 2002; Volpe et al., 2002; Zilberman et al., 2003). At present, it remains unclear how cells discriminate which dsRNA triggers elicit which responses.

In addition to dicer and Ago-related proteins, some invertebrate systems contain pathways that amplify and/or transport guide RNA sequences to other parts of the organism. In *C. elegans* and plants, amplification of the dsRNA signal is thought to initially be mediated by RNA-dependent RNA polymerases (RdRPs). An RNA degradation product (e.g., a guide RNA) may prime RdRPs along the mRNA template, resulting in the production of dsRNA homologous to sequences 5′ (i.e., upstream) of the initially targeted sequence (Sijen et al., 2001). When combined with transport, amplification results in a self-propagating silencing effect throughout the organism. *C. elegans* appears to stand alone among metazoans, however, in regard to the conservation of RdRPs, and thus amplification of RNAi. One possibility is that *C. elegans* acquired RdRPs through horizontal gene transfer, for example, from RNA viruses (C. Mello, personal communication).

Mammalian and *Drosophila* cells apparently lack any evidence of an amplification step (Scharwz et al., 2002) and, at least in cultured cells, any indication of transport of triggers of gene silencing (A. Caudy, et al., personal communication). While this lack of potency may seem a hindrance, it may well be a blessing; for the absence of amplification actually expands the potential utility of RNAi as a genetic tool. Without "transitive" silencing effects, gene silencing can potentially be carried out in an "allele" or "snp" dependent fashion and, in the very least, in an exon-specific manner. When tackling genomes full of multiple mRNA isoforms, exon-specific silencing may demonstrate the true power of RNAi as a genetic tool in mammals.

The RNAi pathway likely arose early during eukaryotic evolution as a mechanism of cell-based immunity direct against viral and genetic parasites. dsRNA viruses and mobile genetic elements with the potential to form dsRNA structures are ubiquitous in nature and can be subject to RNAi-dependent gene silencing in *C. elegans,* plants, *Drosophila,* yeast, and mammals (Hannon, 2002). In addition, elements of the RNAi pathway are also used for regulation of endogenous genes (e.g., during metazoan development), where endogenous, noncoding RNAs are processed and used to seek out targets (e.g., miRNAs).

Endogenously expressed small hairpin RNAs regulate gene expression through the RNAi pathway during *C. elegans* development (Grishok et al., 2001; Hannon, 2002; Hutvagner et al., 2001; Ketting et al., 2001; Knight and Bass, 2001; Reinhart et al., 2000). These small hairpin RNAs ($\sim$70 nt) are processed into a 21- to 22-nt mature form by dicer and then used to seek out mRNA targets of similar sequence (generally via imperfect base-pairing interactions). For the two prototypes of this family, *C. elegans* lin-4 and let-7, silencing occurs at the level of protein synthesis (Bernstein, 2001b). The first small hairpin RNAs were dubbed small temporal RNAs (stRNAs), owing to their role in developmental timing (Ha et al., 1996; Lee et al., 1993; Slack et al., 2000; Wightman et al., 1993). More recently, dozens of orphan hairpins have been identified in *C. elegans, Drosophila,* mouse, and humans, which are collectively referred to as microRNAs (miRNAs) (Lagos-Quintana et al.,

2001; Lau et al., 2001; Lee and Ambros, 2001; Mourelatos et al., 2002; Pasquinelli et al., 2000).

4.2 RNA Interference in Mammals

The first evidence that dsRNA could evoke gene silencing in mammals came from studies using long dsRNA in mouse oocytes, preimplantation embryos (Svoboda et al., 2000; Wianny and Zernicka-Goetz, 2000), and embryonal cell lines (Billy et al., 2001; Paddison et al., 2002a; Yang et al., 2001). In these contexts, cells lack the prominent antiviral responses found in most somatic cells. Such responses include double-stranded RNA-activated protein kinase (PKR) and RNAseL pathways, which are triggered by dsRNA > 30 bp and result in nonspecific translational repression and apoptosis (Baglioni and Nilsen, 1983; Gil and Esteban, 2000; Williams, 1997). These initial glimpses of gene silencing, combined with the strong conservation of key players in the RNAi pathway such as dicer and argonaute (Carmell et al., 2002), suggested that silencing phenomena might be available in somatic cell types if the nonspecific dsRNA responses could be circumvented. However, even when nonspecific dsRNA responses are removed from somatic cells, by either viral inhibitors or targeted disruption, long dsRNA still triggers a residual nonspecific repression of gene expression (Abraham et al., 1999; Paddison et al., 2002a).

Another way around these nonspecific dsRNA responses is to simply reduce the size of the dsRNA trigger of RNAi to < 30 nt to duck the size threshold of PKR and RNAseL. In the past 2 years, two short RNA structures have emerged, which provoke sequence specific gene silencing without activating antiviral responses. These are the siRNA and the shRNA. Both are modeled after biologically active structures in the RNAi pathway: dicer cleavage products and small temporal RNAs or miRNAs, respectively. The first published indication that small dsRNA could trigger RNAi in mammals came from Tuschl's group, which demonstrated that short RNA duplexes resembling the cleavage products of dicer could trigger sequence-specific silencing in mammalian cell lines (Elbashir et al., 2001). These siRNAs contain 21 nt of identity to a homologous mRNA target, 19 nt of dsRNA, and 3′ overhangs of 2 nt. siRNAs presumably bypass the requirement for dicer and enter the silencing pathway by incorporation into RISC complexes (Fig. 4.1). The use of siRNAs has been recently reviewed in detail (Elbashir et al., 2002; McManus and Sharp, 2002), and resources for the design and use of siRNAs are available online (www.mpibpc.gwdg.de/abteilungen/100/105/sirna.html).

As an alternative strategy, we and others have developed in vivo expression constructs for small dsRNA triggers in mammalian cells, which resemble endogenously expressed hairpin RNAs (Brummelkamp et al., 2002a; McManus et al., 2002; Paddison et al., 2002b; Paul et al., 2002; Sui et al., 2002; Yu et al., 2002; Zeng et al., 2002). We have dubbed these shRNAs since, unlike siRNAs, they have an optimal RNA duplex of 23–29 nt, contain a loop structure that joins both strands of the duplex, and require processing by dicer to gain admittance to the RNAi pathway (Fig. 4.1). Figure 4.1 shows the various strategies that have been used to generation and deliver siRNAs (Calegari et al., 2002; Caplen et al., 2001; Elbashir et al., 2001; Kawasaki et al., 2003; Myers et al., 2003; Yang et al., 2002) and shRNAs (Brummelkamp et al., 2002a; McManus et al., 2002; Paddison et al., 2002b; Paul et al., 2002; Sui et al., 2002; Yu et al., 2002; Zeng et al., 2002) in mammalian systems.

Most shRNA silencing strategies rely on RNA polymerase III (pol III) promoter to drive expression in vivo (either human or mouse U6-snRNA or human RNase P (H1) RNA promoters), though some have used RNA polymerase II (pol II) promoters (Paddison and Hannon, 2002). We have compared RNA pol III promoters, including H1, U6, and tRNA(Val) and RNA pol II promoters, and found that RNA pol III promoters (including mouse and human H1, U6, and tRNA) work similarly and are in general more effective than pol II promoters when expressing shRNAs. In regard to structural elements of the hairpins themselves, there is some in vitro biochemical evidence that suggests that RNAseIIIs (e.g., dicer) might have loop structure or sequence preferences (Lamontagne et al., 2003). However, in our laboratory, we have not seen this trend for RNAi in mammals. Comparing multiple shRNAs containing different lengths of dsRNA stems (e.g., 19–29 nt) and loop structures (e.g., 4–14 nt, miRNA styled), we have found that 29-nt hairpins containing a simple loop structure are most effective.

In addition, the incorporation of the U6 snRNA 27-nt "leader" sequence appears to increase the potency of hairpins with suboptimal targeting efficiency (Fig. 4.2). The U6 snRNA leader transcript is a small RNA hairpin that directs the addition of a γ-monomethyl-phosphate guanosine cap (Singh and Reddy, 1989). No work has yet suggested the mechanism that leads to improved shRNA efficacy, but increased stability, more effective transport, or a different localization are possibilities. For 29-nt hairpins that work well, the leader sequence is neutral. Paul et al. (2002) first suggested the use of the U6 leader sequence for 19-nt hairpins. These results suggest that 29-nt hairpins containing the U6 leader sequence and a simple loop structure will be the most effective. Somewhat surprisingly, shRNAs modeled directly after miRNAs, which contain bulges and unique loop structures in general, are less effective than simpler hairpins (Paddison et al., 2002b). This scenario might suggest that miRNAs are not necessarily designed for optimal efficacy or that they require specific cellular contexts or cooperation with additional RNA binding proteins (e.g., fragile X mental retardation protein (FMRP)) to work well. Thus there may be no "magic" variables contained in miRNAs.

The main limitation of siRNAs and transiently transfected shRNA vectors is the inability to evoke stable or inducible gene silencing in mammals. In mammalian cell systems, transient transfection of RNAi triggers (e.g., long dsRNA, siRNAs, or shRNAs) results in a transient effect, lasting 2–7 days due to lack of prominent amplification steps available in other systems. Thus siRNAs by definition have half-lives and are diluted by cell division and turnover of the RISC complex. However, a number of well-characterized stable expression technologies have now been used in combination with shRNA expression to evoke stable gene silencing in mammals both in vitro and in vivo. Among recent reports, stable RNAi has been demonstrated using random plasmid integration (Brummelkamp et al., 2002a; Carmell et al., 2003;

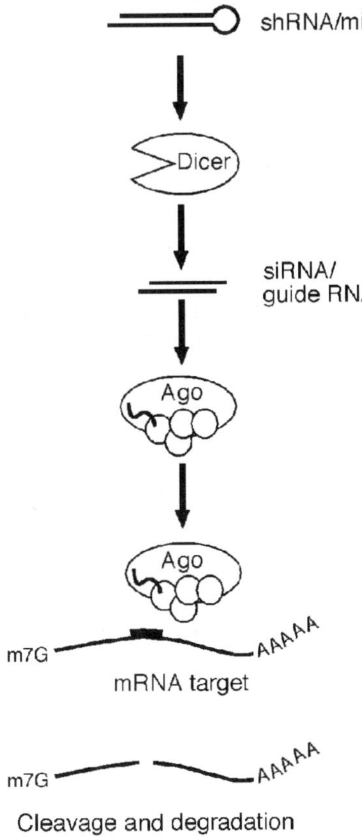

Figure 4.2 An ideal case for using RNAi to map drug activity response for a hypothetic molecular pathway.

Paddison et al., 2002b), episomal plasmid maintenance (Miyagishi and Taira, 2002), and retroviral delivery (Barton and Medzhitov, 2002; Blummelkamp et al., 2002b; Devroe and Silver 2002; Hemann et al., 2003; Paddison and Hannon, 2002; Qin et al., 2003; Tiscornia et al., 2003). In particular, delivery strategies involving retroviruses, adenovirus, or adeno-associated virus are attractive for exploring RNAi in primary cells, which are particularly difficult to manipulate in vitro.

4.3 Journey to the Center of Human Disease

From the standpoint of molecular medicine, RNAi have at least three immediate applications:

- Validation of activities for drugs currently in development.
- Identification of new drug targets.
- Investigation of the underlying biology of diseases for which cellular or rodent models exist.

More long-term developments will likely include therapeutic applications of RNAi triggers, which may effectively replace antisense oligonucleotides currently being used in clinical trials. Only time will tell whether RNAi represents a miracle tool for disease research or merely a step beyond antisense technologies.

Currently, the most common use of RNAi in mammalian cell biology is to demonstrate that a gene is required for a particular molecular process and/or pathway of interest. RNAi experiments enable researchers to assert more rigorous claims as to whether a gene is "necessary and sufficient" for a process by the simple fact that researchers can reproducibly remove gene products from cells product and show that a cellular process and/or phenotype does not occur in the absence of the gene product. For the biotechnology and pharmaceutical industries, RNAi should similarly enable experimenters to determine if removing a cellular activity affected by a drug is compound-mimetic (Fig. 4.2). For example, drugs that have penetrant and specific cellular phenotypes, such as hydroxyurea or aphidicolin, could readily be modeled with RNAi by simply knocking down their cellular targets, ribonucleotide reductase, and DNA polymerase α, respectively. A more realistic case of validating drug activity response would be a compound that was developed to inhibit a particular enzymatic activity in vitro. Using RNAi it could, in theory, be demonstrated that the removal of that enzyme from cells in vitro or in vivo would have the intended effect. Researchers would start by picking a transient or stable RNAi strategy (Table 4.1), testing individual silencing triggers (i.e., siRNAs or shRNAs) for knock-down efficacy (i.e., in a cell type that expresses the gene of interest), and using an appropriate biological assay in cells knocked down for a particular gene (Fig. 4.2).

For many compounds, however, the total picture of a drug's activity may not be as simple as removing one gene product. Many small molecule inhibitors affect multiple pathways, are metabolized into multiple forms, and/or inhibit a broad range of similar gene products. Moreover, not all drug activities can be modeled genetically. Drugs that malign a protein's function by creating a lethal by-product (e.g., camptothecin, Ganciclovir) would fit into this category. In the former case, it may be possible to knock down multiple suspected genes using a combination of siRNAs or shRNAs in parallel to mimic the total effect of the compound. In the latter case, however, the best RNAi or any genetic approach could offer would be enhancer screens to identify secondary drug targets, which might act synergistically with the primary drug to kill diseased cells (i.e., enhancement) or prevent diseased cells from dying (i.e., suppression).

In addition to validating the biological activities of various classes of compounds already in existence, RNAi can be used to identify new putative drug targets via genomewide screens for desirable disease-related phenotypes. For cancer-related research, one area of interest is the search for cancer lethal genes or genes, which, when specifically removed from a transformed cell, result in lethality. Through the use of genomewide RNAi libraries, similar to those used for screens in *C. elegans,* each gene and mRNA isoform can be interrogated for specific lethality in tumor cell lines or in vitro transformed cells. The hope is that such screens will reveal new drug targets or new biology that will make applications of drugs now is use more effective. There are no doubt many theoretical examples of the application of RNAi to human disease. The reality is that such applications (i.e., genetic screens) will require the construction of RNAi libraries for mammalian genomes, just as it has for *C. elegans*. Our group has concentrated constructing human and mouse RNAi libraries using shRNAs rather than siRNAs, largely due to cost (siRNAs = \$100–300/duplex) but also for the added capacity of creating stable and inducible silencing constructs. Moreover, since shRNA constructs are stored as bacterial archives, they can readily be broken up into functional sets (i.e., kinases, G. protein coupled receptors (GPCRs), checkpoints, proteolysis) or collectively pooled and used in forward genetic screens (see below). In the next few years, RNAi libraries will be used in the first genomewide screens in mammals in both in vitro and in vivo formats.

This brings us the perhaps the biggest question in regard to RNAi screens in mammalian systems: whether to use forward or reverse genetic approaches. Genomewide RNAi screens in *C. elegans,* for example, have been carried out in a reverse genetic fashion in which only one to a few different RNAi constructs are introduced to worms contained within a single well of a multiwell plate. Thus affected worms and their progeny can be screened for phenotypic differences caused by the removal of a single gene. In mammalian systems, so far, the most successful genetic screens have involved introducing "gain of function" genetic lesions into cells. Such screens generally consist of expressing, in mass, cDNAs or genomic fragments in receipt cell populations and screening for a positively selectable phenotype (Deiss and Kimchi, 1991; Gudkov and Roninson, 1997; Maestro et al., 1999; Wong et al., 1994). The best example of this type of approach was perhaps among the first, with the cloning of the *ras* oncogene from genomic libraries in rodent cells (Goldfarb et al., 1982; Shih and Weinberg, 1982).

For RNAi-based approaches in mammals, randomized libraries of RNAi constructs can be used in a similar fashion to find genes that might act as tumor suppressors, inhibitors of cell cycle progression, inducers of cell death and/or senescence, etc., basically any gene or pathway whose removal allows cells to proliferate in the screening context (Fig. 4.3). Such screening scenarios would allow for positive selection for clonal cells harboring a single shRNA, while the rest of the population would be blocked from growing by the constraints of the screening assay (e.g., escape from growth arrest). The major advantages of these forward genetic approaches, when applicable, are that the library starting material will have better and more normalized representation for each gene in the genome when compared to random cDNA libraries made from cellular mRNA, and that the screening process selects for functional RNAi triggers. The major drawback is that this approach will apply to only a limited number of pathways and genes that can produce appropriate arrest–growth phenotypes.

More refined genetic systems allow for genetic screens in pathways that might otherwise be neutral for cell growth. For example, forward genetic screening strategies in yeast involve imposing genetic schemes based on the retention of a plasmid bearing a gene of interest such that it complements a genomic mutation in the same gene. Inducing random genomic mutations

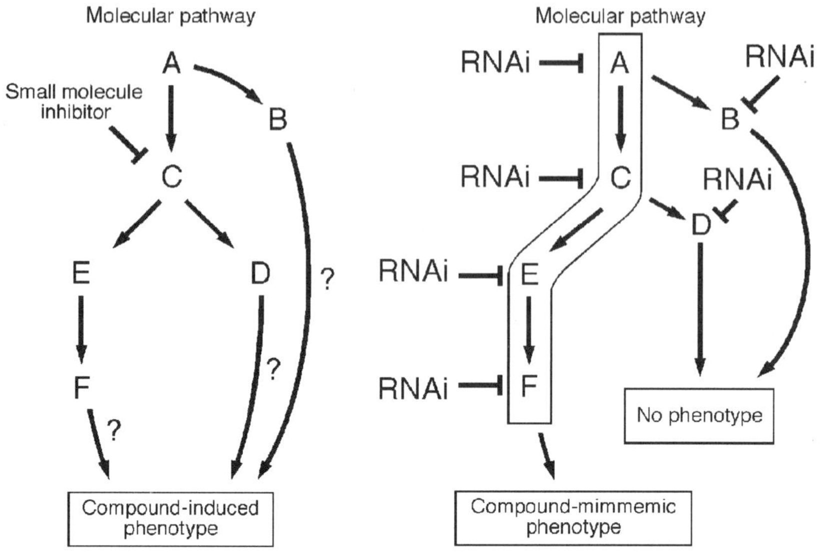

Figure 4.3 Possible forward genetic approaches using genomewide RNAi libraries in mammalian cells.

in such a population of cells can give rise to secondary mutations that genetically interact with the primary lesion and cause an absolute requirement for retention of the complementing plasmid. With the appropriate phenotypic markers, these genetic interactions can be scored as clonal colonies that have near 100% retention of the complementing plasmids. In mammals, similar screens are possible in theory, as many cell lines maintain good viability in colony formation assays, and episomal vectors are available with $< 100\%$ mitotic transmission (allowing for both retention and loss) along with phenotypic and drug selection markers (e.g., green fluorescent protein, 6-thioguanine (hprt−), HATr (hprt+)). In practice though, mammalian cells are more problematic for clonal outgrowth, as doubling time are much longer, drug selections tend to be more variable, and "replica printing" putative positives on large scales would be difficult.

It is conceivable, however, that for certain forward genetic schemes a cell-based technology can be used in place of the Petri dish, making forward genetic screens for neutral or even deleterious phenotypes workable. Fluorescence-activated cell sorting (FACS) represents one such cell-based technology. Using FACS, random populations of even live cells can be "gated" based on the presence or absence of a fluorescent cellular reporter using. In such screens, cells could harbor a green fluorescent protein (GFP) reporter vector or be stained with a fluorescent dye or antibody (e.g., a marker of differentiation of a stem cell). Upon introduction of random pools of RNAi constructs (e.g., retroviral transduction of shRNAs), cells can be collected, sorted, and putative positive RNAi constructs could be identified and rescreened (Fig. 4.3). A FACS-based approach would require that sorting is highly efficient or that the shRNA pool sizes are sufficiently small to accommodate suboptimal gating efficiency.

Other technologies might enable screening for particular phenotypes directly on tissue culture dishes. For example, the ability to PCR amplify shRNAs from single cells may be achievable through the use of ultra-processive DNA polymerases for an initial linear amplification of the entire genome or circularized region (e.g., bacteriophage phi29 DNA polymerase) (Dean et al., 2001). In this case, screens could consist of scanning populations of fixed or live cells directly on the growth plate (e.g., using high-content screening methodologies or a few graduate students and a microscope), then picking cells and recovering shRNAs through PCR and sequence analysis (Fig. 4.3). A simpler approach might involve using pools of 96 or 384 RNAi constructs, associating a phenotype of interest with a pool, and rescreening using an ever smaller pool until the responsible shRNA is found. Regardless of whether such approaches are feasible, the point to be made here is that the combination of molecular and cell-based technologies with forward RNAi genetics may ultimately win the day over well-to-well, reverse genetic approaches, since forward genetic approaches will be both more cost-effective and manageable for individual researchers.

The alternative to the forward genetic approaches is a well-to-well approach, in which individual RNAi constructs are arrayed in single wells of 96- or 384-well plates (Fig. 4.4). The major advantage of a well-to-well approach is that neutral or negatively selected phenotypes (e.g., apoptosis, growth arrest) can be scored in each well for single and multiple gene targeting events, ensuring that each construct is scored independently and, as a result, possibly capturing subtler phenotypes than those captured in forward genetic schemes. The major down-sides of well-to-well approaches are the overall costs involved in the delivery of constructs (i.e., transfection reagents) and reporter assay reagents, not to mention the use of robotic-assisted work stations and the limitations of using 96- and 384-well plate assays.

In *C. elegans,* genomewide and chromosome-wide RNAi screens have probed phenotypes ranging from genome instability (Pothof et al., 2003) to fat regulation (Ashrafi et al., 2003) to longevity (Lee et al., 2003). Similar RNAi screens have now been carried out in cultured *Drosophila* cells (Lum et al., 2003) and are currently under way in plants (D. Baulcombe, personal communication; Waterhouse and Helliwell, 2003). The applications of

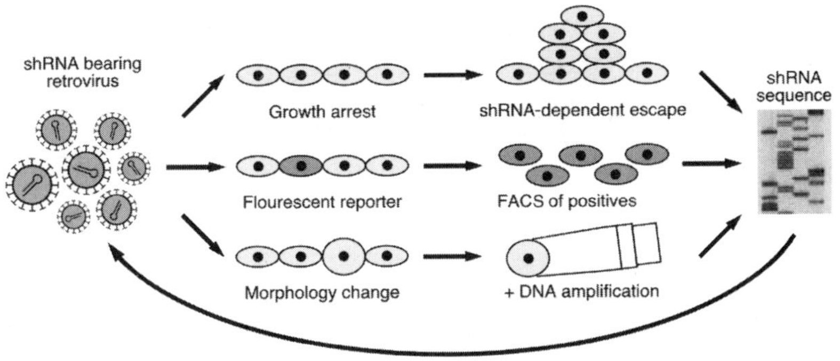

Figure 4.4 RNAi reverse genetics using in vitro cultured mammalian cells.

genomewide RNAi libraries in mammals will likely be as varied as those seen in invertebrate systems. Much of the initial work in mammals will likely explore many of the biological concepts derived from model systems, for example, cell cycle progression, programmed cell death, synthetic lethality (Paddison and Hannon, 2002). However, the crowning achievement of RNAi in mammals may be the identification and validation of putative therapeutic targets in cell culture and in vivo rodent models.

4.4 Using RNA Interference in Animal Models for Human Disease

The ability to trigger RNAi in somatic cells using expressed shRNAs immediately raised the possibility that these RNAi constructs could be used in animals as dominant transgene suppressors of a target gene. To this end, several groups, including our own, have demonstrated shRNA mediated gene silencing in transgenic mice (Carmell et al., 2003; Rubinson et al., 2003), in transplanted mouse hematopoetic stem cells (Hemann et al., 2003; Qin et al., 2003), and in the adult mouse liver (McCaffrey et al., 2002; Song et al., 2003) (Fig. 4.5).

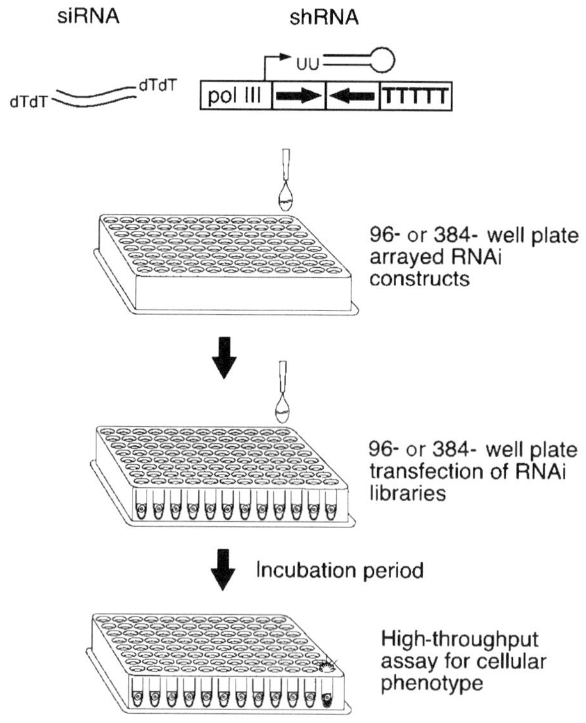

Figure 4.5 In vivo applications of siRNAs and shRNAs in mammals?

Of particular interest in rodent models is the ability to create "epi-allelic" series using RNAi constructs with different silencing efficacies against a gene of interest. Other genetic systems (e.g., budding yeast) are defined by the ability to create loss of function mutations that are either similar or lesser in effect than a genetic null. Obviously, if a gene of interest is essential to an organism the creation of hypomorphic allele is one of the few options available. The fact that not all RNAi triggers work effectively suddenly becomes a boon for this technology, since triggers with different efficacies could, in theory, be made into "allelic" series. With this in mind, Hemann et al. (2003) demonstrated that shRNAs of different knockdown efficacies targeting mouse p53, initially gauged by western blot analysis, can translate into highly reproducible biological phenotypes of corresponding penetrance. In this case, the penetrance of loss of p53 function in vitro, gauged by the level of stimulation of colony formation in mouse embryo fibroblasts, correlated in vivo with the on-set time of Eu-*myc*-driven mouse lymphomas. While this approach has yet to be tried on essential genes, these results suggest that as long as some degree of loss of a gene's function is tolerated that RNAi can be used in this manner.

For the creation of RNAi transgenic mice, so far two approaches have succeeded. Carmell et al. (2003) used random integration of an shRNA construct against Neil1, a putative DNA glycosylase, and tested shRNA-resistant clones for the ability to silencing the gene via RT-PCR, picking the best clones for injections into blastocytes. Rubison et al. (2003) infected embryonic stem (ES) cells with lentiviruses bearing shRNAs constructs and were also successful generating transgenic knock-down mice from transduced ES cells. Both techniques suggest that mammalian development is compatible with using RNAi for gene silencing and that the RNAi construct can essentially be treated as any other transgene (Fig. 4.6).

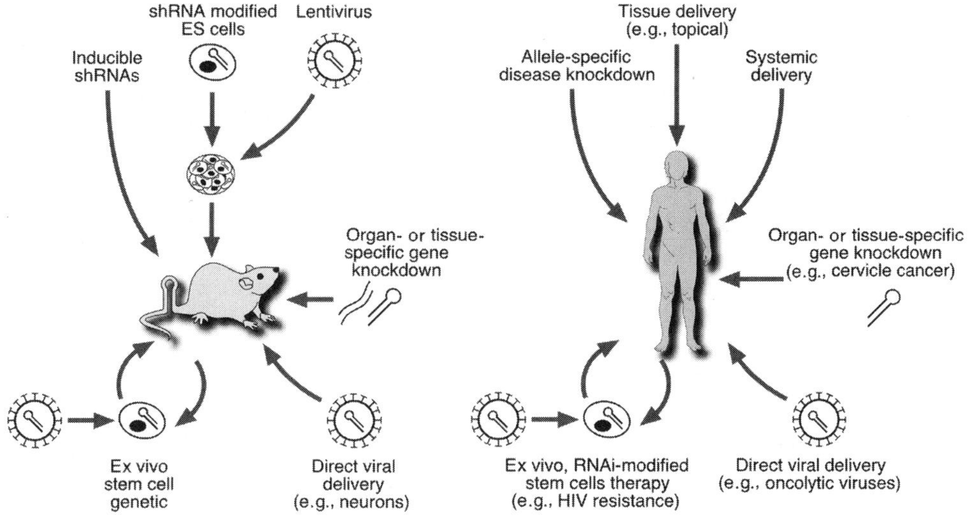

Figure 4.6 In vivo applications of RNAi in rodents models of human disease and potential RNAi-medicated therapies in the clinic.

As for refinements of RNAi in rodents, as with expression of any transgene, copy number and expression level of shRNAs are likely to be of key importance in the reproducibility and penetrance of RNAi generated rodent phenotypes. For example, one of the p53 epi-alleles from Hemann et al. (2003) was generated by simply moving the shRNA cassette to a different portion of a retroviral vector (e.g., long terminal repeat (LTR)), which likely affected the expression levels of the hairpin. The choice of promoter and expression context may thus be key in determining the optimal RNAi strategy for in vivo experiments. However, many unknowns still remain for RNAi in rodents. Are all tissues susceptible to RNAi? Will the effect be uniform in different tissues, developmental compartments, etc? Does co-opting the RNAi pathway effect pathways that are normally regulated by the RNAi machinery? Only with more experience will we be able to answer some of these answers.

Regardless of these lingering questions, the overall point is to be made is that RNAi appears to be remarkably compatible with most gene delivery techniques in mammals. Given the breadth of cell types in which RNAi is available in human and mouse cells, it is likely that absence of RNAi in mammalian tissues will be the exceptional case rather than the norm. As such, the applications for RNAi will likely be as varied and creative as those currently used for expression of transgenes in vivo (Fig. 4.6).

4.5 RNA Interference in the Clinic

It should not be overlooked that small dsRNA triggers of gene silencing might themselves be attractive as small molecule inhibitors of gene activity for the treatment of certain human diseases. Both siRNA and shRNAs could, in theory, occupy special niches in the clinic for genetic targets that are considered undrugable or that require allele-specific or exon-specific targeting events (Blummelkamp et al., 2002b). As with current antisense therapies, the overall hindrance is delivery and uptake of the RNAi trigger itself. In *C. elegans,* due to the efficient amplification and transport of dsRNA, worms can simply be fed bacteria-harboring plasmids that express dsRNA to obtain organismwide gene silencing (Timmons and Fire, 1998). This surprising method of delivery is unlikely to be available for use in humans, however, given the lack of evidence of transport and amplification of dsRNA. However, there are examples of certain strains of bacteria that might be employed for the cytosolic translocation of either dsRNA or shRNA vectors in vivo (Dietrich et al., 1998; Krusch et al., 2002). Most antisense oligo delivery strategies use various combinations of cationic lipids, cell target ligands and/or translocating peptides to help specify cellular addresses and enable uptake (Opalinska and Gewirtz, 2002). Certain tissues seem to be particular promiscuous at uptaking even naked nucleic acids. In mouse models, naked siRNAs and shRNAs have been efficiently delivered to the liver by tail vein injection (McCaffery et al., 2002; Song et al., 2003).

For certain applications the techniques seem to be already in place for RNAi-mediated therapies. For example, it has been suggested from the study of HIV-resistant populations that removal of the CCR5 and CXCR4 co-receptors may confer resistance to HIV infection (Doms and Trono, 2000). The use of self-inactivating retroviruses expressing shRNAs targeting these receptors could in theory cure this disease, at least during the early to middle stages when stromal support cells are not ravaged, if shRNAs were incorporated into hematopoetic stem cells ex vivo and then reintroduced into patients.

RNAi could also serve to target HIV directly where siRNAs or shRNAs are used to target viral transcripts to reduce viral loads. Such strategies have been demonstrated in vitro for inhibiting HIV replication (Jacque et al., 2002; Novina et al., 2002) and several other human viruses, including hepatitis C (Kapadia et al., 2003; Randall et al., 2003), rotavirus (Dector et al., 2002), γ-herpes virus (Jia and Sun, 2003), and influenza (Ge et al., 2003).

Another example for RNAi clinical intervention is the treatment of cervical cancer. The intiating event for cervical cancer is the genomic integration of portions of human papilloma viral genome coding for the E6 and E7 genes, which act to down regulate tumor suppressors p53 and Rb, respectively (Galloway and McDougall, 1989; Helt and Galloway, 2003). Targeting E6 and E7 in the early stages of the disease may help prevent further progression, obviating the need for surgery. Therapies could be tailored to individuals by amplifying and sequencing regions of E6 and E7 directly from Pap smears and designing the appropriate siRNA or shRNAs for topical treatments.

Regardless of whether RNAi constructs will replace the antisense strategies currently in clinical trials, it is likely that RNAi will find its way into the clinic in some capacity.

4.6 Summary

The use of RNAi as a genetic tool has already had a major effect in invertebrate systems like *C. elegans* and *Drosophila*. In the next few years, the first genomewide high-throughput screens in mammals using RNAi will be carried out, along with attempts at using dsRNA triggers of gene silencing to treat certain human diseases. There will no doubt be both notable successes and notable failures as we attempt to apply this genetic tool to various biological problems for the first time in academia and industry. At the very least, with the introduction of RNAi, perhaps mammalian systems will final gain admittance to the pantheon of model genetic systems.

References

Abraham, N., Stojdl, D. F., Duncan, P. I., et al. Characterization of transgenic mice with targeted disruption of the catalytic domain of the double-stranded RNA-dependent protein kinase, PKR. *J. Biol. Chem.* **274,** 5953–5962 (1999).

Ashrafi, K., Chang, F. Y., Watts, J. L., et al. Genome-wide RNAi analysis of *Caenorhabditis elegans* fat regulatory genes. *Nature* **421,** 268–272 (2002).

Baglioni, C., and Nilsen, T. W. Mechanisms of antiviral action of interferon. *Interferon* **5,** 23–42 (1983).

Bargmann C. I. High-throughput reverse genetics: RNAi screens in Caenorhabditis elegans. *Genome Biol.* **2,** REVIEWS1005 (2001).

Barton, G. M., and Medzhitov, R. Retroviral delivery of small interfering RNA into primary cells. *Proc. Natl. Acad. Sci. USA* **99,** 14943–14945 (2002).

Bernstein, E., Caudy A. A., Hammond, S. M., and Hannon, G. J. Role for a bidentate ribonuclease in the initiation step of RNA interference. *Nature* **409,** 363–366 (2001a).

Bernstein, E., Denli, A. M., and Hannon, G. J. The rest is silence. *RNA* **7,** 1509–1521 (2001b).

Billy, E., Brondani, V., Zhang, H., et al. Specific interference with gene expression induced by long, double-stranded RNA in mouse embryonal teratocarcinoma cell lines. *Proc. Natl. Acad. Sci. USA* **98,** 14428–14433 (2001).

Brummelkamp, T. R., Bernards, R., and Agami, R. Stable suppression of tumorigenicity by virus-mediated RNA interference. *Cancer Cell* **2,** 243–247 (2002b).

Brummelkamp, T. R., Bernards, R., and Agami, R. A system for stable expression of short interfering RNAs in mammalian cells. *Science* **296,** 550–553 (2002a).

Calegari, F., Haubensak, W., Yang, D., et al. Tissue-specific RNA interference in postimplantation mouse embryos with endoribonuclease-prepared short interfering RNA. *Proc. Natl. Acad. Sci. USA* **99,** 14236–14240 (2002).

Caplen, N. J., Parrish, S., Imani, F., et al. Specific inhibition of gene expression by small double-stranded RNAs in invertebrate and vertebrate systems. *Proc. Natl. Acad. Sci. USA* **98,** 9742–9747 (2001).

Carmell, M. A., Xuan, Z., Zhang, M. Q., and Hannon, G. J. The argonaute family: Tentacles that reach into RNAi, developmental control, stem cell maintenance, and tumorigenesis. *Genes Dev.* **16,** 2733–2742 (2002).

Carmell, M. A., Zhang, L., Conklin, D. S., et al. Germline transmission of RNAi in mice. *Nature Struct Biol.* **10,** 91–92 (2003).

Dean, F. B., Nelson, J. R., Giesler, T. L., and Lasken, R. S. Rapid amplification of plasmid and phage DNA using phi 29 DNA polymerase and multiply-primed rolling circle amplification. *Genome Res.* **11,** 1095–1099 (2001).

Dector, M. A., Romero, P., Lopez, S., and Arias, C. F. Rotavirus gene silencing by small interfering RNAs. *EMBO J.* **3,** 1175–1180 (2002).

Deiss, L. P., and Kimchi, A. A genetic tool used to identify thioredoxin as a mediator of a growth inhibitory signal. *Science* **252,** 117–120 (1991).

Devroe, E., and Silver, P. A. Retrovirus-delivered siRNA. *BMC Biotechnol.* **2,** 15 (2002).

Dietrich, G., Bubert, A., Gentschev, I., et al. Delivery of antigen-encoding plasmid DNA into the cytosol of macrophages by attenuated suicide Listeria monocytogenes. *Nature Biotechnol.* **16,** 181–185 (1998).

Doms, R. W., and Trono, D. The plasma membrane as a combat zone in the HIV battlefield. *Genes Dev.* **14,** 2677–2688 (2000).

Edgar, R., and Wood, W. Morphogenesis of bacteriophage T4 in extracts of mutant-infected cells. *Proc. Natl. Acad. Sci. USA* **55,** 498–505 (1966).

Elbashir, S. M., Harborth, J., Lendeckel, W., et al. Duplexes of 21-nucleotide RNAs mediate RNA interference in cultured mammalian cells. *Nature* **411,** 494–498 (2001).

Elbashir, S. M., Harborth, J., Weber, K., and Tuschl, T. Analysis of gene function in somatic mammalian cells using small interfering RNAs. *Methods* **26,** 199–213 (2002).

Fire, A., Xu, S., Montgomery, M. K., et al. Potent and specific genetic interference by double-stranded RNA in *Caenorhabditis elegans*. *Nature* **391,** 806–811 (1998).

Galloway, D. A., and McDougall, J. K. Human papillomaviruses and carcinomas. *Adv. Virus Res.* **37,** 125–171 (1989).

Ge, Q., McManus, M. T., Nguyen, T., et al. RNA interference of influenza virus production by directly targeting mRNA for degradation and indirectly inhibiting all viral RNA transcription. *Proc. Natl. Acad. Sci. USA* **100,** 2718–2723 (2003).

Gil, J., and Esteban, M. Induction of apoptosis by the dsRNA-dependent protein kinase (PKR): Mechanism of action. *Apoptosis* **5,** 107–114 (2000).

Goldfarb, M., Shimizu, K., Perucho, M., and Wigler, M. Isolation and preliminary characterization of a human transforming gene from T24 bladder carcinoma cells. *Nature* **296,** 404–409 (1982).

Grishok, A., Pasquinelli, A. E., Conte, D., et al. Genes and mechanisms related to RNA interference regulate expression of the small temporal RNAs that control *C. elegans* developmental timing. *Cell* **106,** 23–34 (2001).

Gudkov, A. V., and Roninson, I. B. Isolation of genetic suppressor elements (GSEs) from random fragment cDNA libraries in retroviral vectors. *Methods Mol. Biol.* **69,** 221–240 (1997).

Ha, I., Wightman, B., and Ruvkun, G. A bulged lin-4/lin-14 RNA duplex is sufficient for *Caenorhabditis elegans* lin-14 temporal gradient formation. *Genes Dev.* **10,** 3041–3050 (1996).

Hall, I. M., Shankaranarayana, G. D., Noma, K., et al. Establishment and maintenance of a heterochromatin domain. *Science* **297,** 2232–2237 (2002).

Hammond, S. M., Bernstein, E., Beach, D., and Hannon, G. J. An RNA-directed nuclease mediates post-transcriptional gene silencing in *Drosophila* cells. *Nature* **404,** 293–296 (2000).

Hannon, G. J. RNA interference. *Nature* **418,** 244–251 (2002).

Helt, A. M., and Galloway, D. A. Mechanisms by which DNA tumor virus oncoproteins target the Rb family of pocket proteins. *Carcinogenesis* **24,** 159–169 (2003).

Hemann, M. T., Fridman, J. S., Zilfou, J. T., et al. An epi-allelic series of p53 hypomorphs created by stable RNAi produces distinct tumor phenotypes in vivo. *Nature Genet.* **33,** 396–400 (2003).

Hutvagner, G., McLachlan, J., Pasquinelli, A. E., et al. A cellular function for the RNA-interference enzyme Dicer in the maturation of the let-7 small temporal RNA. *Science* **293,** 834–838 (2001).

Jacque, J. M., Triques, K., and Stevenson, M. Modulation of HIV-1 replication by RNA interference. *Nature* **418,** 435–438 (2002).

Jia, Q., and Sun, R. Inhibition of gammaherpesvirus replication by RNA interference. *J. Virol.* **77,** 3301–3306 (2003).

Kapadia, S. B., Brideau-Andersen, A., and Chisari, F. V. Interference of hepatitis C virus RNA replication by short interfering RNAs. *Proc. Natl. Acad. Sci. USA* **100,** 2014–2018 (2003).

Kawasaki, H., Suyama, E., Iyo, M., and Taira, K. siRNAs generated by recombinant human dicer induce specific and significant but target site-independent gene silencing in human cells. *Nucleic Acids Res.* **31,** 981–987 (2003).

Ketting, R. F., Fischer, S. E., Bernstein, E., et al. Dicer functions in RNA interference and in synthesis of small RNA involved in developmental timing in *C. elegans*. *Genes Dev.* **15,** 2654–2659 (2001).

Knight, S. W., and Bass, B. L. A role for the RNase III enzyme DCR-1 in RNA interference and germ line development in *Caenorhabditis elegans*. *Science* **293,** 2269–2271 (2001).

Krusch, S., Domann, E., Frings, M., et al. Listeria monocytogenes mediated CFTR transgene transfer to mammalian cells. *J. Gene. Med.* **4,** 655–667 (2002).

Lagos-Quintana, M., Rauhut, R., Lendeckel, W., and Tuschl, T. Identification of novel genes coding for small expressed RNAs. *Science* **294,** 853–858 (2001).

Lau, N. C., Lim, L. P., Weinstein, E. G., and Bartel, D. P. An abundant class of tiny RNAs with probable regulatory roles in *Caenorhabditis elegans*. *Science* **294,** 858–862 (2001).

Lamontagne, B., Ghazal, G., Lebars, I., Yoshizawa, S., Fourmy, D. and Elela, S. A. Sequence dependence of substrate recognition and cleavage by yeast RNase III. *J. Mol. Biol.* 2003 Apr 11; 327(5):985–1000.

Lee, R. C., and Ambros, V. An extensive class of small RNAs in *Caenorhabditis elegans*. *Science* **294,** 862–864 (2001).

Lee, R. C., Feinbaum, R. L., and Ambros, V. The *C. elegans* heterochronic gene *lin-4* encodes small RNAs with antisense complementarity to lin-14. *Cell* **75,** 843–854 (1993).

Lee, S. S., Lee, R. Y., Fraser, A. G., et al. A systematic RNAi screen identifies a critical role for mitochondria in *C. elegans* longevity. *Nature Genet.* **33,** 40–48 (2003).

Lum, L., Yao, S., Mozer, B., et al. Identification of hedgehog pathway components by RNAi in *Drosophila* cultured cells. *Science* **299,** 2039–2045 (2003).

Maestro, R., Dei, Tos, A. P., Hamamori, Y., et al. Twist is a potential oncogene that inhibits apoptosis. *Genes Dev.* **13,** 2207–2217 (1999).

McCaffrey, A. P., Meuse, L., Pham, T. T., et al. RNA interference in adult mice. *Nature* **418,** 38–39 (2002).

McManus, M. T. and Sharp, P. A. Gene silencing in mammals by small interfering RNAs. *Nat. Rev. Genet. 2002* Oct; 3(10):737–747. Review.

McManus, M. T., Petersen, C. P., Haines, B. B., Chen, J. and Sharp, P. A. Gene silencing using micro-RNA designed hairpins. RNA 2002 Jun; 8(6):842–850.

Miyagishi, M., and Taira, K. U6 promoter driven siRNAs with four uridine 3′ overhangs efficiently suppress targeted gene expression in mammalian cells. *Nature Biotechnol.* **20,** 497–500 (2002).

Mourelatos, Z., Dostie, J., Paushkin, S., et al. miRNPs: A novel class of ribonucleoproteins containing numerous microRNAs. *Genes Dev.* **16,** 720–728 (2002).

Myers, J. W., Jones, J. T., Meyer, T., and Ferrell, J. E. Recombinant dicer efficiently converts large dsRNAs into siRNAs suitable for gene silencing. *Nature Biotechnol.* **21,** 324–328 (2003).

Novina, C. D., Murray, M. F., Dykxhoorn, D. M., et al. siRNA-directed inhibition of HIV-1 infection. *Nature Med.* **8,** 681–686 (2002).

Nykanen, A., Haley, B., and Zamore, P. D. ATP requirements and small interfering RNA structure in the RNA interference pathway. *Cell* **107,** 309–321 (2001).

Opalinska, J. B., and Gewirtz, A. M. Nucleic-acid therapeutics: Basic principles and recent applications. *Nature Rev. Drug Discov.* **1,** 503–514 (2002).

Paddison, P. J., Caudy, A. A., Bernstein, E., et al. Short hairpin RNAs (shRNAs) induce sequence-specific silencing in mammalian cells. *Genes Dev.* **16,** 948–958 (2002b).

Paddison, P. J., Caudy, A. A., and Hannon, G. J. Stable suppression of gene expression in mammalian cells by RNAi. *Proc. Natl. Acad. Sci. USA* **99,** 1443–1448 (2002a).

Paddison, P. J., and Hannon, G. J. RNA interference: the new somatic cell genetics? *Cancer Cell* **2,** 17–23 (2002).

Paul, C. P., Good, P. D., Winer, I., and Engelke, D. R. Effective expression of small interfering RNA in human cells. *Nature Biotechnol.* **20,** 505–508 (2002).

Pothof, J., van Haaften, G., Thijssen, K., et al. Identification of genes that protect the *C. elegans* genome against mutations by genome-wide RNAi. *Genes Dev.* **17,** 443–448. (2003).

Qin, X. F., An, D. S., Chen, I. S., and Baltimore, D. Inhibiting HIV-1 infection in human T cells by lentiviral-mediated delivery of small interfering RNA against CCR5. *Proc. Natl. Acad. Sci. USA* **100,** 183–188 (2003).

Randall, G., Grakoui, A., and Rice, C. M. Clearance of replicating hepatitis C virus replicon RNAs in cell culture by small interfering RNAs. *Proc. Natl. Acad. Sci. USA* **100,** 235–240 (2003).

Reinhart, B. J., Slack, F. J., Basson, M., et al. The 21-nucleotide let-7 RNA regulates developmental timing in *Caenorhabditis elegans*. *Nature* **403,** 901–906 (2000).

Rubinson, D. A., Dillon, C. P., Kwiatkowski, A. V., et al. A lentivirus-based system to functionally silence genes in primary mammalian cells, stem cells and transgenic mice by RNA interference. *Nature Genet.* **33,** 401–406 (2003).

Schwarz, D. S., Hutvagner, G., Haley, B., and Zamore, P. D. Evidence that siRNAs function as guides, not primers, in the *Drosophila* and human RNAi pathways. *Mol. Cell* **10,** 537–548. (2002).

Sijen, T., Fleenor, J., Simmer, F., et al. On the role of RNA amplification in dsRNA-triggered gene silencing. *Cell* **107,** 465–476 (2001).

Singh, R., and Reddy, R. Gamma-monomethyl phosphate: A cap structure in spliceosomal U6 small nuclear RNA. *Proc. Natl. Acad. Sci. USA* **86,** 8280–8283 (1989).

Shih, C., and Weinberg, R. A. Isolation of a transforming sequence from a human bladder carcinoma cell line. *Cell* **29,** 161–169 (1982).

Slack, F. J., Basson, M., Liu, Z., et al. The lin-41 RBCC gene acts in the *C. elegans* heterochronic pathway between the let-7 regulatory RNA and the LIN-29 transcription factor. *Mol. Cell* **5,** 659–669 (2000).

Song, E., Lee, S. K., Wang, J., et al. RNA interference targeting Fas protects mice from fulminant hepatitis. *Nature Med.* **9,** 347–351 (2003).

Sui, G., Soohoo, C., Affar el, B., Gay, F., Shi, Y., Forrester, W. C. and Shi, Y. A DNA vector-based RNAi technology to suppress gene expression in mammalian cells. *Proc. Natl. Acad. Sci. USA.* 2002 Apr 16; 99(8):5515–5520.

Svoboda, P., Stein, P., Hayashi, H., and Schultz, R. M. Selective reduction of dormant maternal mRNAs in mouse oocytes by RNA interference. *Development* **127,** 4147–4156 (2000).

Timmons, L., and Fire, A. Specific interference by ingested dsRNA. *Nature* **395,** 854 (1998).

Tuschl, T., Zamore, P. D., Lehmann, R., et al. Targeted mRNA degradation by double-stranded RNA in vitro. *Genes Dev.* **13,** 3191–3197 (1999).

Volpe, T. A., Kidner, C., Hall, I. M., et al. Regulation of heterochromatic silencing and histone H3 lysine-9 methylation by RNAi. *Science* **297,** 1833–1837 (2002).

Waterhouse, P. M., and Helliwell, C. A. Exploring plant genomes by RNA-induced gene silencing. *Nature Rev. Genet.* **4,** 29–38 (2003).

Wianny, F., and Zernicka-Goetz, M. Specific interference with gene function by double-stranded RNA in early mouse development. *Nature Cell. Biol.* **2,** 70–75 (2000).

Wightman, B., Ha, I., and Ruvkun, G. Posttranscriptional regulation of the heterochronic gene lin-14 by lin-4 mediates temporal pattern formation in *C. elegans*. *Cell* **75,** 855–862 (1993).

Williams, B. R. Role of the double-stranded RNA-activated protein kinase (PKR) in cell regulation. *Biochem. Soc. Trans.* **25,** 509–513 (1997).

Yang, D., Buchholz, F., Huang, Z., et al. Short RNA duplexes produced by hydrolysis with *Escherichia coli* RNase III mediate effective RNA interference in mammalian cells. *Proc. Natl. Acad. Sci. USA* **99,** 9942–9947 (2002).

Yang, S., Tutton, S., Pierce, E., and Yoon, K. Specific double-stranded RNA interference in undifferentiated mouse embryonic stem cells. *Mol. Cell. Biol.* **21,** 7807–7816 (2001).

Yu, J. Y., DeRuiter, S. L., and Turner, D. L. RNA interference by expression of short-interfering RNAs and hairpin RNAs in mammalian cells. *Proc. Natl. Acad. Sci. USA* **99,** 6047–6052 (2002).

Zamore, P. D., Tuschl, T., Sharp, P. A., and Bartel, D. P. RNAi: Double-stranded RNA directs the ATP-dependent cleavage of mRNA at 21 to 23 nucleotide intervals. *Cell* **101,** 25–33 (2000).

Zeng, Y., Wagner, E. J., and Cullen, B. R. Both natural and designed microRNAs can inhibit the expression of cognate mRNAs when expressed in human cells. *Mol. Cell* **9,** 1327–1333 (2002).

Zilberman, D., Cao, X., and Jacobsen, S. E. ARGONAUTE4 control of locus-specific siRNA accumulation and DNA and histone methylation. *Science* **299,** 716–719 (2003).

CHAPTER 5

Applications and Issues for Tissue Arrays in Target and Drug Discovery

Eric Jonasch, Kim-Anh Do, Christopher Logothetis, and Timothy J. McDonnell

Molecular Cancer Therapeutics: Strategies for Drug Discovery and Development, edited by George C. Prendergast
ISBN 0-471-43202-4

The past several years have witnessed a rapid evolution in high-throughput tissue analysis techniques applicable to target discovery, biomarker analysis, and drug discovery. The commercial availability of cDNA and oligonucleotide microarrays is enabling the comprehensive assessment of global transcript profiles from tissues and cell types of biologic and clinical interest. In addition, techniques for high-throughput protein analysis have recently become available. These methods include mass-spectroscopy-based techniques such as matrix assisted laser desorption/ionization (MALDI) and surface-enhanced laser desorption/ionization (SELDI) (Hutchens and Yip, 1993) that provide size and functional characteristics of protein extracts (Fetsch et al., 2002). These various high-throughput tissue-based methodologies have necessitated the development and application of sophisticated informatics and biostatistical approaches to enable investigators to analyze and interpret the resultant enormous volume of data.

A stated goal of many studies involving the use of these high-throughput techniques is to establish transcript or proteomic profiles of biologically and/or clinically relevant events, such as disease progression or therapeutic response. To accomplish this goal it is necessary to establish the validity of candidate marker (target) information in a large cohort of corresponding patient tissue samples that are relevant. This validation process is frequently labor intensive and time-consuming using traditional histopathological and light microscopic techniques. An important advance to address this limitation was the development of technical means to produce tissue microarrays (TMAs). The TMA represents a high-throughput platform to accelerate the assessment of candidate marker relevance (Kononen et al., 1998). Arrays can be used to assess a large number of variables within the same specimen or to evaluate the same variable among a large number of different samples. Image-acquisition and data analysis tools are available to enable the generation of a virtual library of images that can be shared between investigators with internet access. The TMA when coupled to structured marker information and computational analysis has already proven to be a valuable tool not only for biomarker validation and assessment but also for hypothesis generation and tissue-based data modeling.

It is becoming increasingly evident that a "multidimensional" approach needs to be taken in data gathering and analysis for cancer research (Duyk, 2002). The TMA is an important component of the spectrum of technologies available for tissue analysis and ideally should be used in the context of a programmatic approach to cancer research. By using an integrated approach, the largest amount of high-quality data can be gathered, interpreted, and understood in the appropriate clinical context. The design, generation, and analysis of TMAs is the focus of this chapter. Specific examples of TMA applications are presented and advantages and limitations are discussed.

TMAs can be constructed to address a variety of issues of clinical and biologic interest. The purpose of the microarray should be defined by the investigator before array generation. TMAs can be constructed to assess the normal tissue distribution of candidate biomarkers. TMAs can represent stage of tumor progression with cores representative of normal, preneoplastic, in situ carcinoma, invasive carcinoma, and metastatic carcinoma for an individual tumor type. Arrays can be designed to enable direct comparisons between the histopathological spectrum of specific tumor types, such as non-small cell lung cancer or non-Hodgkin lymphoma. Arrays can incorporate direct comparisons between neoplastic and matched non-neoplastic tissue of origin for a cohort of patients. Similarly, arrays can be designed for matched primary tumor and metastatic tumor. Alternatively, TMAs may be designed to represent individuals in a patient cohort over time. TMAs lend themselves to DNA expression analysis, using fluorescence in situ hybridization (FISH) techniques. Again, large numbers of samples can be treated in a virtually identical fashion, providing a way to rapidly acquire data on gene amplification, deletion, and mutation (Andersen et al., 2001; Bubendorf et al., 1999; Fuller et al., 2002; Simon et al., 2002; Tzankov et al., 2003).

The availability of tumor tissue with associated clinical annotation is an exceedingly valuable, and limited, resource. These specimens enable the correlation of biomarker expression with individual patient response, or resistance, to therapy. Maximum use of these limited samples can be achieved by incorporating TMA strategies. An additional, and welcome, benefit of using TMA in biomarker assessment is the significant savings in time and reagent cost.

5.1 Construction of Tissue Microarrays

The concept of TMA construction is simple: tissue cores, typically 0.6–2.0 mm in diameter, are obtained from several hundred donor blocks and are arrayed with high precision in a single recipient paraffin block (Fig. 5.1). In this manner, as many as 1000 cores can be arrayed in a single standard block. The key factors that ensure maximum utility of the TMA include identifying appropriate sampling areas on the donor tissue blocks, designing the array pattern in a manner that facilitates interpretation, and applying meticulous technique in the transfer process. The TMA apparatus (Beecher Instruments, Silver Spring, MD) consists of a turret with two attached stylet holders and variably sized stylets (Fig. 5.2). Two precision micrometers with digital displays enable accurate positioning of the stylets and placement of tissue cores into the recipient block. An array block holder holds the recipient block in place during array construction with the aid of two magnets in the base of the stage. A donor block bridge is used to cover the recipient block and holder during core acquisition from individual donor blocks.

To begin, a single hematoxylin and eosin (H&E)-stained slide is obtained from the donor block and is evaluated, preferably by a trained pathologist, for areas of interest. This step ensures that material is taken from areas of viable tissue with the representative histology. The area of interest is outlined on the H&E slide. Donor blocks must be at least 1 mm thick to be suitable for constructing tissue arrays, but blocks should ideally be 3–4 mm or thicker for optimal results. It is feasible to array up to 1200 tissue cores in a 40- by 25-mm block. However, constructing blocks with >700 cores can be technically difficult, and for most purposes, 300–500 cores per slide provides relatively easy sample handling coupled with optimal slide organization. The spacing

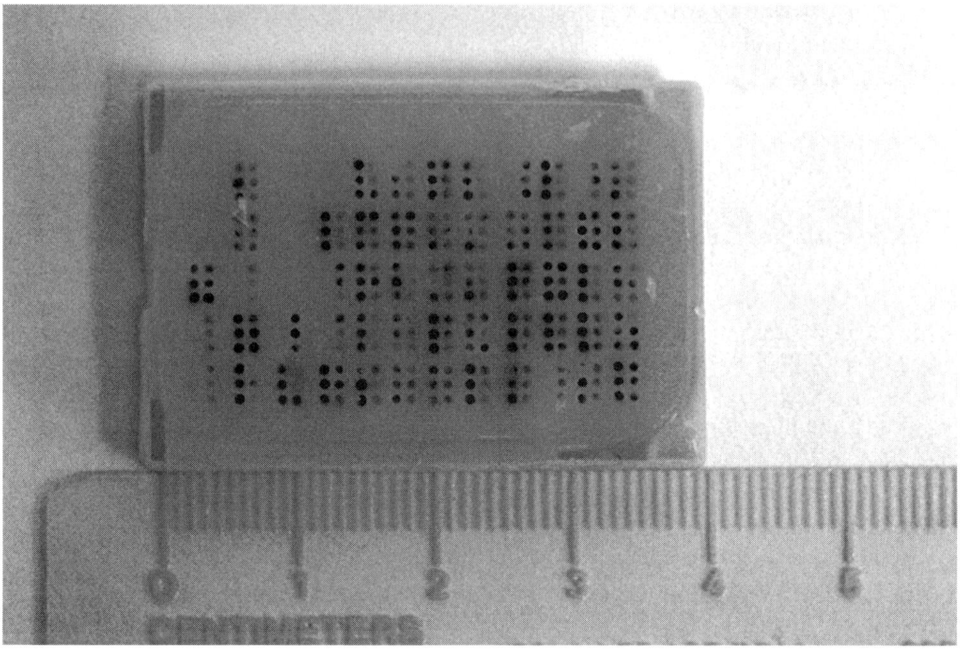

Figure 5.1 The surface of a tissue microarray paraffin block. Each core is 0.6 mm in diameter. Cores are arrayed in triplicates from each donor block.

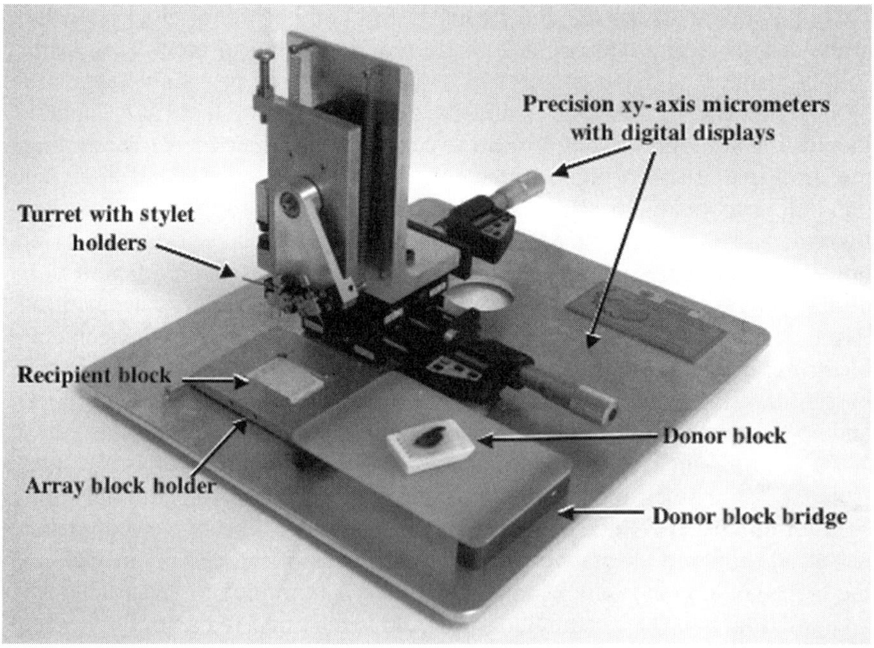

Figure 5.2 The Beecher tissue microarray apparatus. The Recipient block of the TMA under construction (*pink*) is held in place by the array block holder. Cores of tissue are obtained from individual donor blocks and addressed into the recipient blocks using the stylet needles. Precise placement of the cores is enabled by the adjusting the micrometers.

between the centers of two adjacent cores in the recipient array may range from 0.65 to 1 mm (0.6 mm core diameter). Sufficient space at the edges of the TMA block should be designated to avoid cracking of the paraffin. In general, 2.5- to 3-mm margins are usually adequate for this purpose. Initially, the TMA apparatus micrometers are set at 0, and a hole is made in the recipient block using the stylus. The H&E slide and the corresponding tissue block are then aligned; the slide–block complex is moved under the sampling needle that is lowered to retrieve the donor core. The depth of the needle movement is controlled by a manually adjusted depth stop bolt and the upward movement is controlled by a spring attached to the vertical slide on the turret. The donor block bridge is removed and the needle is pushed down until its tip reaches the hole in the recipient array block or is slightly above the surface level. While holding this position, the stylet is used to empty the tissue core into the recipient block hole. The micrometers are then advanced to the next *xy* coordinate and the cycle is repeated until the TMA is completely constructed. An experienced operator can reliably place 30–70 cores per hour.

The array block is removed from the recipient block holder and placed in a warm chamber (37°C) for 10–15 min. This serves to promote adherence of the tissue biopsies to the walls of the holes in the array paraffin block and makes the wax flexible for easier manipulation. After the block has warmed, a glass microscope glass slide or other clean and smooth surface can be used to level the surface of the TMA block. Some recommend the use of adhesive-coated tape sectioning system to cut sections to facilitate transfer of the relatively unstable paraffin sections onto microscope slides (Beecher technical manual), but others recommend against this approach, as it may result in increased sample loss (Hoos and Cordon-Cardo, 2001). These slides can subsequently be manipulated in the same manner as other paraffin tissue section.

The array design should incorporate a strategy to enable the orientation of the TMA section to be easily determined. A perfectly symmetrical array of the same tissue source can result in an inability to orient the corresponding section appropriately and subsequently lead to uncertainty regarding core addresses. This problem may be avoided by interspersing tumor specimens with readily identifiable marker tissue in a preset, non-symmetrical pattern (Hoos and Cordon-Cardo, 2001). Alternatively, the array may incorporate some defined asymmetry in core placement to enable orientation to be defined with certainty.

5.2 Automation and High-Throughput Array Systems

Although manual array generation provides substantial economies of scale and time compared to conventional whole-slide immunohistochemistry, efforts to automate the process can promote the efficiency of array generation. Efforts are now under way to create automated or semiautomated arrayers that will increase productivity of laboratory personnel, allow precision to be increased via an improvement in machine design, as opposed to operator skill level, and markedly shorten the time required to generate high-quality TMAs. Examples include models that incorporate a stereomicroscope that holds a reference slide prepared from the donor block. Movements of the microscope stage are precisely coordinated with the movement of the sampling stylet over the corresponding position in the donor block. Further efforts at automation include an instrument that can hold 27 donor or recipient blocks, allowing rapid transfer of material, in conjunction with dedicated software for specimen tracking. This technology allows the transfer of 120–180 cores per hour, as opposed to 30–70 per hour with a manual arrayer. The additional capital expenditure necessary to acquire automated TMA instruments currently limits their widespread availability.

Automated systems capable of performing most tasks from recipient block generation all the way to slide cutting and processing may become available in the future. Currently, combining semiautomated arrayers with conventional high-throughput slide processing equipment already provides a highly efficient way of producing large numbers of slides in a short time frame.

5.3 Software and Web-Based Archiving Tools

As with other high-throughput platforms, such as oligonucleotide arrays, efficient information management and storage capabilities rapidly become issues with routine use of TMAs. One of the challenges in TMA analysis is efficient image capture, storage, and analysis. The imaging, archiving, and retrieval software and hardware necessary to manage these data is complex and continues to evolve. Critical qualities needed in these systems include image fidelity and flexibility. There are several commercially available systems to address these issues. Common features include a digital image acquisition system and proprietary software to enable image archiving and management. There are several Web-based software applications available (Table 5.1).

Liu et al. (2002) reviewed their system for high-throughput analysis and storage of TMA data. The data are placed into a digital image collection using the Bliss system from Bacus Laboratories, Inc. (Lombard, IL). Each individual array has its own *Excel* scoring workbook, based on the three-dimensional layout of the array, and consists of multiple worksheets. The first worksheet, or master, contains the two-dimensional layout of the TMA. Subsequent worksheets are then used to record the staining interpretations for individual antibodies. Proprietary TMA-deconvoluter software is used to translate the large amounts of three-dimensional TMA data from these raw workbooks into a two-dimensional spreadsheet table format in *Excel.* The structured data are then amenable to further analysis, including hierarchical clustering via "Cluster" and "TreeView," and can be analyzed using other statistical manipulations. This group has also developed "Stainfinder," a Web-based program that permits linkage between a selected row in the "TreeView" graphical output file and the digital images recorded for each core.

Table 5.1 Web-Based Resources for TMA Analysis

Resource	Function	URL
"TMA Deconvoluter"	Translates three-dimensional TMA spreadsheet data into two-dimensional spreadsheet format	genome-www.stanford.edu/TMA
"Stainfinder"	Links "TreeView" graphical output to all immunostaining for one arrayed core	genome-www.stanford.edu/TMA
"Cluster"	Performs heirarchical cluster analysis on array data	rana.lbl.gov/EisenSoftware.htm
"TreeView"	Graphically displays "Cluster" analyses with original output and dendrograms	rana.lbl.gov/EisenSoftware.htm
"Webslide Browser"	Stores TMA image data	bacuslabs.com
"Zoom Viewer"		www.mgisoft.com

The quality of the images and data acquired should be comparable to that acquired via conventional means. Recently, an integrated system for storage and retrieval of prostate TMA data has been developed using a moderately compressed jpeg format with a Web-based retrieval format (Bova et al., 2001). After generation and staining of a 403-core TMA, images were acquired via the Bliss system and were then converted to fpx (flash-pix) format and uploaded into a "LivePicture Image" server (MGI Software, Richmond Hill, ON, Canada). A Web interface and backend database was developed for this system by PELICAN Informatics (Johns Hopkins University, Baltimore). This interface was assessed in a blinded fashion by two pathologists, with overall interpretability of images rated at 99%. The interobserver and intraobserver variation of Gleason grading obtained from the same two pathologists was equal to or better than that reported in the literature for standard microscope-based Gleason grading.

Manley et al. (2001) reviewed their experience with TMA data acquisition and management. The components included the TMA database, the TMA-image database, and the prostate pathology and clinical information databases. One of the useful features of their approach is the generation of an *Access* 2000-based relational database that facilitates data entry and data access.

We recently developed a tissue array database (TAD) for the storage and analysis of the large amounts of information generated from TMA datasets (Coombes et al., 2002). The database consists of a Web-based front-end application and a back-end relational database. Information is entered, edited, retrieved, and visualized through the individual investigator's Web browser. The relational database provides a mechanism to facilitate sharing of data while maintaining data integrity and security. The database consists of *SQL Server* 2000 on a dedicated *Windows* 2000 server (BiostatSQL, Pentium III with 512 MB RAM and 33 GB hard disk space). The Web-based front-end application has been developed in Active Server Pages and Java, and runs on Internet Information Services (IIS) on a similar computer (ACCG, Pentium III with 1 GB RAM and 17 GB hard disk space). After logging into the TAD server, users are presented with a list of menus that direct them through creating or scoring a TMA. The TMA can be created from an existing format or a completely new one by assigning the number of rows, columns, replicates per column and gap size. Once created, users are presented with a virtual TMA that mirrors the TMA slide map (Fig. 5.3). To score a particular core,

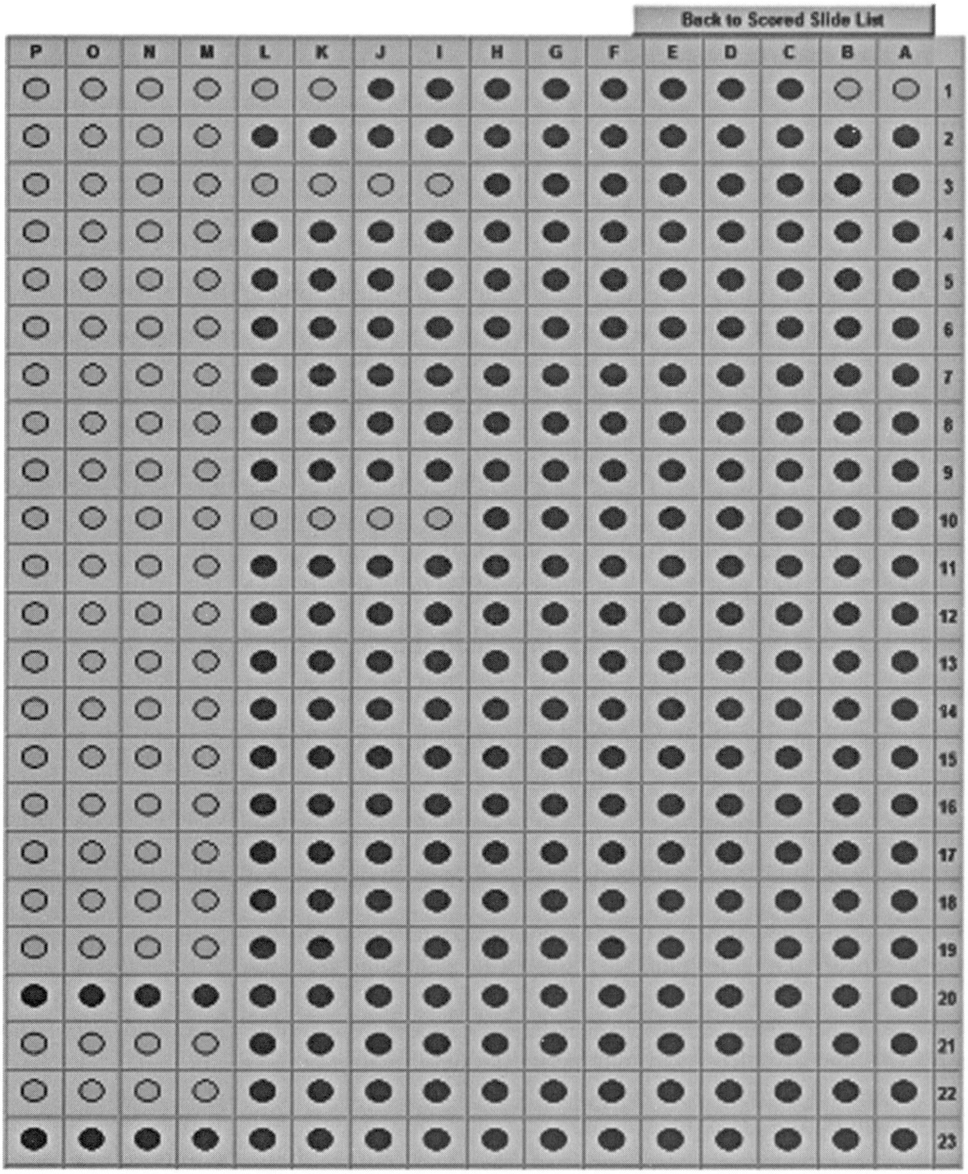

Figure 5.3 A virtual array interface in the TAD developed by investigators at the M. D. Anderson Cancer Center (Coombes et al., 2002). The virtual array is a representation of each addressed cored from a corresponding TMA. Color-coded virtual cores facilitate scoring of the array.

users clicks on one of the cores on the virtual array, which activates a pop-up window with the core location and scoring attributes (intensity, involvement, localization, etc). After the saving the record, the data are stored in cache and the array is saved. When users finish scoring, the virtual array is saved and all of the data entered are written to the database.

A high-resolution digital image of each core for each of the nine biomarkers was created and stored using a Zeiss Axioplan 2 Universal Microscope fitted with a Bacus Laboratories Slide Scanner (BLISS) and integrated with a BLISS image analysis workstation. This system enables the creation of a distributable virtual microscope slide ("Webslide"). The "Webslides" are stored on a dedicated image server using software to share and serve "Webslides" across the Internet, or an intranet, using "WebSlide Browser." To facilitate scoring the TMA, a Bacus Laboratories Active X Application Programming Interface (API) was used to hard link individual TMA core images on the "WebSlide" server to the specific corresponding core in the TAD. This enables users to click on a core on the virtual array and simultaneously activate the scoring pop-up window and the image corresponding to that core in the same window (Fig. 5.4). This magnification of the core image can be changed from 5× to 20× using a mouse click and remain in perfect optical

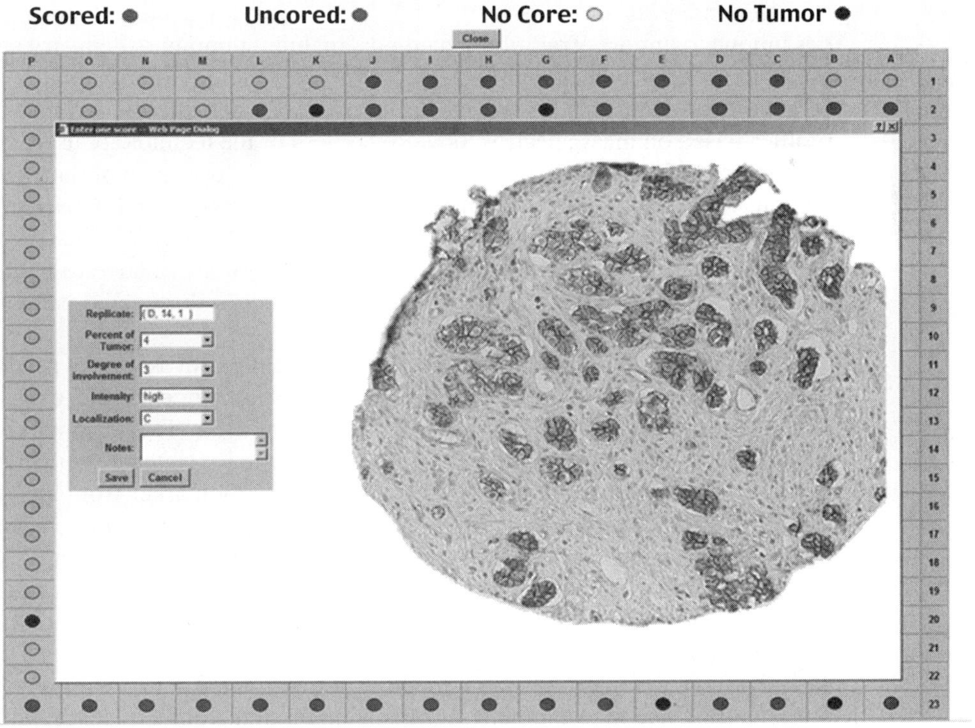

Figure 5.4 A TAD interface used during biomarker scoring of a TMA. By clicking on a virtual core, the image of the core and scoring window appears. Structured data are then entered using the pull-down menu and saved. Magnification of the image can by altered from 5× to 20× by clicking on the core image. Once the core is scored and saved, the color of the virtual core changes appropriately.

resolution. Using the Bacus software in conjunction with the TAD facilitated data integrity by eliminating transposition and clerical errors associated with transferring hand-written scores from paper to computer. Consequently, the time necessary to score each of the slides is lessened. It is important, that the availability of TMA information in this format greatly facilitates data acquisition and interactions between investigators and enables the application of biostatistical strategies not readily amenable to traditional histopathological techniques.

5.4 Statistical Analytic Strategies for TMA-Based Data

Sophisticated statistical methodology and software are crucial in the analysis of TMA data. Developing computational models and methods for the investigation of TMA gene expression patterns has already gained in importance among biostatistical research projects. The increasing specialization of biomedicine is anticipated to further contribute to its prominence. The first step in the search for biomarkers often begins with exploratory preclinical studies, comparing tumor tissue with nontumor tissue, to identify characteristics unique to tumors. Traditional methods, including immunochemistry and western blots, have been supplemented with a technology-driven explosion of gene-expression profiles based on microarrays, and protein expression profiles based on mass spectroscopy. Regardless of the technology used, the requirement of statistical consideration is crucial at every stage of data generation: the experimental design stage, image processing, and subsequent downstream analyses. This applies when identifying genes or clusters of genes that appear to differentially express in tumor tissue relative to nontumor tissue and, subsequently, when using the subset of discovered genes in a classification or predictive tool.

Before embarking on rigorous modeling of TMA data, investigators need to have a clear description of the data structures involved. For example, expression of immunohistochemical staining data arising from TMA biomarkers is often represented by a two-dimensional vector x_{ij} = (intensity, involvement), where x_{ij} represents measurements of the ith biomarker from the jth patient. Both intensity and involvement are often ordered variables, where intensity = 0 (none), 1 (low), or 2 (high) and involvement = 0, 1, 2 or 3.

Statistical analyses of such data may require investigators either to create a new variable that collapses the two measurements and combines them into one or to perform statistical analyses for each measurement separately. Alternatively, more complex statistical methods for correlated data are required.

In contrast, the measurements for gene expression from gene expression microarrays are usually performed on a continuous scale. The resulting data often require a preprocessing step that includes a transformation technique, such as taking the logarithm or the square root of the data, followed by appropriate normalization (Dudoit et al., 2002; Yang et al., 2002). Many statistical methods developed for the analysis of microarray gene expression data are immediately applicable to the analysis of TMA data.

Often a scientific study may involve a mixture of data types. For example, researchers studying prostate cancer at M.D. Anderson Cancer Center collected data on nine biomarkers of interest: Bax, Bcl-2, Bcl-xL, Fas, Mdm2, p21WAF1, p53, NFkB, and Bin1. They characterized the expression of these biomarkers by both involvement and intensity, with the aim of describing co-expressions of the variables with Gleason grade (i.e., tumor grade). Subsequent research focused on constructing models linking biomarker expression with important clinical end points such as disease stage (an ordered categorical data type), prostate-specific antigen (PSA; continuous data type), or time-to-event end points with possible censoring, (e.g., survival time or time to progression).

5.5 Correlative and Association Studies

Standard statistical methods such as Kendall's Tau-b or Pearson correlation coefficients can be employed to investigate co-expression between pairs of biomarkers. Contingency table analysis such as Fisher's exact and χ^2 tests can uncover associations between a particular biomarker expression (measured as an ordered variable) with a binary variable (e.g., case/control or gender), or with another ordered categorical variable (e.g., stage; grade). For comparing

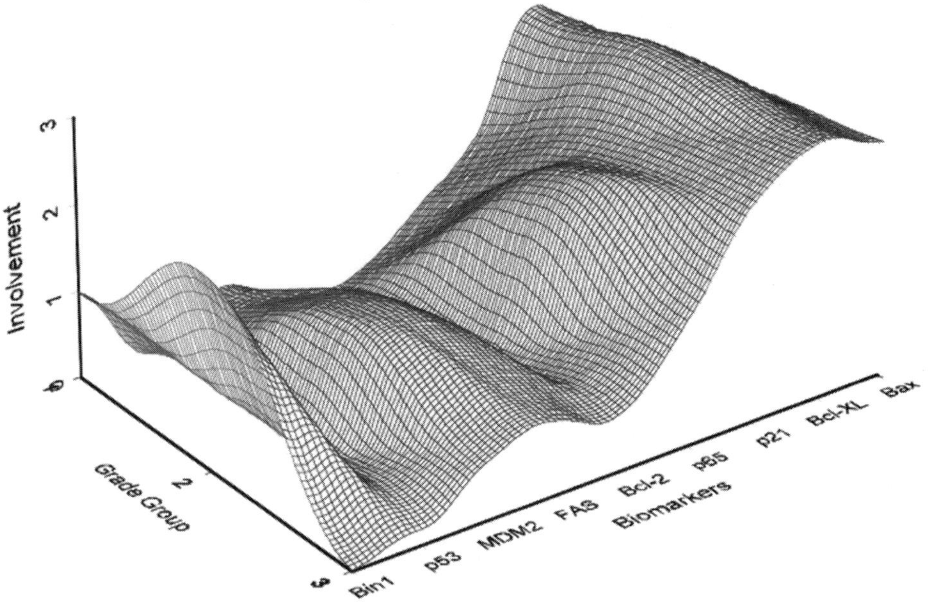

Figure 5.5 A three-dimensional plot produced by SPLUS, depicting biomarker measurements (involvement) for a collection of nine different biomarkers over three Gleason grade groups (grade group = 1 for Gleason score < 7, grade group = 2 for Gleason score = 7, and grade group = 3 for Gleason score > 7).

continuous measurements between groups, ANOVA and the equivalent Mann-Whitney nonparametric methods are popular statistical tools. For multivariable analyses, preliminary interactive three-dimensional plots that allow rotation around one of the axes can aid in uncovering specific structures in the data and are easily implemented in SPLUS (1988–2000) or MATLAB (1992–2003) (Fig. 5.5). Many traditional data-reduction techniques can be used to select a more limited set of biomarkers for future exploration. One can start by from simply ranking the biomarkers on the basis of a summary statistics, such as by the p-values obtained from the repeated application of the Student t or the nonparametric Wilcoxon test, or by the area under the receiver operating curve, and selecting the ones with the highest ranks. More complex techniques may involve hierarchical clustering and principal components analysis.

5.6 Classification and Predictive Studies

Many investigators are interested in building specific models, for example, to define a panel of biomarkers that can be used to predict an endpoint of interest. For a binary or categorical endpoint (e.g., case/control; tumor stage), logistic regression or an equivalent recursive partitioning tree method (e.g., CART, Classification and rebression trees) (Breiman et al., 1984) is often employed to build a predictive tool, from which one can perform additional bootstrap resampling, Monte Carlo simulations, or cross-validation for validation purposes (Fig. 5.6). For a general continuous end point, linear and nonlinear least-squares regression or regression trees are applicable. A more flexible procedure is multivariate adaptive regression splines (MARS). This method models relationships that are nearly additive or that involve interactions with a small number of variables (Friedman and Roosen, 1995). Analyses of time-to-event

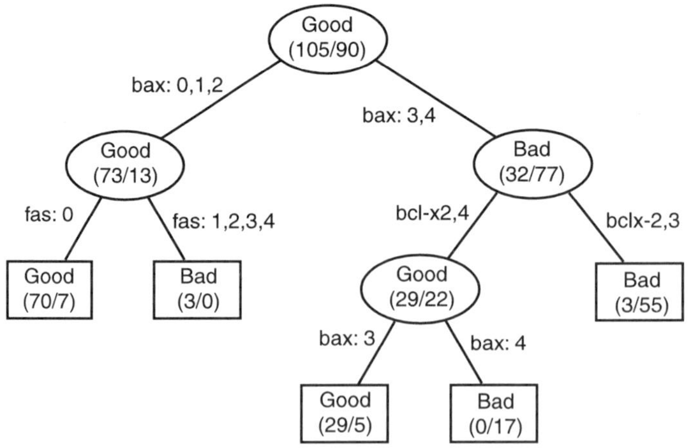

Figure 5.6 The results of a classification tree. The dependent variable is categorized into two groups: good = {Gleason grade = $2+3, 3+2, 3+3$} versus bad = {Gleason grade = $3+5, 4+4, 4+5, 5+3$}. The significant predictors are the expression measurements (involvement) for three biomarkers: *Bax, Fas,* and *Bcl-xL.* This tree may be employed as a classification tool for all tumor tissues with Gleason grade = $3+4$ or $4+3$.

end points (survival, time to progression) require Kaplan-Meier curves for graphical display and Cox proportional-hazards regression or survival tree methods (e.g., RPART, Recursive Partitioning) (Therneau and Atkinson, 1998). When a biomarker result can take many ordered values, for example, when larger values indicate a strong association with disease in a case-control study, a receiver operating characteristic (ROC) curve can be used that simultaneously depicts the relationship between sensitivity and specificity (Pepe, 2001). Sensitivity defines the true positive rate, – that is, the proportion of case subjects who are biomarker positive – whereas specificity defines the true negative rate – that is, the proportion of control subjects who are biomarker negative.

5.7 Issues on Dependent Data and Multiple Comparisons

Most standard statistical methods require the assumption of independence in the data. Often in practice, numerous interdependent observations are measured on the same subject. For example, several biopsy cores can be taken from a specific tumor tissue or repeated measurements can be performed on the same subject over time in a longitudinal study. Appropriate statistical methods – for example, mixed-effects models (Pinheiro, 2000) and Bayesian hierarchical models (Gelman et al., 1995) – that can incorporate the covariance structure of the data into the model have been developed and are continually improved. In summary, investigators interesting into pursuing high-throughput analysis of TMA results should contact collaborator-consultants who have biostatistical expertise.

5.8 The Search for Significant Biomarkers Involves Multiple Comparisons

Formal evaluation of differential expression may be approached as a collection of tests for each biomarker of the null hypothesis of no difference or as estimating the probability that a biomarker shows differential expression. Testing raises the need to account for multiple comparisons, by which one needs to adjust the test level of each biomarker to have a suitable overall level of significance. Biomarkers with significant differential expression are often reported in order of increasing p-value. A well-recognized problem of this multiplicity is that the chances of obtaining a positive result become high even if all null hypotheses are true. There are many existing methods to adjust the derived p-values for multiple comparisons, including the traditional Bonferroni approach, a common approach to control the familywise error rate. The tests are done at a level of stringency so that the probability of making one or more type I errors is smaller than some nominal α level. Many scientists find this kind of control to be overly conservative for microarray studies. A less conservative adjustment is the Holm method, which orders the

p-values and makes successively smaller adjustments. An additional method for bioinformaticians that accounts for the dependence structure that may exist between certain groups of biomarkers has been proposed (Westfall and Young, 1993).

Alternatively, if the goal of an experiment is to generate a list of interesting biomarkers, a certain number of false-positive results may be tolerable. The false discovery rate (FDR) was introduced as an alternative error measure for multiple-hypothesis testing and provided a sequential controlling p-value method. (Benjamini and Hochberg, 1995). The FDR represents the expected proportion of false positive findings among all the rejected hypotheses and, therefore, leads to an increase in power. A modified version of the FDR, the positive false discovery rate (pFDR), and an analogue of the p-value, called the q-value, a hypothesis testing error measure for each observed statistic with respect to the pFDR, have recently been developed (Storey, 2002). The major requirement of the general methodology is that the null versions of the test statistics can be simulated by permuting the data while preserving the experimental design. The FDR method of adjustment is particularly useful when an investigator is concerned with examining the true significance of a small subset of biomarkers in the presence of hundreds or thousands of other biomarkers. Furthermore, the investigator has the flexibility to calibrate the sample size, in terms of the number of biomarkers or number of replicate assays, to correspond to an appropriate FDR for each specific experiment.

5.9 Consideration of Heterogeneity in the Use of TMAs

Trained pathologists are, more than other investigators involved in tissue-based research, familiar with the issue and implications of heterogeneity. By definition, tissues consist of heterogeneous cell populations. In obtaining multiple samples of the same tissue the relative proportions of the individual cell populations making up the tissue can vary considerably. These variations, obviously, have the potential to affect the interpretation of corresponding results. Further complexity is encountered in the heterogeneity and variation imparted by disease states, such as cancer. For example, an investigator interested in whether a candidate biomarker is associated with malignant lymphoma would be well advised to become familiar with the histological and clinical diversity of these neoplasms. Investigators involved in the field of genomics research frequently minimize these issues, perhaps because they are defined by something as "crude" as the human eye and light microscopic.

It seems remarkable, with these issues in mind, that useful information can be derived from cores of tissue 0.6 mm in diameter. In this regard, it needs to be appreciated that heterogeneity and potential sampling bias is not limited to TMA applications but is, in fact, inherent in any and all routine histopathological or biomarker assessment. Routine tissue samples are fixed and processed in volumes of approximately 1 cm^3. Thus the standard 5-μm-thick tissue section represents 0.05% of the tumor. When a tissue microarray is generated, this sample size is further reduced to about 0.3% of the amount usually evaluated. The question may, therefore, be stated in terms of the reliability of a TMA strategy to "represent," with statistical validity, the individual donor blocks used to construct the TMA.

Several studies have addressed the reliability of TMAs for biomarker assessment. In one series, 2317 histologically characterized tissue samples of urinary bladder cancer from 1849 patients were used to construct four replica TMAs with 0.6-mm cores (Nocito et al., 2001). Proliferative indices were scored for each core in each of the four arrays and compared with that obtained from each of the 2317 donor blocks of origin using Ki67 immunohistochemistry. Not only was each TMA highly similar to the data obtained from sections from the individual donor blocks ($p < 0.0001$) but the associations between tumor grade, Ki67 labeling index, tumor stage, and prognosis were also maintained in each of the four replicate TMAs. A similar strategy was used to assess the validity of TMA versus "large" tissue sections in a series of 553 breast carcinomas assessed for estrogen receptor, progesterone receptor, and p53 expression (Torhorst et al., 2001). The investigators determined that a single core was sufficient to establish correlations between alterations in marker expression and clinical outcome ($p < 0.0015$). Other studies arrive at similar conclusions regarding the validity of tissue microarrays compared to standard approaches (Camp et al., 2000; Hoos et al., 2001).

In summary, these studies show that TMAs provide relevant and reproducible histological information despite the small amount of tissue being analyzed. Although the concerns arising out of tissue heterogeneity and the need to take this into consideration when designing tissue-centric experiments remains, TMA samples are equal in quality to those provided by conventional histological slides. Indeed, because of their small size, it is possible to generate arrays from a larger number of sites within the total tumor mass, accelerating the analysis of intratumor heterogeneity. Certainly it can be anticipated that disease and tissue heterogeneity will on occasion result in discrepancies between samples represented on a tissue microarray and corresponding donor blocks for individual biomarker expression. Nonetheless, there is now consistent evidence supporting the power, advantages, and statistical validity of the TMA strategy.

5.10 Tissue Microarray Applications

The use of TMA enables the rapid screening of candidate biomarkers, or therapeutic targets, for their frequency of expression. Correlation with disease progression and therapeutic response can be evaluated to the extent that the samples are linked to clinical annotation. This constitutes a critical step in the "validation" or "credentialing" of candidate markers of interest. This clinical annotation could include, but is not limited to, clinical stage, pathologic stage, tumor grade, therapeutic interventions, time to progression, and disease free survival. TMAs are now commonly integrated into global genomic strategies, such as cDNA or oligonucleotide "chip" arrays, to identify and characterize clinically relevant differentially expressed gene products (Bubendorf et al., 1999; Mousses et al., 2001) . The TMA, in this context, provides the means to accelerate the throughput of candidate biomarker evaluation and maximizes the use of, the frequently limiting, relevant tissue resources.

The increasing commercial availability of antibodies that are specific for the activation state, or conformation, of proteins of interest is enabling the interrogation of cell-signaling pathways in formalin-fixed and paraffin-embedded tissues. These reagents are refining our ability not only to obtain an assessment of relative protein levels but to provide a means to determine whether these proteins are actively mediating a signaling event. This strategy was recently applied to assess the state of the tumor growth factor β (TGFβ) signaling pathway in primary breast carcinomas using TMAs (Xie et al., 2002). The investigators determined that the majority of breast cancers possessed an intact TGFβ signaling axis, as evidenced by the presence of the downstream mediators of TGFβ signaling, Smad2, Smad4, and phospho-Smad2 (active). It is important that they determined that loss of phospho-Smad2 expression in stage II breast cancers was associated with a reduction in overall survival. This finding was independent of other known prognostic markers of breast cancer progression.

In addition to the predictive information obtained from these types of studies, the use of conformation-dependent antibodies may also be anticipated to provide information regarding the efficacy of therapeutic agents intended to selectively disrupt specific signaling pathways in human disease states. In this regard it is noteworthy that the feasibility to correlate disruption of epidermal growth factor receptor signaling by selective tyrosine kinase inhibitors has recently been demonstrated using immunohistochemical techniques (Baker et al., 2002; Kim et al., 2003). Cell microarrays, similar to TMAs, are also valuable tools that can be used to assess the effect of the expression of specific proteins as well as to examine drug–target interactions (Ziauddin and Sabatini, 2001).

5.11 Summary

The past 5 years have experienced significant advances in our ability to apply global genomic technologies to obtain comprehensive information of human disease states. The ability to convert this deluge of information into useful knowledge will depend on the extent that these strategies are integrated into the overall clinical context. The challenge will be to apply these emerging technologies to rigorously annotated and meticulously obtained human tissue samples, to identify and characterize the consistent molecular features of diseased tissue that may serve as the basis for effective therapy development and appropriate clinical application. The use of tissue microarrays will continue to be an important component in this process of discovery and validation.

References

Andersen, C. L., Hostetter, G., Grigoryan, A., et al. Improved procedure for fluorescence in situ hybridization on tissue microarrays. *Cytometry* **45,** 83–86 (2001).

Baker, C. H., Solorzano, C. C., and Fidler, I. J. Blockade of vascular endothelial growth factor receptor and epidermal growth factor receptor signaling for therapy of metastatic human pancreatic cancer. *Cancer Res.* **62,**1996–2003 (2002).

Benjamini, Y., and Hochberg, Y. Controlling the false discovery rate: A practical and powerful approach to multiple testing. *J. R. Stat. Soc. B* **57,** 289–300 (1995).

Bova, G. S., Parmigiani, G., Epstein, J. I., et al. Web-based tissue microarray image data analysis: initial validation testing through prostate cancer Gleason grading. *Hum. Pathol.* **32,** 417–427 (2001).

Breiman, L., Friedman, J. H., Olshen, R. A., and Stone, C. J. *Classification and Regression Trees.* Belmont, CA, Wadsworth International (1984).

Bubendorf, L. M., Kolmer, A., Kononen, J., et al. Hormone therapy failure in human prostate cancer: Analysis by complementary DNA and tissue microarrays. *J. Natl. Cancer Inst.* **91,**1758–1764 (1999).

Camp, R. L., Charette, L. A., and Rimm, D. L. Validation of tissue microarray technology in breast carcinoma. *Lab Invest.* **80,** 1943–1949 (2000).

Coombes, K. R., Zhang, L., Bueso-Ramos, C., Brisbay, S., Logothetis, C., Roth, J., Keating, M. J., McDonnell, T. J. TAD: A web interface and database for tissue microarrays. *Applied Bioinformatics.* **1,** 155–158 (2002).

Dudoit, S., Yang, Y. H., Speed, T. P., and Callow, M. J. Statistical methods for identifying differentially expressed genes in replicated cDNA microarray experiments. *Statistica Sinica* **12,** 111–139 (2002).

Duyk, G. M. Sharper tools and simpler methods. *Nature Genet.* **32 (suppl),** 465–468 (2002).

Fetsch, P. A., Simone, N. L., Bryant-Greenwood, P. K., et al. Proteomic evaluation of archival cytologic material using SELDI affinity mass spectrometry: Potential for diagnostic applications. *Am. J. Clin. Pathol.* **118,** 870–876 (2002).

Friedman, J. H., and Roosen, C. B. An introduction to multivariate adaptive regression splines. *Stat. Methods Med. Res.* **4,** 197–217 (1995).

Fuller, C. E., Wang, H., Zhang, W., et al. High-throughput molecular profiling of high-grade astrocytomas: The utility of fluorescence in situ hybridization on tissue microarrays (TMA-FISH). *J. Neuropathol. Exp. Neurol.* **61,** 1078–1084 (2002).

Gelman, A., Carlin, J. B., Stern, H. S., and Rubin, D. B. *Bayesian Data Analysis.* New York, Chapman & Hall, (1995).

Hoos, A., and Cordon-Cardo, C. Tissue microarray profiling of cancer specimens and cell lines: opportunities and limitations. *Lab. Invest.* **81,** 1331–1338 (2001).

Hutchens, T. W., and Yip, T. T. New desorption strategies for the mass spectrometric analysis of macromolecules. *Rapid Commun. Mass Spectrom.* **7,** 576–580 (1993).

Kim, S. J., Uehara, H., Karashima, T., et al. Blockade of epidermal growth factor receptor signaling in tumor cells and tumor-associated endothelial cells for therapy of androgen-independent human prostate cancer growing in the bone of nude mice. *Clin. Cancer Res.* **9,** 1200–1210 (2003).

Kononen, J., Bubendorf, L., Kallioniemi, A., et al. Tissue microarrays for high-throughput molecular profiling of tumor specimens. *Nature Med.* **4,** 844–847 (1998).

Liu, C. L., Prapong, W., Natkunam, Y., et al. Software tools for high-throughput analysis and archiving of immunohistochemistry staining data obtained with tissue microarrays. *Am. J. Pathol.* **161,** 1557–1565 (2002).

Manley, S., Mucci, N. R., De Marzo, A. M., and Rubin, M. A. Relational database structure to manage high-density tissue microarray data and images for pathology studies focusing on clinical outcome: The prostate specialized program of research excellence model. *Am. J. Pathol.* **159,** 837–843 (2001).

Mousses, S., Kallioniemi, A., Kauraniemi, P., Elkahloun, A., and Kallioniemi, O. P. Clinical and functional validation using tissue and cell microarrays. *Curr. Opin. Chem. Biol.* **6,** 97–101 (2001).

Nocito, A., Bubendorf, L., Maria Tinner, E., et al. Microarrays of bladder cancer tissue are highly representative of proliferation index and histological grade. *J. Pathol.* **194,** 349–357 (2001).

Pepe, M. S. Receiver operating characteristic methodology. *J. Am. Stat. Assoc.* **95,** 308–311 (2001).

Pinheiro, J. C. *Mixed Effects Models in S and S-Plus.* Springer-Verlag, NY, 552 (2000).

Simon, R., Struckmann, K., Schraml, P., et al. Amplification pattern of 12q13-q15 genes (MDM2, CDK4, GLI) in urinary bladder cancer. *Oncogene* **21,** 2476–2483 (2002).

Storey, J. D. A direct approach to false discovery rates. *J R. Stat. Soc. B* **64,** 289–300 (2002).

Therneau, T. M., and Atkinson, E. J. An introduction to recursive partitioning using the RPART routines. [Technical Report #61]. Available at www.mayo.edu/hsr/techrpt.html (1998).

Torhorst, J. C., Bucher, J., Kononen, P., et al. Tissue microarrays for rapid linking of molecular changes to clinical endpoints. *Am. J. Pathol.* **159,** 2249–2256 (2001).

Tzankov, A., Zimpfer, A., Lugli, A., et al. High-throughput tissue microarray analysis of G1-cyclin alterations in classical Hodgkin's lymphoma indicates overexpression of cyclin E1. *J. Pathol.* **199,** 201–207 (2003).

Westfall, P. H., and Young, S. S. *Resampling-Based Multiple Testing: Examples and Methods for p-value Adjustment.* New York, Wiley & Sons (1993).

Xie, W., Mertens, J. C., Reiss, D. J., et al. Alterations of Smad signaling in human breast carcinoma are associated with poor outcome: A tissue microarray study. *Cancer Res.* **62,** 497–505 (2002).

Yang, Y. H., Dudoit, S., Luu, P., et al. Normalization for cDNA microarray data: A robust composite method addressing single and multiple slide systematic variation. *Nucleic Acids Res.* **30,** e15 (2002).

Ziauddin, J., and Sabatini, D. M. Microarrays of cells expressing defined cDNAs. *Nature* **41,** 107–110 (2001).

CHAPTER 6

Protein Transduction Strategies for Target and Mechanism Validation

Sergei A. Ezhevsky and Steven F. Dowdy

Molecular Cancer Therapeutics: Strategies for Drug Discovery and Development, edited by George C. Prendergast
ISBN 0-471-43202-4

The common sense of cancer therapy is similar to that of military action: Destroy the enemy while leaving your friends unharmed. The objective of targeting tumor cells while minimizing harm to the surrounding tissues has inspired the development of numerous experimental approaches and the discovery of a plethora of tumor-specific genetic alterations. However, the question remains how to convert the specific features of tumor cells into effective antitumor therapies. The design of modern anticancer drugs may demand a high level of specificity to target certain protein–protein interactions that are altered in tumor cells. This high level of specificity is achievable by introduction of peptides, full-length proteins or functional domains of proteins into tumor cells. A major obstacle toward such specific anticancer therapies is the natural size restriction imposed on the delivery of macromolecules across cellular membranes. To remove this obstacle, new and more sophisticated strategies that allow for unrestricted delivery of biologically active molecules are needed. One such strategy that has recently been developed toward these ends is protein transduction.

6.1 What Is Protein Transduction?

Protein transduction is defined as a receptor-independent and transporter-independent translocation of macromolecules, including peptides, proteins, and siRNAs, across the cellular membrane. Two pioneering papers published in 1988 ignited the field of protein transduction by offering the unexpected observation that the TAT protein from HIV can pass through the cell membrane barrier, reach the nucleus, and transactivate an long terminal repeat (LTR)-driven viral promoter (Frankel and Pabo, 1988; Green and Loewenstein, 1988). Interestingly, TAT protein not only retained enzymatic activity upon intracellular accumulation but was subsequently shown to deliver heterologous proteins into the cell (Fawell et al., 1994). Thus a novel field of delivering macromolecules into cells was born.

The protein transduction domain (PTD) of TAT maps to a short basic region (YGRKKRRQRRR) that is involved in binding the RNA stem loop of the nascent HIV transcript (Green and Loewenstein, 1988; Vives et al., 1997). Fusion of the TAT PTD to the protein is sufficient to promote translocation across the cellular membrane (Vives et al., 1997). The PTD can be either chemically cross-linked to purified proteins or, more efficiently, expressed as an in-frame fusion protein with the gene of interest followed by purification from bacterial cultures (Ezhevsky et al., 1997; Nagahara et al., 1998).

Although TAT was the first PTD characterized, subsequent studies have identified several other strong PTDs. Transduction properties were found in the *Drosophila* Antennapedia (Antp) homeodomain transcription factor (Derossi et al., 1994). The minimal Antp PTD region was narrowed down to 16 residues (RQIKIWFQNRRMKWKK). In addition, poly-arginine has also been shown to promote efficient transduction of organic compounds and peptides (Rothbard et al., 2002). These three "premier" PTDs have been extensively reviewed elsewhere (Ford et al., 2001; Lindgren et al., 2000; Lindsay, 2002; Snyder and Dowdy, 2001; Wadia and Dowdy, 2002). The mechanism of TAT-mediated cell membrane penetration or "transduction" is still a subject of some controversy. Early observations suggested that TAT enters cells by adsorptive endocytosis (Mann and Frankel, 1991). If so, the basic PTD in this case must possess at least three features: binding to acidic moieties on the cell membrane, promoting endocytosis, and facilitating escape from endosomal vesicles.

Another study showed that protein transduction was energy independent, based on its occurrence at low temperature and under ATP-depleted conditions (Vives et al., 1997). Transduction in this study was confirmed by immunostaining of TAT on the fixed cells. This report suggested that PTDs could directly penetrate through lipid bilayers and that endocytosis was not required for protein transduction. However, while this working model has prevailed in the literature, recently the same group has reevaluated it and drawn an opposite conclusion (Richard et al., 2003). Due to extremely tight binding of the TAT peptide to the cell surface, conventional wash-out procedures fail to remove all bound peptides from the cell surface. Consequently, fixation of cells with membrane bound PTDs results in an artificial redistribution of the PTD from the membrane into the cytosol, and even into the nucleus, potentially giving false-positive results that PTDs directly "penetrate" the cellular membrane (Leifert et al., 2002; Richard et al., 2003). Thus the conclusion regarding the energy-independency of transduction was based on this artificial staining, and therefore it needs to be reevaluated by phenotypic assays in living cells. While these observations favor an endocytotic mechanism of transduction, they do not exclude other physiological ways of penetrating the cell membrane. Regardless of what the mechanism(s) may ultimately be, certain precautions have to be applied when interpreting data on protein transduction obtained from the immunostaining or from the other methods of visualization. Although these observations question previous results based on visualization, they should not distract from the plethora of live cell phenotypes generated by transduction. Consequently, this review will concentrate exclusively on phenotypic changes in live cells and *in vivo* animal models resulting from protein transduction.

6.2 Advantages and Disadvantages

The field of protein transduction has expanded remarkably over recent years from in vitro observations to in vivo delivery of biologically active compounds (Jo et al., 2001; Schwarze et al., 1999; Xia et al., 2001). Although a role for protein transduction in cancer therapy has yet to be established, it is clear that protein transduction has tremendous potential to deliver biologically active cargo to specifically kill tumor cells.

The most impressive feature of protein transduction is its broad flexibility (Fig. 6.1). The classes of cargo molecules that have been delivered represent

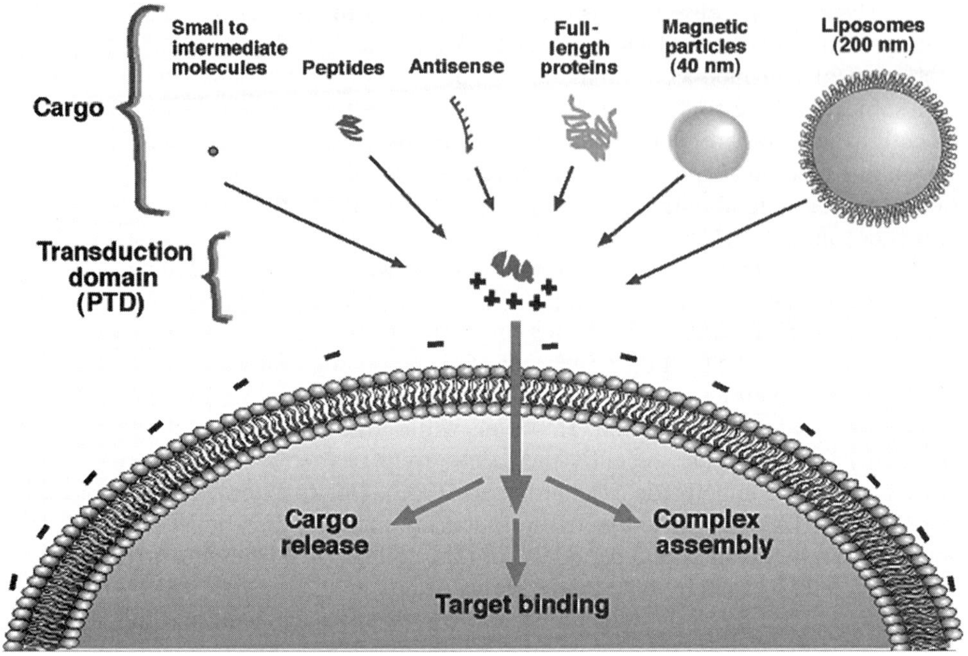

Figure 6.1 Protein transduction is defined as a receptor-independent and transporter-independent translocation of macromolecules across the cellular membrane. The protein transduction domain is a short, positively charged amino acid sequence that when linked to "cargo" promotes its delivery into living cells. Delivered classes of cargo molecules represent a wide range of sizes and biophysical properties from small molecules to peptides to proteins to PNA to DNA to phage particles to magnetic nanoparticles and liposomes.

a wide range of sizes and biophysical properties, including small molecules, peptides, proteins, peptide nucleic acid (PNA), DNA, phage particles, magnetic nanoparticles, and liposomes (Torchilin et al., 2001). Due to the absence of the size limitations, multifunctional protein domains can be genetically combined into one single polypeptide chain, including, but not limited to, delivery tag plus recognition domain and an action site, the latter of which can be either in the "on" or "off" mode, pending the cellular environment, specific cell type, or addition of a second activating molecule or substrate (Vocero-Akbani et al., 1999). The need to target tumor cells in a highly precise manner can be accommodated in the complexity of the protein transduced, dramatically increasing the range of the prospective intracellular interactions and mechanisms targeted.

Many intracellular differences between normal and tumor cells could conceivably be converted into potent tools to restrict tumor growth. The conversion of a pro-drug into the active drug with the help of protein transduction might rely on minor alterations in the tumor-specific phenotype due to improved precision of the recognition "device." For example, similar to pathogen-specific proteases, a tumor-specific protease (if found) would cleave a transduced, chimeric protein, constructed with an appropriate cleavage site and release a killing part of the construct (endonuclease, caspase) that would be harmful for the tumor cells only. It is important that those alterations might not be relevant to tumor growth per se. Some of the potential major advantages of protein transduction to cancer therapy are listed below.

- **Transduction is not cell-type restricted.** Tumor and normal cells are equally transduced. Hence one can find the cellular sensitivity to the treatment from direct comparison of tumor cells and normal counterparts in vitro and, in pending positive outcomes, the treatment can be directly applied to the appropriate model in vivo. This approach could minimize the time gap between bench observations and clinical trials.
- **Cell-permeable polypeptides bypass transcriptional and translational controls.** This feature offers an advantage of protein transduction over gene therapy, in which gene expression and percentage of cells expressing the gene remain unpredictable and potentially problematic.
- **Cell-permeable polypeptides are not susceptible to the most common mechanism of multidrug resistance.** Protein transduction across the cell membrane occurs in a manner that is independent of multidrug resistance (gene) 1/P-glycoprotein activity. Consequently, cell-permeable peptides linked to conventional drugs can overcome multidrug resistance of tumor cells (Rousselle et al., 2000). This ability offers an potentially important advantage to combination therapy of many cancers where multidrug resistance commonly arises.
- **High levels of mechanism specificity are achievable.** The high level of specificity of cell-permeable compounds (enzymes, substrates, inhibitors) allows precise targeting, therefore minimizing unpredictable side effects of the treatment.
- **The intracellular level of transduced proteins is adjustable.** Similarly to small, diffusible molecules, the intracellular concentration of the drug can be regulated by added amount at least in vitro. The half-life of the penetrating proteins is also adjustable (see below), which makes the treatment reversible and prevents any irreversible genetic alterations.
- **Stability of transduced proteins/peptides.** To minimize the degradation of the transduced peptides it may be useful to employ D-isomer retro-inverso peptides that have similar surface topology and cell penetrable qualities relative to L-enantiomers (Bonny et al., 2001) but are not susceptible to endopeptidases. As a result of substantial stabilization, the reduction in the effective amount of the transduced peptide can be achieved.

The disadvantages of protein transduction are dialectically related to the advantages; in other words, we can convert positive attributes of transduction into negative ones under the appropriate conditions. For example, lack of receptor dependency results in transduction into most, if not all, cell types, which increases the demand for the specificity of transduced proteins. While posttranslational modifications have been shown to occur on transduced proteins, co-translational protein modifications are sometimes absolutely necessary to exert appropriate functions and would thereby be excluded. Pharmacokinetic studies are limited to one paper, which demonstrates fast blood clearance of PTD peptides (Lee and Pardridge, 2001), questioning how high the concentration of TAT-chimeric proteins could be achieved in any given organ on systemic delivery. Last, although it has not been yet extensively tested, the immunogenicity of the cell-permeable proteins is a potential concern.

6.3 Applications in Signal Transduction

The mounting body of knowledge about cell signaling provides ample opportunities for therapeutic intervention. More specifically, targeting of short protein domains, involved in protein–protein interactions, can interrupt the chain reaction of cell signaling and stop the cascade at any point from the membrane receptor to the nuclear messenger. Peptide-based modulators of such interactions have already been shown to be powerful tools for basic research and have the capacity for future clinical applications. In this regard, the conjunction of protein–protein recognition sequences with protein transduction domains holds a great deal of promise to specifically treat tumors in vivo (Dunican and Doherty, 2001).

Receptor tyrosine kinase (RTK) signal transduction pathways play an essential role in growth factor-dependent cell proliferation and thereby possess strong oncogenic potential (Garbay et al., 2000). The first member of the RTK pathway is the receptor-bound protein 2 (Grb2), a small adaptor polypeptide that interacts with phospho-tyrosine of tyrosine kinase receptors through its SH2 domain (Tari and Lopez-Berestein, 2001). Receptor-bound Grb2 results in the recruitment and complex formation with of SOS, a *ras* GEF protein that exchanges GDP for GTP to activate *ras* (Fig. 6.2). Activated *ras* initiates the mitogen-activated protein kinase (MAPK) cascade by recruiting *raf*, which triggers an extracellular signal-regulated kinase (ERK). ERK, in turn, translocates to the nucleus and stimulates transcription of early genes (Chang et al., 2003). Thus blocking the formation of Grb/Sos complexes has great potential to avert a vast downstream signaling network.

Blocking Grb2/Sos interactions relies on peptides that mimic the proline-enriched protein–protein contact domain of SOS that bind the SH3 domain of Grb2. SOS contains four proline-enriched regions and each of them shows low affinity for the SH3 domain of Grb2 (Cussac et al., 1994). However, linking two SOS proline-enriched domains together (peptidimer) yields a protein with a 400-fold higher affinity for Grb2 compared to the monomer (Garbay et al., 2000). The peptidimer has no effect on cell growth when added to cell culture media on its own. However, when linked to the Antp PTD, it disrupts Grb2/SOS complex formation and inhibits the phosphorylation and activation of MAP kinase that is induced by EGF addition. The proliferation of transformed cells overexpressing an oncogenic analog of the EGF receptor (*her*-2) was also dramatically inhibited by the peptide. It is important that Antp-peptidimer conjugates did not cease proliferation of normal cells, suggesting that other signaling pathways compensate for the disrupted EGF signaling (Cussac et al., 1999). Thus interruption of *ras* signaling pathway by a transducible peptide was shown to be sufficient in this study for blocking cell proliferation of transformed cells.

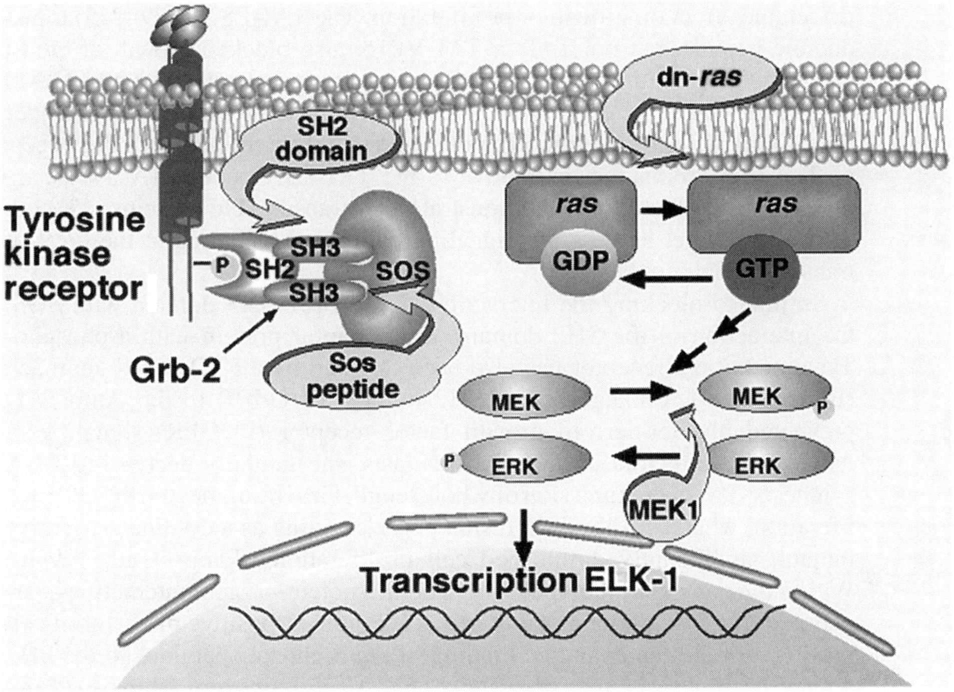

Figure 6.2 The protein transduction strategy has been used to target receptor tyrosine kinases to signal transduction pathways at different levels. SOS peptide linked to the Antp-PTD disrupts the formation of Grb2/SOS complexes and inhibits EGF-induced MAP kinase phosphorylation. Antp-SH2 fusions acting as a dominant-negative mutant prevents the interaction of an adaptor protein's SH2 domain with the phospho-Tyr of kinase receptor and blocks signaling to the downstream effectors. Introduction of dominant-negative *ras* fused to the TAT PTD blocks all *ras*-mediated cellular responses and at the same time, MEK1-derived peptide cross-linked to a PTD prevents ERK-mediated activation of the transcriptional activity of ELK.

The differential reaction of normal and transformed cells to the permeable Antp-peptidimer, which disrupts SOS/Grb2 interaction, offers an illustration of how an key protein–protein interaction can be targeted for cancer therapy. A second example shows that targeting the oncogene *her*-2 may be beneficial. *her*-2 is a clinically validated receptor target that promotes aggressive, highly metastatic breast tumors with increased resistance to chemotherapy. Thus methods that directly target expression of *her*-2 would be expected to offer a clinical benefit. The transcriptional factor ESX, in complex with nuclear cofactor DRIP130, binds and strongly activates the *her*-2 promoter (Chang et al., 1997). Disruption of the interaction between ESX and DRIP130 impairs the expression of the *her*-2 gene, ceases the proliferation and viability of *her*-2-expressing breast cancer cells. Asada et al. (2002) determined the critical region for the DRIP130-ESX interaction, and designed a TAT cell-permeable peptide with competitive binding capabilities. When added to the culture medium, the TAT-ESX peptide reduced the protein level of *her*-2 in cell lines overexpressing the oncogene. In contrast, the control TAT transduction domain or irrelevant chimeric proteins TAT-VP16 did not have a detectable effect on growth or proliferation. The TAT-ESX (129–145) transducible peptide, but not TAT or TAT-VP16, also blocked growth of breast cancer cells and induced apoptosis. It is important that cells with a low level of endogenous *her*-2 expression and with ectopic expression of *her*-2 under the control of a heterologous promoter, and therefore refractory to the effects of TAT-ESX, remained insensitive to the TAT-ESX peptide (Asada et al., 2002). These observations demonstrate that transducible proteins can both selectively target and disrupt signaling pathways, including at the level of transcriptional regulation.

Similar to blocking the interaction of the Grb2 SH3 domain with SOS, the interaction of the SH2 domain of an adaptor protein with a phospho-Tyr moiety on a receptor has also been targeted by a PTD-based approach (Buday, 1999). Linkage of the SH2 domain of Grb10 to the Antp PTD prevented platelet-derived growth factor receptor (PDGFR) signaling to downstream effectors in normal fibroblasts, substantially decreasing DNA synthesis. The overexpression of whole length Grb10 augmented the cell proliferation, whereas Antp-SH2 fusion peptide, acting as a dominant-negative mutant, significantly diminished cell proliferation (Wang et al., 1999b). Within the context of manipulating specific protein–protein interactions, targeting of the SH2 domain may serve as a dominant-positive mediator of cell proliferation. As an example, binding of Antp-phosphopeptides to the SH2 domain of the p85 regulatory subunit of phosphotidylinositol 3-kinase (PI3K) can activate the enzyme in vitro and stimulate a mitogenic response in muscle cell lines. Remarkably, this peptide is as effective as serum epidermal growth factor (EGF) and fibroblast growth factor (FGF) in promoting entry into S-phase (Derossi et al., 1998). Although this example does not directly relate to cancer therapy, it illustrates the capability and potential of protein transduction as a powerful dissecting tool for specific protein-protein interactions. Such approaches may promote the discovery of novel modulators for RTKs, including inhibitors of EGF receptors (e.g., vascular endothelial growth factor receptor 2; VEGFR2), which may serve as antiangiogenic drugs.

One well-studied anticancer strategy is to block *ras* signaling by introduction of a dominant negative form of this small GTPase (Baldari et al., 1993). Dominant negative *ras* (dn-*ras*) competes with the wild-type *ras* for binding to the GTP exchange factor, which forms unproductive complexes with dn-*ras,* preventing the binding and activation of wild-type *ras*. Dominant-negative *ras* fused to the TAT PTD was efficiently transduced into isolated nondividing, human blood eosinophils, where it prevented interleukin 5 (IL-5)-dependent activation of ERK-1 and -2 (Hall et al., 2001). Transduction of the TAT-dn-*ras* chimera into eosinophils also blocked all *ras*-mediated cellular responses on activated cytokine-, chemokine-, and G-protein-coupled receptors (Myou et al., 2002). It is worth mentioning that eosinophils are poorly transfected or infected and that transduction of TAT-dn-*ras* is the only way to introduce this protein into the vast majority of the cell population. The applicability of PTD-based strategies in cells that are poorly transfected by other methods represents an additional advantage of this technology.

The significance of a tractable method to target *ras* is high: Oncogenic mutations of *ras* occur in one third of human malignancies, leading to constitutively active *ras* and promoting growth factor–independent cell proliferation (Adjei, 2001). Blocking of farnesylation of *ras* by targeting farnesyl protein transferase was strategized as one way to inhibit *ras*-transforming activity (Cox, 2001; Rane and Prendergast, 2001). Unfortunately, K-*ras,* which accounts for the majority of all *ras* mutations, can escape inhibition of farnesyl protein transferase, due to alternate geranylgeranylation in the presence of farnesyl (protein) transferase inhibitor (FTIs), although these inhibitors can still target other proteins, such as RhoB, which is is sufficient to induce cell cycle arrest and apoptosis in neoplastically transformed cells (Cox, 2001; Prendergast, 2001). Another way to reverse the *ras*-dependent phenotype is by abolishing function of downstream intermediates, such as the MEK/ERK interaction. The final member of the pathway, ERK, has been targeted by cell-permeable inhibitors (Kelemen et al., 2002). A MEK1-derived peptide, fused to membrane-translocating PTD, inhibits ERK activation in vitro and prevents ERK activation in TPA-stimulated 3T3 cells in a concentration-dependent manner (Kelemen et al., 2002). The ERK cell-permeable peptides also blocked ERK-mediated activation of the transcriptional activity of ELK1 (Kelemen et al., 2002).

Targeting signaling cascades by PTDs confirms the precision that is achievable by protein transduction. With the knowledge of protein interacting domains, it is feasible in principle to design PTD inhibitors that will break signaling chains at any link. The broadening or narrowing possible set of effectors depends on how close the impact is to the membrane receptor or nuclear targets. As an example, the NF-κB pathway plays an important role in the regulation of immune and inflammatory responses (Fig. 6.3). Recently discovered NF-κB affiliations with apoptosis, differentiation, and cell migration have brought more attention to the possible oncogenic implications of this transcription factor (Baldwin, 2001). The maintenance of certain levels of transcriptionally active NF-κB is required for normal cell proliferation, cytokine production, and self-defense surveillance. However, constitutively high levels of NF-κB activity promote unrestricted cell growth with less susceptibility to pro-apoptotic treatments (Baldwin, 2001). Therefore, targeting NF-κB requires the ability to regulate levels of this transcription factor in a provisional manner. This is an ideal assignment for modulation by PTD peptides.

IκB is an inhibitor of NF-κB that works by binding to the nuclear localization domain of NF-κB and preventing nuclear accumulation (Richmond, 2002). Extracellular stimuli induce *N*-terminal phosphorylation of IκB, resulting in its degradation, followed by NF-κB translocation into the nucleus and NF-κB-dependent transcriptional activation. However, the critical event in NF-κB activation is IκB phosphorylation by IKKα and IKKβ kinases (Karin and Ben-Neriah, 2000). Mutant forms of IκB with alanine substitutions at positions Ser-32 and Ser-36 are not phosphorylated by inhibitor of κB-kinase (IKK) and thereby avoid degradation. Adenoviral delivery of mutant IκB represses tumor necrosis factor α (TNF-α) induced activation of NF-κB in carcinoma cells in vitro and reduces tumor growth in SCID mice in vivo (Wang et al., 1999a). Thus IκB mutant acts as a superrepressor by sequestering NF-κB and avoiding the transcription of NF-κB-dependent genes (Yamamoto and Gaynor, 2001).

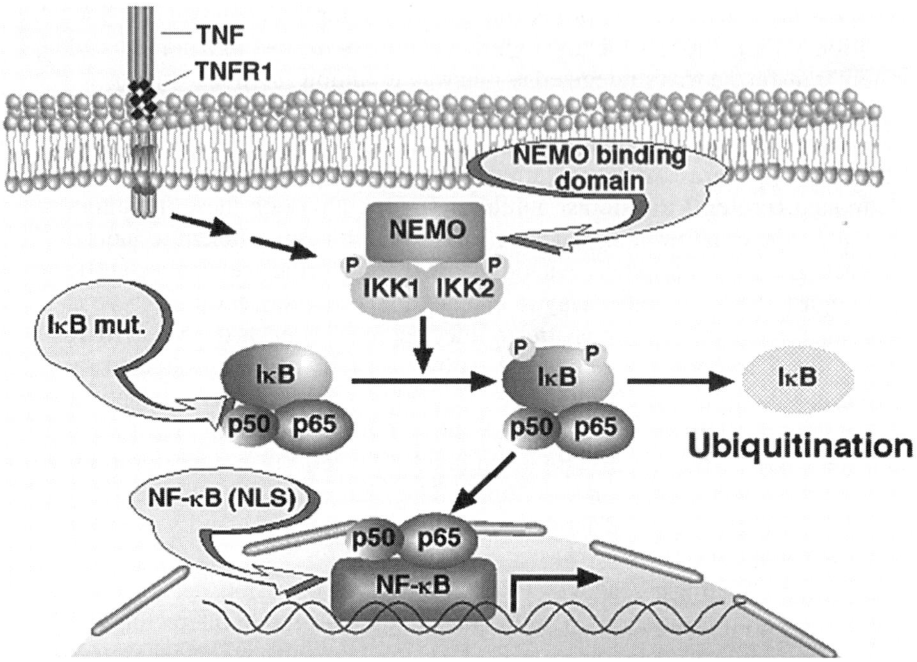

Figure 6.3 The key element in the activation of NF-κB-mediated transcription is phosphorylation and degradation of IκB. The nonphosphorylatable, nondegradable superrepressor srIκBα linked to the TAT PTD inhibits TNF-α-induced NF-κB activation and thereby prevents NF-κB-mediated transcription. Phosphorylation of IκB requires cognate action of the enzymatic unit IKK and the regulatory protein NEMO. NBD on IKK has been delivered via the Antp PTD. Membrane transducible peptides abolish IκBα phosphorylation and inhibite cytokine-induced NF-κB-dependent transcription. Last, cell-permeable peptides linked to the nuclear localization sequences of NF-κB prevent nuclear localization of NF-κB, and block NF-κB signaling partway and T cell proliferation.

The above powerful features of mutant IκB have prompted researchers to use it in PTD transduction experiments. The nonphosphorylatable, nondegradable superrepressor IκBα(srIκ Bα) linked to the TAT PTD inhibits TNF-α- and IL-1-induced NF-κB activation in HeLa cells and thereby prevents NF-κB-mediated transcription at concentrations as low as 600 nM (Kabouridis et al., 2002). The inhibitory effect of TAT-srIκBα depends on the duration of pre-incubation before TNF-α addition and reflects the rate of intracellular accumulation of the transduced chimeric protein. (Kabouridis et al., 2002). In contrast, treatment with control TAT-GFP or TAT-$\tilde{\beta}$Galactosidase (TAT-$\tilde{\beta}$Gal) did not have any effect on NF-κB activity (Kabouridis et al., 2002), confirming the absence of reactivity of the TAT PTD alone (11 amino acids in length). This is an important observation, since the whole length TAT protein causes the activation of the NF-κB pathway and phosphorylation of ERK1 and ERK2 (Badou et al., 2002; Bruce-Keller et al., 2001).

Differentially targeting the same IκBα provides further evidence of the specificity and precision of PTD-mediated transduction. For instance, IκBα in bone marrow macrophages is not phosphorylated on Ser32 and Ser36 but on Tyr residues. Accordingly, the IκBα mutant with Tyr-42 substitution serves as a super-repressor of the NF-κB pathway in bone marrow macrophages.

Indeed, TAT-IκBα (Y42F) construct prevents nuclear translocation of NF-κB and inhibits the differentiation of macrophages into osteoclasts at concentration as low as 100 nM, whereas the TAT PTD alone does not affect the differentiation of macrophages (Abu-Amer et al., 2001). These examples clearly demonstrate how various modifications of the same transducible inhibitor affect its cell type specificity and compensate for the indiscriminative TAT-mediated protein delivery to all cell types.

Cell-permeable PTD-containing peptides have also been used to directly target NF-κB. Phosphorylation of IκB requires cognate action of the enzymatic unit IKK and the regulatory protein NF-κB essential modifier (NEMO). The NEMO binding domain (NBD) on IKK is a short sequence within the carboxy-terminus. NBD peptide delivery via the Antp PTD blocks association of NEMO with the IKK complex, abolishing IκBα phosphorylation and inhibiting cytokine-induced NF-κB-dependent transcription (May et al., 2000). In another example, cell-permeable peptide consisting of the tandem of two nuclear localization sequences of NF-κB, acting as a dominant negative inhibitor, prevented nuclear localization of NF-κB and blocked NF-κB signaling partway and T cell proliferation (Fujihara et al., 2000). Although these two peptides restrain NF-κB signaling, their working concentrations range (1–100 μM) are 100-fold higher than the reported effective concentrations of the TAT-IκBα superrepressor.

Thus the application of protein transduction technology to the accumulated knowledge of signaling pathways has great potential for preclinical studies and confirms both the specificity and flexibility of protein transduction as a platform delivery technology.

6.4 Applications to Cell Cycle Regulation

The conversion of normal cells to malignant cells requires the genetic alteration of proto-oncogenes (positive regulators), tumor-suppressor genes (negative regulators), and DNA damage repair genes (Hanahan and Weinberg, 2000). A key aspect of the modification of these genes is the destruction of G_1 phase cell cycle regulation that results in uncontrolled proliferation.

The retinoblastoma tumor-suppressor gene product (pRb) is a key negative regulator of transition from early G_1 phase, across the restriction point into late G_1 phase (Ho and Dowdy, 2002). pRb is an active transcriptional repressor when bound to transcription factors, such as members of the E2F family (Fig. 6.4). Inactivation of pRb by hyperphosphorylation by cyclin-dependent kinases (CDK) results in the release of E2F, allowing for the coordinated transcription of late G_1 phase specific genes important for DNA synthesis and S phase entry. CDKs are activated by complex formation with cyclins (A, B, D, E) and negatively regulated by kinase inhibitors, $p16^{INK4a}$, p21, and p27 (Lee and Yang, 2001; Sherr and Roberts, 1999). Genetic alteration of this pathway, such as inactivation of either $p16^{INK4a}$, amplification of cyclin D1 or CDK4, or loss or mutation of *RB,* occurs in the vast majority of human malignancies (Sherr, 2001; Sherr and McCormick, 2002). Therefore,

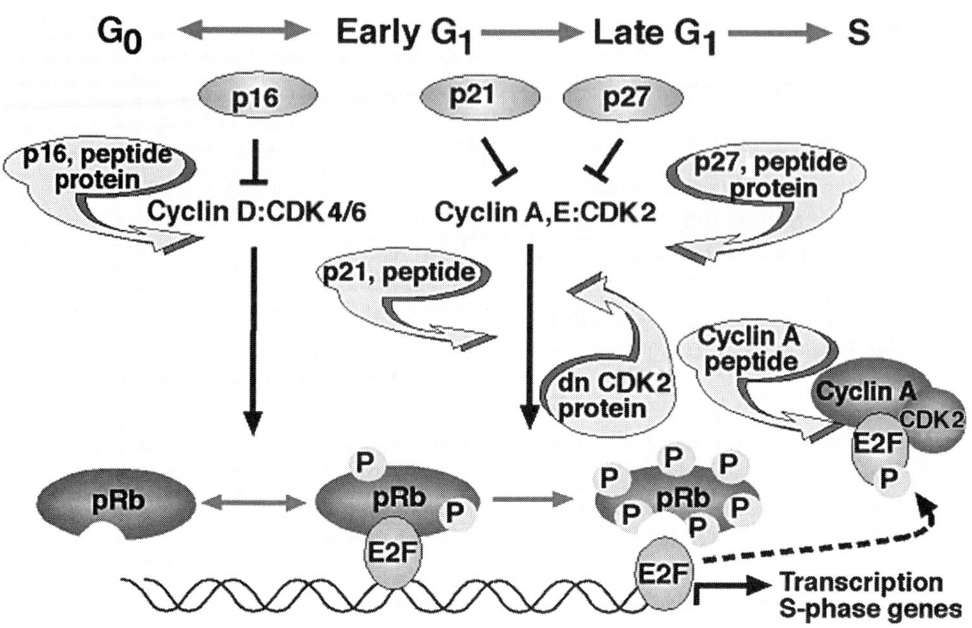

Figure 6.4 The major player in cell cycle regulation is the product of retinoblastoma tumor-suppressor gene, pRb. pRb binds transcription factors, such as members of the E2F family and histone deacetylases and thereby represses specific gene expression. pRB phosphorylation by CDKs results in the release of E2F, which promotes transcription of late G_1 phase specific genes important for S phase entry. CDKs are activated by complex formation with cyclins (A, B, D, E) and negatively regulated by kinase inhibitors, $p16^{INK4A}$, p21, and p27. Both p16 peptides and full-length p16 protein fused to either TAT or Antp PTDs efficiently transduce into cells, prevent pRb phosphorylation, and elicit an early G_1 cell cycle arrest. As expected, universal inhibitors p21 and p27 also block pRb phosphorylation when introduced in membrane-permeable forms. The dominant-negative mutant of CDK2 (dn-CDK2), linked to the TAT PTD, efficiently and reversibly blocks the proliferation of human fibroblasts and maintains pRb in its active, growth inhibitory state. Fusion of the cyclin A recognition site on E2F1 to the TAT PTD results in cell transducible peptides that disrupt the substrate–kinase interaction and prevent from entering the S phase. Cell-permeable kinase inhibitors are valuable tools in the study of the negative control of G_1 phase of the cell cycle.

epigenetic reconstitution of tumor-suppressor function represents a specific manner in which to selectively target tumor cells versus surrounding normal cells.

The $p16^{INK4}$ tumor-suppressor gene is frequently altered in human tumors by point mutation, deletion, or silencing due to promoter methylation, all of which results in its functional inactivation (Rocco and Sidransky, 2001; Ruas et al., 1999). Consequently, p16 is a major target of cancer therapies, perhaps second only to the p53 tumor-suppressor gene. $p16^{INK4}$ binds to monomeric CDK4 or CDK6 and thereby prevents the formation of active cyclin D:CDK4/6 complexes. Inhibition of cyclin D:CDK4/6 activity leads to a G_0/G_1 phase cell cycle arrest (Shapiro et al., 2000).

Fahraeus et al. (1996) showed that the third ankyrin-like repeat of p16 is responsible for CDK4/6 binding. A 20 amino acid p16 peptide (residues 84–103) derived from this domain is sufficient for binding to CDK4/6 and inhibiting cyclin D:CDK4/6-dependent phosphorylation of pRb in vitro

(Fahraeus et al., 1996). It is significant that when the p16 peptide was coupled to the Antp PTD, it blocked S phase entry in nonsynchronized human keratinocytes by ~90% at a concentration of 24 μM (Fahraeus et al., 1996, 1998).

The p16 peptide cross-linked to the TAT PTD also had the same effect, namely an acute inhibition of cyclin D:CDK4/6 activity followed by a G_1 cell cycle arrest (Gius et al., 1999). These observations provide experimental evidence that both the Antp and TAT PTDs are functionally equivalent in the performing delivery of this peptide. However, the TAT domain can promote not only peptide transduction but also that of the full-length p16 protein transduction (Ezhevsky et al., 1997). Full-length p16 protein fused to the TAT PTD blocked cyclin D:CDK4/6 kinase activity and elicited a G_1 cell cycle arrest at the concentration of 300 nM (Ezhevsky et al., 1997). Note that the effective concentration of the transducible full-length p16 was ~ 100 times lower then the concentration of p16 peptide.

These observations suggest that larger peptides and proteins have increased specificity for their cognate intracellular targets and may thereby ultimately result in a substantial decrease in the effective concentration.

p21 and p27 are so-called universal inhibitors, which bind to and inhibit preexisting, active cyclin D:CDK4/6, cyclin E:CDK2, and cyclin A:CDK2 complexes (Sherr and Roberts, 1999). Overexpression of p21 or p27 leads to an early G_1 phase cell cycle arrest. In addition, the cyclin-dependent kinase inhibitor p21 is a major mediator of the p53-dependent growth-arrest pathway (el-Deiry et al., 1994).

Treatment of cells with a 20 amino-acid peptide based on the carboxy-terminal CDK-binding domain of p21 coupled to the Antp PTD inhibited pRb phosphorylation and induced a strong G_1 phase cell cycle arrest (Ball et al., 1997). Bonfanti et al. (1997) also used two peptides corresponding to p21 CDK-binding domains, residues 17–33 and 63–77, and fused them to the Antp PTD, resulting in prevention of cell growth in two human ovarian cancer cell lines, while the same peptides minus the PTD were inactive. Similar to p21 peptides, treatment of human hepatocytes with a full-length p27 protein fused to the TAT PTD resulted in a cell cycle arrest (Nagahara et al., 1998). In addition, treatment of human pre-B cell lymphomas with TAT-p27 protein resulted in cell cycle arrest and induction of apoptosis (Banerji et al., 2001). In contrast, treatment of these cells with control TAT-eGFP protein or mutant TAT p27 protein, which cannot bind CDK2, had no effect.

The dominant-negative form of CDK2 (dn-CDK2) that sequesters cyclin from the endogenous wild-type CDK is a functional analog of p21 and p27. However, it exclusively targets cyclin E and A. TAT-dn-CDK2 fusion proteins efficiently and reversibly block the proliferation of human fibroblasts and maintain pRb in its active, growth inhibitory state (Ezhevsky et al., 2001). These observations demonstrate that both transducible peptides and proteins are capable of targeting active cyclin:CDK complexes.

Cell cycle control genes are deregulated in the vast majority of human tumors (Sherr and McCormick, 2002). Targeting of these genes and/or their products is now under investigation as potential cancer therapies (Ortega et al., 2002). Unfortunately, the inhibitors of cell cycle progression are almost equally efficient at blocking proliferating of normal cells and tumor cells, as might be predicted. Therefore, the therapeutic window for a tumor-specific treatment may be unacceptably narrow. One potential benefit of transducible cell cycle inhibitors relies on the high adsorptive capabilities of the cell-permeable proteins. As a result, these agents, when administrated locally near the tumor mass, may not spread systemically. However, this assumption still needs to be checked experimentally.

Another theoretical possibility is to use transducible cell cycle inhibitors for temporally restraining normal cell proliferation in the course of cancer therapy. Pretreatment of the normal cells (epithelial, bone marrow, etc.) with cell cycle inhibitors will impose a temporal cell cycle arrest and may protect normal cells from the damage of chemotherapy (Blagosklonny and Pardee, 2001).

Activation of cyclin A:CDK2 complexes is a necessary and critical step for entering S phase (Sherr and Roberts, 1999). Therefore, the substrate-docking site located on cyclin A:CDK 2 complexes is a good target for disruption of substrate–kinase interaction. One cellular target of cyclin A:CDK2 kinase is the E2F1 transcriptional factor (Krek et al., 1994). Chen et al. (1999) fused the cyclin A recognition site on E2F1 to the TAT PTD and demonstrated that transformed cells treated with the peptide undergo apoptosis and it is important that it did not affect normal cells. This group speculated that deregulation of E2F transcription factors occurs frequently during transformation, and impairing E2F functioning led to the tumor selective sensitization to cyclin:CDK inhibitors. These results lay down the foundation for development of cell-permeable inhibitors of CDKs as anticancer agents (Chen et al., 1999).

The von Hippel-Lindau (VHL) tumor-suppressor gene is functionally inactivated in the patients with sporadic renal cell carcinomas (RCCs) (Zbar, 1995). The growth of RCC depends on insulin-like growth factor 1 (IGF1) activation of the IGF1 receptor, which activates protein kinase C-δ and promotes cell proliferation (Datta et al., 2000). A small region of VHL binds to the cytoplasmic domain of the IGF1 receptor and thereby interrupts IGF1 signaling pathway (Datta et al., 2000). This region of VHL is often mutated in RCC and results in unrestricted signaling from the IGF1 receptor to protein kinase C (PKC) and supports cell proliferation (Datta et al., 2001).

An epigenetic compensation is a reasonable way to intervene into the malfunctioning of the VHL tumor suppressor. Treatment of RCC cells with a transducible TAT-VHL peptide that binds the IGF1 receptor inhibited cell proliferation (Datta et al., 2001). In vivo treatment of subcutaneous RCC tumors in nude mice resulted in tumor growth retardation. In addition, treatment with TAT-VHL peptide in cell culture retarded migration potential across a matrigel barrier (Datta et al., 2001). It is impressive that TAT-VHL treatment of mice harboring RCC tumors dramatically reduced the tumor invasiveness into the muscle wall. The authors concluded that TAT-mediated transduction of active peptides has merit to treat RCCs, either alone or in conjunction with other therapies. These results strengthen the notion that TAT-mediated transduction of peptides and proteins are capable of penetrating and distributing relatively homogeneously within solid tumors.

The neurofibromatosis type 2 (NF2) or merlin tumor suppressor gene is mutated in the majority of schwannomas (Merel et al., 1995). Loss of merlin function causes the alteration of cell shape and of cell–cell communication and contributes to tumor formation. It is remarkable that addition of TAT-merlin protein to schwannoma cells in culture reverses cytoskeletal defects due to loss of merlin and restores the cell to a near normal type (Bashour et al., 2002). These observations suggest that epigenetic reconstitution of merlin function by TAT-mediated protein transduction can complement aberrant merlin functioning and may reverse cell growth and tumor formation in vivo (Bashour et al., 2002).

In conclusion, targeting proteins that have a validated and reproducible biologic end point is a promising application for protein transduction technology. In this regard, the restoration or reconstitution of tumor-suppressor function that results in a permanent downstream consequence will likely yield the most productive in vivo results. However, temporary cytostatic effects by cyclin:CDK inhibitors may both lower the threshold of combination chemotherapy treatment and add increased selectivity.

6.5 Induction of Apoptosis

The fields of apoptosis and cancer genetics have been linked recently (Huang and Oliff, 2001). First, oncogenic alterations often lead to the disruption of apoptotic pathway. Second, most conventional cytotoxic anticancer drugs are initiators of apoptosis. Therefore, defects in apoptotic programs contribute significantly to cancer treatment failure (Lowe and Lin, 2000). Fortunately, tumor-selected antiapoptotic mutations generally affect the initial steps of the apoptotic pathway, leaving the execution machinery undamaged. Consequently, two basic strategies have emerged: Either restore the missing sensor functions at the beginning of the cell death program (e.g., p53) or directly trigger the last irreversible stage(s) of apoptosis exclusively in tumor cells.

6.5.1 *Bcl*-2 FAMILY

Proteins belonging to the *Bcl*-2 family play a crucial role in the regulation of programmed cell death (Adams and Cory, 1998; Gross et al., 1999). This family makesup antiapoptotic proteins, such as Bcl-2 and Bcl-X(L), and pro-apoptotic proteins, such as Bax and Bak. Heterodimerization of proteins from these two polar groups modulates cellular response to environmental cues, resulting in life or death of the cell (Cheng et al., 2001). The domains involved in heterodimerization are the *Bcl*-2 homology domains 1–4 (BH1–4) (Lutz, 2000). Specific mutations in these regions disrupt protein–protein interactions and either increase or decrease antiapoptotic activity. In addition, upregulation of antiapoptotic proteins such as *Bcl*-2 and *Bcl*-X(L) contributes to the tumorigenesis and resistance to drug treatments in certain types of

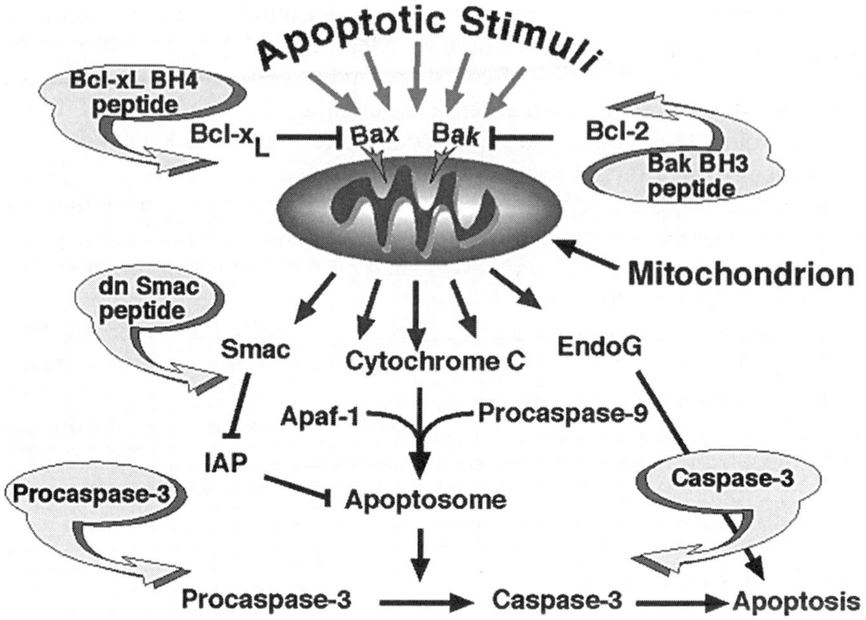

Figure 6.5 Bcl-2 family members play a crucial role in the regulation of apoptosis and make up both antiapoptotic proteins, such as Bcl-2 and Bcl-X(L), and pro-apoptotic proteins, such as Bax and Bak. Fusion of the BH3 death-promoting domain of Bak and Bax to the Antp PTD induces apoptosis. In contrast, fusion of the Bcl-X(L) BH4 domain to the TAT PTD prevents apoptosis. $Smac^{WT}$-Antp peptides have also been used to override the mitochondria-dependent activation step of apoptosis, resulting in enhanced apoptotic capabilities of chemotherapeutic agents. Last, the transduction of caspase 3 has been proven to target tumor cells.

cancer, including follicular lymphoma, breast, prostate, lung, ovary, and colon carcinomas (Zornig et al., 2001). Antagonizing these death suppressors is an attractive target for combination anticancer therapy. Hence, small peptides mimicking BH domains may be valuable tools in promoting or preventing cell death, and may serve as potential drugs to modulate the cell's susceptibility to apoptosis (Fig. 6.5).

The BH3 domain is involved in the death-promoting functions of Bak and Bax (Lutz, 2000). The Bak BH3 peptide is believed to bind *Bcl*-xL and competitively abrogate *Bcl*-xL/Apaf-1 heterodimerization, resulting in Apaf-1-dependent activation of caspases and consequently apoptosis (Cosulich et al., 1997). Fusion of the BH3 domain peptide to the Antp PTD (Antp-BH3) resulted in efficient transduction into HeLa cells and induced apoptosis, including caspase-dependent cleavage of poly (ADP-ribose) polymerase, cytoplasmic contraction, membrane blebbing, and the formation of apoptotic bodies (Holinger et al., 1999). Morphological changes take place within two to three hours of the addition of the Antp-BH3 peptide and cell viability dropped dramatically within six hours. In contrast, Antp PTD or BH3 peptides alone were ineffective. Importantly, the mutant Ant-BH3-L78A peptide with a single amino acid substitution and aberrant binding activity to *Bcl*-xL failed to initiate apoptosis, implying that the intact binding site and the transduction domain comprise a truly pro-apoptotic internalized drug (Holinger et al., 1999). Consistent with these observations, Letai et al. (2002)

further refined these observations by utilizing the poly-Arg PTD to deliver BH3 domain peptides from multiple pro-apoptotic family members. These observations also demonstrate that transducible peptides preserve their 3D structure and biological function, even after passing through cellular and/or intracellular membranes.

Looking beyond the BH3 domain, the BH4 domain is present only among antiapoptotic *Bcl*-2 family members and is absolutely required for the prevention of cell killing (Reed et al., 1996). This domain plays an important role in preventing the loss of mitochondria potential and subsequent cytochrome-c release after pro-apoptotic stimulus. However, the question is, can the BH4 domain act on its own, as a single peptide? Fusion of the *Bcl*-xL BH4 domain peptide to the TAT PTD resulted in efficient transduction into HeLa cells (Shimizu et al., 2000). The TAT-BH4 peptide, but not the TAT or BH4 peptides alone, significantly prevented VP-16-induced apoptosis in a concentration-dependent manner (Shimizu et al., 2000). Taken together, these observations demonstrate that PTD delivery of specific pro-apoptotic peptides that exceed the bioavailability delivery size results in specific modulation of the cellular apopotic machinery.

6.5.2 CASPASE-3

The activation of pro-caspase 3, an effector caspase, is the final step in the apoptotic pathway (Zimmermann et al., 2001). Proteolytic cleavage of the pro-caspase 3, stimulated by pro-apoptotic stimuli, generates two subunits that join to form an active, heterotetrameric enzyme. Activated caspase 3 in turn triggers activation of caspase-activated DNAse (CAD) by cleavage of the inhibitor of CAD (ICAD) (Nagata, 2000; Vaughan et al., 2002). With the aim of introducing functionally active caspase 3 into the cell, the activating protease needs to be supplied to cleave pro-caspase 3 into the active form. To do so, one approach is to substitute the natural cleavage sites for novel protease ones. Vocero-Akbani et al. (1999) have devised such an approach by substituting HIV protease cleavage sites for the endogenous ones. The resultant TAT-Caspase 3^{HIV} protein was specifically activated only in HIV-infected cells and induced cell-specific apoptosis.

This strategy has also been applied to the culture of cardiomyocytes (Wu et al., 2000). The addition of TAT-pro-Caspase 3^{HIV} along with TAT-HIV protease to the culture of cardiomyocytes causes significant loss of cell viability within 16 h. Introduction of the TAT-pro-caspase 3^{HIV} alone or the TAT-pro-caspase 3^{HIV} (mutant) with the HIV protease did not have an effect. The level of induced cytotoxicity increased with increasing concentrations of the TAT-wild-type pro-caspase 3 (Wu et al., 2000).

This experimental model proves that the activation of caspase 3 is sufficient to promote apoptosis in cardiomyocytes. It is important that, it also represents the proof of concept that PTD-mediated intracellular delivery of a caspase can promote the irreversible steps of programmed cell death. The next obvious move is to make caspase activation tumor-dependent. This can be achieved by inserting the cleavage site for either tumor-specific proteases or at least for proteases with augmented activities in tumor cells. Although there are currently no identified intracellular tumor-specific proteases, several extracellular proteases are upregulated during oncogenesis, including metalloproteases and PSA (Frankel et al., 2002).

The validity of applying this approach to cancer therapy has recently been confirmed by creating a TAT–oxygen-dependent degradation (ODD)–caspase-3 fusion protein (Harada et al., 2002). The hypoxia-inducible factor (HIF-1α) transcription factor regulates multiple genes involved in responding to changes in the intracellular oxygen levels, including upregulation of vascular endothelial growth factor (VEGF) (Semenza, 2000). Activation of HIF-1α is regulated by the inverse of oxygen concentration; hypoxia causes stabilization of the protein. The ODD sequence within central region of the HIF-1α promotes ubiquitination and proteasomal degradation of the protein under normoxic conditions. Low oxygen pressure (hypoxia) in solid tumors results in the inactivation of ODD-dependent degradation, stabilization of HIF-1α and transcriptional induction of target genes to stimulate angiogenesis (Maxwell and Ratcliffe, 2002). Thus the ODD serves to act as a biochemical sensor of intracellular oxygen.

Harada et al. (2002) tested the functionality of tethering the ODD to heterologous proteins by linking it to a transducible TAT-β-galactosidase (β-Gal), originally characterized by Schwarze et al. (1999). Injection of TAT-ODD-β-Gal into mice bearing solid tumors resulted in intensive β-Gal activity in the hypoxic core of the tumor with low oxygen tension. In contrast, injection of the control TAT-β-Gal protein (minus the ODD) resulted in equal distribution throughout the tumor. In addition, injection of TAT-β-Gal protein, but not TAT-ODD-β-Gal protein, led to positive β-Gal activity in normal tissue (liver), implying the rapid degradation of TAT-ODD-β-Gal under normoxic conditions. Harada et al. (2002) then tested a modified form of this protein that included caspase 3, TAT-ODD-Casp3^{WT} by injecting this protein into tumor-bearing mice. It was impressive, that treatment of mice with TAT-ODD-Casp3^{WT} protein resulted in reduced tumor masses, whereas treatment with an inactive, mutant TAT-ODD-Casp3^{MUT} protein had no effect. As predicted, caspase 3 stabilization in hypoxic tumor tissue activated the transduced caspase 3 and induced tumor-specific apoptosis in vivo.

The advantage of this Trojan horse strategy relies on the virtually unlimited possibilities for converting a latent form of a cytotoxic enzyme into an active one, exclusively in the tumor cells. As a hypothetical possibility, one can consider cyclin E-specific protease. There are multiple truncated forms of cyclin E in breast cancer cells that originate from proteolytic cleavage. These forms are found exclusively in tumor cells, apparently due to activity of tumor-specific protease(s) (Porter et al., 2001). Porter et al. (2001) mapped the cyclin E cleavage sites and proved the necessity of truncated cyclin E protein for tumor cell growth. These observations suggest that the design of inhibitors of the cyclin E-specific protease may prevent formation of superactive cyclin E. Although this is an attractive idea, it will require additional investment in the identification of the protease and drug search. However, the cleavage site(s) could be introduced into either the TAT-caspase 3 or TAT-ODD-caspase 3 fusion proteins and then assayed for tumor selectivity.

The activation of caspases does not necessarily need to be tumor specific to have some therapeutic benefit. Mai et al. (2001) fused small KLAKLAK peptides to a PTD and triggered apoptotic induction activation of pro-caspase 3. Direct injection of the PTD- KLAKLAK peptide into solid tumors led to a mix of in vivo responses from a reduction to a complete halt in tumor growth rates. However, due to the nonspecific activation of caspase 3, this approach could be used clinically only by direct administration. Nevertheless, this group is confident that the drug will serve best in the treatment of a surgically inaccessible site or as an adjuvant therapy in conjunction with other conventional therapies and surgical debulking (Mai et al., 2001).

6.5.3 PRO-APOPTOTIC SMAC PEPTIDE

Inhibitors of apoptosis proteins (IAPs), including XIAP and survivin, regulate the activity of caspases by blocking caspase-active sites (Yang and Li, 2000). IAPs are often overexpressed in malignant tissues and are cited as one basis for chemoresistance or radioresistance. Consequently, selective inactivation of IAPs is a promising strategy for defeating the resistance of a tumor cell to a pro-apoptotic therapy. The mitochondrial protein Smac/Diablo is a natural inhibitor of IAPs that binds these proteins and disrupts their ability to sequester and inactivate caspases (Holcik et al., 2001; Shi, 2002). The IAP binding site on Smac resides in the N-terminal four amino acids, which are critical and sufficient for binding (Liu et al., 2000). One feasible strategy to promote caspase-dependent apoptosis is to introduce the Smac peptide directly into the cytoplasm, thereby overriding mitochondrial steps needed to activate apoptosis.

To explore a Smac-based strategy to promote apoptosis, Arnt et al. (2002) fused four to eight residues of the Smac N-terminus that is sufficient to disrupt IAP–caspase interaction to the Antp PTD. Treatment of human breast cancer cells with the $Smac^{WT}$-Antp peptide significantly enhanced drug-induced apoptosis (Arnt et al., 2002). By itself, the $Smac^{WT}$-Antp peptide did not induce apoptosis; it merely sensitized the cells to the proapoptotic stimulus triggered by the chemotherapeutic drug. Moreover, while subtherapeutic concentrations of pro-apoptotic drugs, such as SN-38 and paclitaxel, failed to induce apoptosis on their own, they synergized with the $Smac^{WT}$-Antp peptide to kill cells. In contrast, treatment with a mutated $Smac^{MUT}$-Antp peptide did not promote apoptosis in the cell lines studied. It is important that the $Smac^{WT}$-Antp peptide not only enhanced apoptotic induction by the chemotherapeutic agents but also provided a long-term antiproliferative effect of the combined treatment, as has been demonstrated by colony-forming assays (Arnt et al., 2002).

One drawback of such an approach for therapeutic purposes is the nonspecific delivery of a transducible Smac-PTD peptide to all cell types, normal as well as tumorigenic. Normal cells may be as vulnerable to this treatment as tumorigenic cells. Hence the approach may be limited to combination with chemotherapeutic agents that show tumor selectivity but limited efficacy.

Further refinements in selective tumor transduction may aid with the development of this strategy.

One well-known candidate for selective anticancer therapy is the TNF-related apoptosis-inducing ligand (TRAIL) (Wajant et al., 2002). TRAIL induces apoptosis by activating the TRAIL-R1 and TRAIL-R2 death receptors. Many normal cells express "decoy" receptors that compete for TRAIL binding (but do not transmit death signals), thereby preventing activation of the TRAIL receptors. However, tumor cells fail to express decoy TRAIL-R1 and are, therefore, more susceptible to TRAIL-induced apoptosis. Recombinant epitope-nontagged version of TRAIL derivatives have been shown to induce apoptosis in a broad range of tumor cells and a limited number of nontransformed, normal cell lines (Wajant et al., 2002). However, not all tumor cells respond robustly to TRAIL; some neuroblastoma, melanoma, and pancreatic carcinoma cells possess defects in apoptotic signaling that render them nonresponsive to TRAIL (Eggert et al., 2001; Hersey and Zhang, 2001; Ibrahim et al., 2001). It is notable, however, that treatment of these transformed cells with a Smac^{WT}-TAT PTD chimera can sensitize these cells to TRAIL, causing a remarkable decline in tumor cell viability on its administration (Fulda et al., 2002). Taken together, these observations offer of a proof of concept for the utility of combining sensitizing transducible peptides with tumor-selective anticancer agents.

A major remaining challenge is to expand these cell culture observations to in vivo mouse tumor models to assess the potential efficacy and specificity of tumor cell death. Fulda et al. (2002) locally administered TRAIL with Smac-TAT peptide into gliomas established by implanting tumor cells into the striatum of athymic mice. It is striking that treated mice showed complete tumor regression in the absence of clinical symptoms after receiving this combination treatment. In contrast, mice treated with vehicle, Smac-TAT peptide alone, or TRAIL alone developed an increased tumor burden and died within 30 days. It is notable that co-injection of Smac-TAT peptide and TRAIL into normal mouse brain also did not result in any detectable neurotoxicity (Fulda et al., 2002). The action of the transducible Smac-TAT peptide alone was insufficient to induce apoptosis. Thus this preclinical study further validated the potential of protein transduction technology to selectively target tumor cell biology, while leaving surrounding normal tissues relatively unharmed. It is important that, because this is an epigenetic manipulation, cells are sensitized only to apoptosis for as long as the transducible protein–peptide exists within the cell. Similar to a loaded and cocked gun, the cell's suicide protocol is ready to be fired. The administration of a subthreshold, tumor-specific proapoptotic compound then "pulls the trigger" exclusively in transformed cells, leaving normal cells unharmed.

6.5.4 p53 TUMOR SUPPRESSOR

The p53 tumor suppressor gene, a DNA damage sensor, is mutated in 50% of all human tumors and the loss of normal p53 or the p53 pathway increases the resistance of cancer cells to therapy (Sherr and McCormick, 2002). Due to mutations and alterations in the DNA damage repair machinery, tumor cells are thought to undergo continuous DNA damage, whereas normal cells maintain the ability to repair damaged DNA. Consequently, in theory and in practice, reconstitution of p53 function restores the link between proapoptotic stimuli and the apoptotic execution machinery, resulting in cell death in a DNA damage-dependent manner. Thus many investigators have sought to reactivate p53 as a strategy for anticancer therapy.

Hupp et al. (1995) identified a C-terminal peptide of p53 that activates both wild-type p53 and several mutant forms of p53 that have deficiencies in DNA binding. Selivanova et al. (1997) linked the C-terminal p53 peptide to the Antp PTD and found that it induced p53-dependent apoptosis in several tumor cell lines. It is important that normal cells harboring wild-type p53 were resistant to the p53-Antp peptide (Selivanova et al., 1997). These observations demonstrate that activation of p53 can selectively kill tumor cells; however, the potential in vivo efficacy of this approach remains to be assessed.

Takenobu et al. (2002) generated a genetic fusion of the full-length p53 protein with a C-terminal poly-Arg PTD. Treatment of p53 null osteosarcoma cells resulted in activation of the promoter of the p21WAF1 gene, a transcriptional target of p53, and inhibition of cell growth. The p53-PTD fusion also sensitized tumor cells to cisplatin-induced apoptosis. While still at an early point, these observations present evidence that favors the concept that a fully functional p53 protein can be reconstituted intracellularly by protein transduction (Takenobu et al., 2002).

p53 is negatively regulated by the human homologul of MDM2 (HDM2) protein, which binds to the N-terminal transactivation domain of p53 and inhibits its transcriptional activity as well as promotes its degradation (Wu et al., 1993). It is not surprising that HDM2 is overexpressed in certain tumors that contain wild-type p53, resulting in its functional inactivation. Harbour et al. (2002) generated a transducible N-terminal p53 peptide fused to the TAT PTD and found that it disrupted p53/HDM2 binding and resulted in the liberation of p53. Treatment of cells in culture with the TAT-N-terminal p53 peptide bypassed HDM2 regulation of p53 and induced p53-dependent apoptosis. It is important that this same peptide had minimal effects on normal cells. Moreover, treatment retinoblastoma xenograft tumors generated in a rabbit eye model induced tumor-specific apoptosis, resulting in a dramatic tumor volume reduction (Harbour et al., 2002). Together, these observations are among the first in vivo studies with transducible peptides that demonstrate selective tumor killing.

6.6 Applications in Cancer Vaccines

Cytotoxic T lymphocytes (CTL) recognize tumor-associated antigens (TAAs) bound to MHC class I molecules and elicit cytotoxic responses that cause the development of immune memory cells that may safeguard against recurrent tumorigenesis (Jager et al., 1999). TAAs in the form of exogenous peptides have been shown to induce a systemic immune response on proper administration to the tumor-bearing patient (Eisenbach et al., 2000). However, the low immunogenicity of natural epitopes expressed by tumor cells, impairing the capabilities of the host's T-cells to trigger a sufficient immune response (Meng and Butterfield, 2002). Currently, there are two well-known methods to override the immunogenicity barrier: modifying TAA peptides to augment their interaction with HLA and T-cell receptors and exploiting tumor antigen-loaded dendritic cells (DCs) (the most efficient type of antigen-presenting cell) as enhancers of an antitumor immune response (Mayordomo et al., 1995). DCs proteolytically degrade antigens into peptides and present processed peptides in complexes with MHC class I and II; therefore, DCs are key cells for initiating an immune reaction in naive T cells. DCs loaded with tumor antigens in vitro induce immune responses in vivo after reimplantation into the host (Tarte and Klein, 1999). To load DCs one can either transfect in cDNAs or infect them with various viruses encoding tumor-specific antigens. However, the problem of how to efficiently pulse sufficient numbers of DCs with antigens for clinical practice remains unknown.

PTD peptides offer one way to deliver TAAs to sufficient numbers of DCs. Attachment of the antigen to a PTD to facilitate direct delivery of antigens to the cytosolic compartment of DCs is an attractive method for enhancing antigen presentation, while avoiding potential problems of DNA integration associated with genetic approaches. Indeed, linkage of the Antp PTD to an otherwise nonimmunogenic peptide promotes internalization, processing, and presenting of the epitope by immature DCs, efficiently enough to activate antigen-specific CTLs (Chikh et al., 2001). Encapsulation of Antp-fusions into liposomes, as a protective measure against extracellular protein degradation, can further improve the response of CTLs in vivo. These observations imply that successful conversion of an otherwise nonimmunogenic antigen into an immunogenic antigen by delivery via a PTD is a suitable model for weak tumor epitopes, perhaps useful for the development of more robust cancer vaccines (Chikh et al., 2001).

Shibagaki and Udey (2002) have further refined the use of protein transduction to present tumor antigens in vivo. Ovalbumin (OVA) is a convenient and standard model antigen that is recognized by MHC class I molecules and allows for a straightforward comparison of DCs transduced with recombinant PTD-OVA with DCs loaded with OVA peptide alone. It is interesting that on injection in mice, DCs transduced with TAT-OVA elicited a strong OVA-specific CTL response in vitro (Shibagaki and Udey, 2002). It is important that both control TAT PTD peptide fused to irrelevant proteins and OVA peptide alone (minus the TAT PTD) were ineffective in activating a CTL response. Mice engrafted with thymoma cells constitutively expressing OVA quickly develop solid tumors (Shibagaki and Udey, 2002). It is impressive that immunization of tumor-bearing mice with DCs transduced with TAT-OVA peptide showed a marked reduction in tumor volume. In contrast, DCs pulsed with the control OVA peptide were significantly less effective in tumor protection.

The advantage of loading DCs with PTD–antigen peptides is futher illustrated by studies with tyrosinase-related protein 2 (TRP2), a tumor-rejection antigen for human B16 melanoma (Wang et al., 2002). DCs loaded with TRP2 peptide can initiate CTL response and promote B16 tumor rejection. However, the efficiency of TRP2 pulsed DCs is unacceptably low. In contrast, TRP2 peptide covalently linked to the TAT PTD dramatically potentiated the

DCs ability to present MHC-loaded antigen (peptide) to T cells and increased antigen immunogenicity (Wang et al., 2002). Immunization of mice with DCs pretreated with TAT-TRP2 peptides during 2 weeks resulted in complete protection from B16 tumor challenge as measured by decreased number of lung metastases. Moreover, the survival rate of mice immunized with DCs loaded with TAT-TRP2 was also significantly increased.

There may be significant advantages to priming DCs with full-length antigens instead of antigen peptides. First, full-length antigens are processed intracellularly and displayed on the DC surface as peptides bound to MHC class I and class II. Those complexes present peptides to both $CD8^+$ CTL and $CD4^+$ Th cells. Second, antigen-transduced DCs display greater numbers of epitopes recognized by CTL than peptide-pulsed DCs do. In terms of potential clinical applications, protein-transduced DCs are superior to peptide-pulsed DCs, because predetermination of MHC class I and class II binding peptides is necessary. Transduction appears to be a more convenient and efficient way to load DCs compared to viral infection or transfection (Shibagaki and Udey, 2002). PTD-based strategies to load DCs with full-length antigens may, therefore, offer superior utility.

TAT-mediated antigen delivery into DCs enhances protective immunity and therapeutic immunity against B16 tumors: Immunization of mice with loaded DCs after intravenous injection of B16 melanoma cells notably reduced the number of lung metastases (Wang et al., 2002). It is interesting that both $CD4^+$ and $CD8^+$ T cells were activated on mice immunization with TAT-peptide transduced DCs, possibly explaining the improved antitumor immunity versus the poorly immunogenic antigen (Wang et al., 2002).

Due to rapid MHC turnover and peptide degradation, the half-life of peptide-MHC class I complexes is limited. Consequently, DCs loaded in vitro with peptides will have a limited time to present antigen to CTLs in vivo. Fusion of antigen peptides to PTDs increases the intracellular pool of antigen and may prolong the time of DC-dependent antigen presentation. Thus loading DCs with PTD–antigen peptides offers an attractive strategy for improving the immunogenicity of tumor antigens and thereby the efficacy of anticancer vaccines. In addition, another obvious advantage of transduction is the fast and reliable validation of novel putative tumor antigens. Any protein that is differentially displayed on the surface of tumor cells can be defined, cross-linked to a PTD domain, transduced into DCs and tested on CTL response without even knowing the biochemical structure of the protein. With currently available methods of fast protein separation, creating a library of transducible antigens is a realistic consideration. In summary, the possible applications of protein transduction technology to anticancer vaccines further highlight the flexibility and utility of this technology.

6.7 Summary

The extensive development of protein transduction over the past several years has opened a variety of new opportunities for anticancer therapeutic strategies. By overcoming the problem of delivering large molecules across plasma membrane, the spectrum of potential targets has been significantly increased. This strategy does not stand alone; its applications in combinatorial cancer therapy may offer one of the stronger demonstrations of its utility and flexibility. However, in the laboratory, protein transduction has the potential in the short-term to rapidly increase the time needed to validate targets, potentially speeding the translation of experimental therapies to clinical trials.

References

Abu-Amer, Y., Dowdy, S. F., Ross, F. P., et al. TAT fusion proteins containing tyrosine 42-deleted I kappa B alpha arrest osteoclastogenesis. *J. Biol. Chem.* **276,** 30499–30503 (2001).

Adams, J. M., and Cory, S. The Bcl-2 protein family: Arbiters of cell survival. *Science* **281,** 1322–1326 (1998).

Adjei, A. A. Blocking oncogenic *ras* signaling for cancer therapy. *J. Natl. Cancer Inst.* **93,** 1062–1074 (2001).

Arnt, C. R., Chiorean, M. V., Heldebrant, M. V., et al. Synthetic Smac/DIABLO peptides enhance the effects of chemotherapeutic agents by binding XIAP and cIAP1 in situ. *J. Biol. Chem.* **277,** 44236–44243 (2002).

Asada, S., Choi, Y., Yamada, M., et al. External control of HER2 expression and cancer cell growth by targeting a *ras*-linked coactivator. *Proc. Natl. Acad. Sci. USA* **99,** 12747–12752 (2002).

Badou, A., Bennasser, Y., Moreau, M., et al. Tat protein of human immunodeficiency virus type 1 induces interleukin-10 in human peripheral blood monocytes: Implication of protein kinase C-dependent pathway. *J. Virol.* **74,** 10551–10562 (2002).

Baldari, C. T., Heguy, A., and Telford, J. L. *Ras* protein activity is essential for T-cell antigen receptor signal transduction. *J. Biol. Chem.* **268,** 2693–2698 (1993).

Baldwin, A. S. Control of oncogenesis and cancer therapy resistance by the transcription factor NF-kappaB. *J. Clin. Invest.* **107,** 241–246 (2001).

Ball, K. L., Lain, S., Fahraeus, R., et al. Cell-cycle arrest and inhibition of Cdk4 activity by small peptides based on the carboxy-terminal domain of p21WAF1. *Curr. Biol.* **7,** 71–80 (1997).

Banerji, L., Glassford, J., Lea, N. C., et al. BCR signals target p27(Kip1) and cyclin D2 via the PI3-K signalling pathway to mediate cell cycle arrest and apoptosis of WEHI 231 B cells. *Oncogene* **20,** 7352–7367 (2001).

Bashour, A. M., Meng, J. J., Ip, W., et al. The neurofibromatosis type 2 gene product, merlin, reverses the F-actin cytoskeletal defects in primary human Schwannoma cells. *Mol. Cell. Biol.* **22,** 1150–1157 (2002).

Blagosklonny, M. V., and Pardee, A. B. Exploiting cancer cell cycling for selective protection of normal cells. *Cancer Res.* **61,** 4301–4305 (2001).

Bonfanti, M., Taverna, S., Salmona, M., et al. p21WAF1-derived peptides linked to an internalization peptide inhibit human cancer cell growth. *Cancer Res.* **57,** 1442–1446 (1997).

Bonny, C., Oberson, A., Negri, S., et al. Cell-permeable peptide inhibitors of JNK novel blockers of beta-cell death. *Diabetes* **50,** 77–82 (2001).

Bruce-Keller, A. J., Barger, S. W., Moss, N. I., et al. Pro-inflammatory and pro-oxidant properties of the HIV protein TAT in a microglial cell line: attenuation by 17 beta-estradiol. *J. Neurochem.* **78,** 1315–1324 (2001).

Buday, L. Membrane-targeting of signalling molecules by SH2/SH3 domain-containing adaptor proteins. *Biochim Biophys Acta* 1422:187–204 (1999).

Chang, C. H., Scott, G. K., Kuo, W. L., et al. ESX: A structurally unique ETS overexpressed early during human breast tumorigenesis. *Oncogene* **14,** 1617–1622 (1997).

Chang, F., Steelman, L. S., Shelton, J. G., et al. Regulation of cell cycle progression and apoptosis by the Ras/Raf/MEK/ERK pathway. *Int. J. Oncol.* **22,** 469–480 (2003).

Chen, Y. N. P., Sharma, S. K., Ramsey, T. M., et al. Selective killing of transformed cells by cyclin/cyclin- dependent kinase 2 antagonists. *Proc. Natl. Acad. Sci. USA* **96,** 4325–4329 (1999).

Cheng, E. H., Wei, M. C., Weiler, S., et al. BCL-2, BCL-X(L) sequester BH3 domain-only molecules preventing BAX- and BAK-mediated mitochondrial apoptosis. *Mol. Cell* **8,** 705–711 (2001).

Chikh, G. G., Kong, S., Bally, M. B., et al. Efficient delivery of antennapedia homeodomain fused to CTL epitope with liposomes into dendritic cells results in the activation of CD8(+) T cells. *J. Immunol.* **167,** 6462–6470 (2001).

Cosulich, S. C., Worrall, V., Hedge, P. J., et al. Regulation of apoptosis by BH3 domains in a cell-free system. *Curr. Biol.* **7,** 913–920 (1997).

Cox, A. D. Farnesyltransferase inhibitors: Potential role in the treatment of cancer. *Drugs* **61,** 723–732 (2001).

Cussac, D., Frech, M., and Chardin, P. Binding of the Grb2 SH2 domain to phosphotyrosine motifs does not change the affinity of its SH3 domains for SOS proline-rich motifs. *EMBO J.* **13,** 4011–4021 (1994).

Cussac, D., Vidal, M., Leprince, C., et al. A SOS-derived peptidimer blocks the *ras* signaling pathway by binding both Grb2 SH3 domains and displays antiproliferative activity. *FASEB J.* **13,** 31–38 (1999).

Datta, K., Nambudripad, R., Pal, S., et al. Inhibition of insulin-like growth factor-I-mediated cell signaling by the von Hippel-Lindau gene product in renal cancer. *J. Biol. Chem.* **275,** 20700–20706 (2000).

Datta, K., Sundberg, C., Karumanchi, S. A., and Mukhopadhyay, D. The 104–123 amino acid sequence of the beta-domain of von Hippel-Lindau gene product is sufficient to inhibit renal tumor growth and invasion. *Cancer Res.* **61,** 1768–1775 (2001).

Derossi, D., Joliot, A. H., Chassaing, G., and Prochiantz, A. The third helix of the Antennapedia homeodomain translocates through biological membranes. *J. Biol. Chem.* **269,** 10444–10450 (1994).

Derossi, D., Williams, E. J., Green, P. J., et al. Stimulation of mitogenesis by a cell-permeable PI 3-kinase binding peptide. *Biochem. Biophys. Res. Commun.* **251,** 148–152 (1998).

Dunican, D. J., and Doherty, P. Designing cell-permeant phosphopeptides to modulate intracellular signaling pathways. *Biopolymers* **60,** 45–60 (2001).

Eggert, A., Grotzer, M. A., Zuzak, T. J., et al. Resistance to tumor necrosis factor-related apoptosis-inducing ligand (TRAIL)-induced apoptosis in neuroblastoma cells correlates with a loss of caspase-8 expression. *Cancer Res.* **61,** 1314–1319 (2001).

Eisenbach, L., Bar-Haim, E., and El-Shami, K. Antitumor vaccination using peptide based vaccines. *Immunol. Lett.* **74,** 27–34 (2000).

el-Deiry, W. S., Harper, J. W., O'Connor, P. M., et al. WAF1/CIP1 is induced in p53-mediated G1 arrest and apoptosis. *Cancer Res.* **54,** 1169–1174 (1994).

Ezhevsky, S. A., Ho, A., Becker-Hapak, M., et al. Differential regulation of retinoblastoma tumor suppressor protein by G(1) cyclin-dependent kinase complexes in vivo. *Mol. Cell. Biol.* **21,** 4773–4784 (2001).

Ezhevsky, S. A., Nagahara, H., Vocero-Akbani, A. M., et al. Hypo-phosphorylation of the retinoblastoma protein (pRb) by cyclin D:Cdk4/6 complexes results in active pRb. *Proc. Natl. Acad. Sci. USA* **94,** 10699–10704 (1997).

Fahraeus, R., Lain, S., Ball, K. L., and Lane, D. P. Characterization of the cyclin-dependent kinase inhibitory domain of the INK4 family as a model for a synthetic tumour suppressor molecule. *Oncogene* **16,** 587–596 (1998).

Fahraeus, R., Paramio, J. M., Ball, K. L., et al. Inhibition of pRb phosphorylation and cell-cycle progression by a 20-residue peptide derived from p16CDKN2/INK4A. *Curr. Biol.* **6,** 84–91 (1996).

Fawell, S., Seery, J., Daikh, Y., et al. TAT-mediated delivery of heterologous proteins into cells. *Proc. Natl. Acad. Sci. USA* **91,** 664–668 (1994).

Ford, K. G., Souberbiele, B. E., Darling, D., and Farzaneh, F. Protein transduction: An alternative to genetic intervention? *Gene Therapy* **8,** 1–4 (2001).

Frankel, A. D., and Pabo, C. O. Cellular uptake of the TAT protein from human immunodeficiency virus. *Cell* **55,** 1189–1193 (1998).

Frankel, A. E., Bugge, T. H., Liu, S., et al. Peptide toxins directed at the matrix dissolution systems of cancer cells. *Protein Pept. Lett.* **9,** 1–14 (2002).

Fujihara, S. M., Cleaveland, J. S., Grosmaire, L. S., et al. A D-amino acid peptide inhibitor of NF-kappa B nuclear localization is efficacious in models of inflammatory disease. *J. Immunol.* **165,** 1004–1012 (2000).

Fulda, S., Wick, W., Weller, M., and Debatin, K. M. Smac agonists sensitize for Apo2L/TRAIL- or anticancer drug-induced apoptosis and induce regression of malignant glioma in vivo. *Nature Med.* **8,** 808–815 (2002).

Garbay, C., Liu, W. Q., Vidal, M., and Roques, B. P. Inhibitors of *Ras* signal transduction as antitumor agents. *Biochem. Pharmacol.* **60,** 1165–1169 (2000).

Gius, D. R., Ezhevsky, S. A., Becker-Hapak, M., et al. Transduced p16INK4a peptides inhibit hypophosphorylation of the retinoblastoma protein and cell cycle progression prior to activation of Cdk2 complexes in late G1. *Cancer Res.* **59,** 2577–2580 (1999).

Green, M., and Loewenstein, P. M. Autonomous functional domains of chemically synthesized human immunodeficiency virus TAT trans-activator protein. *Cell* **55,** 1179–1188 (1998).

Gross, A., McDonnell, J. M., and Korsmeyer, S. J. BCL-2 family members and the mitochondria in apoptosis. *Genes Dev.* **13,** 1899–1911 (1999).

Hall, D. J., Cui, J., Bates, M. E., et al. Transduction of a dominant-negative H-*ras* into human eosinophils attenuates extracellular signal-regulated kinase activation and interleukin-5-mediated cell viability. *Blood* **98,** 2014–2021 (2001).

Hanahan, D., and Weinberg, R. A. The hallmarks of cancer. *Cell* **100,** 57–70 (2000).

Harada, H., Hiraoka, M., and Kizaka-Kondoh, S. Antitumor effect of TAT-oxygen-dependent degradation-caspase-3 fusion protein specifically stabilized and activated in hypoxic tumor cells. *Cancer Res.* **62,** 2013–2018 (2002).

Harbour, J. W., Worley, L., Ma, D. D., and Cohen, M. Transducible peptide therapy for uveal melanoma and retinoblastoma. *Arch. Ophthalmol.* **120,** 1341–1346 (2002).

Hersey, P., and Zhang, X. D. How melanoma cells evade TRAIL-induced apoptosis. *Nature Rev. Cancer* **1,** 142–150 (2001).

Ho, A., and Dowdy, S. F. Regulation of G(1) cell-cycle progression by oncogenes and tumor suppressor genes. *Curr. Opin. Genet. Dev.* **12,** 47–52 (2002).

Holcik, M., Gibson, H., and Korneluk, R. G. XIAP: Apoptotic brake and promising therapeutic target. *Apoptosis* **6,** 253–261 (2001).

Holinger, E. P., Chittenden, T., and Lutz, R. J. Bak BH3 peptides antagonize Bcl-xL function and induce apoptosis through cytochrome c-independent activation of caspases. *J. Biol. Chem.* **274,** 13298–13304 (1999).

Huang, P., and Oliff, A. Signaling pathways in apoptosis as potential targets for cancer therapy. *Trends Cell. Biol.* **11,** 343–348 (2001).

Hupp, T. R., Sparks, A., and Lane, D. P. Small peptides activate the latent sequence-specific DNA binding function of p53. *Cell* **83,** 237–245 (1995).

Ibrahim, S. M., Ringel, J., Schmidt, C., et al. Pancreatic adenocarcinoma cell lines show variable susceptibility to TRAIL-mediated cell death. *Pancreas* **23,** 72–79 (2001).

Jager, E., Jager, D., and Knuth, A. CTL-defined cancer vaccines: Perspectives for active immunotherapeutic interventions in minimal residual disease. *Cancer Metastasis Rev.* **18,** 143–150 (1999).

Jo, D. W., Nashabi, A., Doxsee, C., et al. Epigenetic regulation of gene structure and function with a cell-permeable Cre recombinase. *Nature Biotechnol.* **19,** 929–933 (2001).

Kabouridis, P. S., Hasan, M., Newson, J., et al. Inhibition of NF-kappa B activity by a membrane-transducing mutant of I kappa B alpha. *J. Immunol.* **169,** 2587–2593 (2002).

Karin, M. and Ben-Neriah, Y. Phosphorylation meets ubiquitination: The control of NF-[κ]B activity. *Ann. Rev. Immunol.* **18,** 621–663 (2002).

Kelemen, B. R., Hsiao, K., and Goueli, S. A. Selective in vivo inhibition of mitogen-activated protein kinase activation using cell-permeable peptides. *J. Biol. Chem.* **277,** 8741–8748 (2002).

Krek, W., Ewen, M. E., Shirodkar, S., et al. Negative regulation of the growth-promoting transcription factor E2F-1 by a stably bound cyclin A-dependent protein kinase. *Cell* **78,** 161–172 (1994).

Lee, H. J., and Pardridge, W. M. Pharmacokinetics and delivery of TAT and TAT-protein conjugates to tissues in vivo. *Bioconjugate Chem.* **12,** 995–999 (2001).

Lee, M. H., and Yang, H. Y. Negative regulators of cyclin-dependent kinases and their roles in cancers. *Cell. Mol. Life Sci.* **58,** 1907–1922 (2001).

Leifert, J. A., Harkins, S., and Whitton, J. L. Full-length proteins attached to the HIV tat protein transduction domain are neither transduced between cells, nor exhibit enhanced immunogenicity. *Gene Ther.* **9,** 1422–1428 (2002).

Letai, A., Bassik, M. C., Walensky, L. D., et al. Distinct BH3 domains either sensitize or activate mitochondrial apoptosis, serving as prototype cancer therapeutics. *Cancer Cell* **2,** 183–192 (2002).

Lindgren, M., Hallbrink, M., Prochiantz, A., and Langel, U. Cell-penetrating peptides. *Trends Pharm. Sci.* **21,** 99–103 (2000).

Lindsay, M. A. Peptide-mediated cell delivery: Application in protein target validation. *Curr. Opin. Pharm.* **2,** 587–594 (2002).

Liu, Z., Sun, C., Olejniczak, E. T., et al. Structural basis for binding of Smac/DIABLO to the XIAP BIR3 domain. *Nature* **408,** 1004–1008 (2000).

Lowe, S. W., and Lin, A. W. Apoptosis in cancer. *Carcinogenesis* **21,** 485–495 (2000).

Lutz, R. J. Role of the BH3 (Bcl-2 homology 3) domain in the regulation of apoptosis and Bcl-2-related proteins. *Biochem. Soc. Trans.* **28,** 51–56 (2000).

Mai, J. C., Mi, Z. B., Kim, S. H., et al. A proapoptotic peptide for the treatment of solid tumors. *Cancer Res.* **61,** 7709–7712 (2001).

Mann, D. A., and Frankel, A. D. Endocytosis and targeting of exogenous HIV-1 TAT protein. *EMBO J.* **10,** 1733–1739 (1991).

Maxwell, P. H., and Ratcliffe, P. J. Oxygen sensors and angiogenesis. *Semin. Cell. Dev. Biol.* **13,** 29–37 (2002).

May, M. J., D'Acquisto, F., Madge, L. A., et al. Selective inhibition of NF-kappaB activation by a peptide that blocks the interaction of NEMO with the IkappaB kinase complex. *Science* **289,** 1550–1554 (2000).

Mayordomo, J. I., Zorina, T., Storkus, W. J., et al. Bone marrow-derived dendritic cells pulsed with synthetic tumour peptides elicit protective and therapeutic antitumour immunity. *Nature Med.* **1,** 1297–1302 (1995).

Meng, W. S., and Butterfield, L. H. Rational design of peptide-based tumor vaccines. *Pharm. Res.* **19,** 926–932 (2002).

Merel, P., Hoang-Xuan, K., Sanson, M., et al. Predominant occurrence of somatic mutations of the NF2 gene in meningiomas and schwannomas. *Genes Chromosomes Cancer* **13,** 211–216 (1995).

Myou, S., Zhu, X. D., Boetticher, E., et al. Blockade of focal clustering and active conformation in beta(2)-integrin-mediated adhesion of eosinophils to intercellular adhesion molecule-1 caused by transduction of HIV TAT-dominant negative *ras*. *J. Immunol.* **169,** 2670–2676 (2002).

Nagahara, H., Vocero-Akbani, A. M., Snyder, E. L., et al. Transduction of full-length TAT fusion proteins into mammalian cells: TAT-p27Kip1 induces cell migration. *Nature Med.* **4,** 1449–1452 (1998).

Nagata, S. Apoptotic DNA fragmentation. *Exp. Cell Res.* **256,** 12–18 (2000).

Ortega, S., Malumbres, M., and Barbacid, M. Cyclin D-dependent kinases, INK4 inhibitors and cancer. *Biochim. Biophys. Acta* **1602,** 73–87 (2002).

Porter, D. C., Zhang, N., Danes, C., et al. Tumor-specific proteolytic processing of cyclin E generates hyperactive lower-molecular-weight forms. *Mol. Cell. Biol.* **21,** 6254–6269 (2001).

Prendergast, G. C. Actin' Up: RhoB in cancer and apoptosis. *Nature Rev. Cancer* **1**, 162–168 (2001).

Rane, N., and Prendergast, G. C. Farnesyltransferase inhibitors: Mechanisms and clinical applications. *Exp. Opin. Invest. Drugs* **10**, 2105–2116 (2001).

Reed, J. C., Zha, H., Aime-Sempe, C., et al., Structure-function analysis of Bcl-2 family proteins. Regulators of programmed cell death. *Adv. Exp. Med. Biol.* **406,** 99–112 (1995).

Richard, J. P., Melikov, K., Vives, E., et al. Cell-penetrating peptides. A reevaluation of the mechanism of cellular uptake. *J. Biol. Chem.* **278,** 585–590 (2003).

Richmond, A. NF-kappa B, chemokine gene transcription and tumour growth. *Nature Rev. Immunol.* **2,** 664–674 (2002).

Rocco, J. W., and Sidransky, D. p16(MTS-1/CDKN2/INK4a) in cancer progression. *Exp. Cell Res.* **264,** 42–55 (2001).

Rothbard, J. B., Kreider, E., Vandeusen, C. L., et al. Arginine-rich molecular transporters for drug delivery: Role of backbone spacing in cellular uptake. *J. Med. Chem.* **45,** 3612–3618 (2002).

Rousselle, C., Clair, P., Lefauconnier, J. M., et al. New advances in the transport of doxorubicin through the blood-brain barrier by a peptide vector-mediated strategy. *Mol. Pharmacol.* **57,** 679–686 (2000).

Ruas, M., Brookes, S., McDonald, N. Q., and Peters, G. Functional evaluation of tumour-specific variants of p16INK4a/CDKN2A: Correlation with protein structure information. *Oncogene* **18,** 5423–5434 (1999).

Schwarze, S. R., Ho, A., Vocero-Akbani, A. and Dowdy, S. F. In vivo protein transduction: Delivery of a biologically active protein into the mouse. *Science* **285,** 1569–1572 (1999).

Selivanova, G., Iotsova, V., Okan, I., et al. Restoration of the growth suppression function of mutant p53 by a synthetic peptide derived from the p53 C-terminal domain. *Nature Med.* **3,** 632–638 (1997).

Semenza, G. L. HIF-1: Using two hands to flip the angiogenic switch. *Cancer Metastasis Rev.* **19,** 59–65 (2000).

Shapiro, G. I., Edwards, C. D., and Rollins, B. J. The physiology of p16(INK4A)-mediated G1 proliferative arrest. *Cell Biochem. Biophys.* **33,** 189–197 (2000).

Sherr, C. J. The INK4a/ARF network in tumour suppression. *Nature Rev. Mol. Cell. Biol.* **2,** 731–737 (2001).

Sherr, C. J., and McCormick, F. The RB and p53 pathways in cancer. *Cancer Cell* **2,** 103–112 (2002).

Sherr, C. J., and Roberts, J. M. CDK inhibitors: Positive and negative regulators of G1-phase progression. *Genes Dev.* **13,** 1501–1512 (1999).

Shi, Y. Mechanisms of caspase activation and inhibition during apoptosis. *Mol. Cell* **9,** 459–470 (2002).

Shibagaki, N., and Udey, M. C. Dendritic cells transduced with protein antigens induce cytotoxic lymphocytes and elicit antitumor immunity. *J. Immunol.* **168,** 2393–2401 (2002).

Shimizu, S., Konishi, A., Kodama, T., and Tsujimoto, Y. BH4 domain of antiapoptotic Bcl-2 family members closes voltage-dependent anion channel and inhibits apoptotic mitochondrial changes and cell death. *Proc. Natl. Acad. Sci. USA* **97,** 3100–3105 (2000).

Snyder, E. L., and Dowdy, S. F. Protein/peptide transduction domains: Potential to deliver large DNA molecules into cells. *Curr. Opin. Molec. Therapeutics* **3,** 147–152 (2001).

Takenobu, T., Tomizawa, K., Matsushita, M., et al. Development of p53 protein transduction therapy using membrane-permeable peptides and the application to oral cancer cells. *Mol. Cancer Ther.* **1,** 1043–1049 (2002).

Tari, A. M., and Lopez-Berestein, G. GRB2: A pivotal protein in signal transduction. *Semin. Oncol.* **28,** 142–147 (2001).

Tarte, K., and Klein, B. Dendritic cell-based vaccine: A promising approach for cancer immunotherapy. *Leukemia* **13,** 653–663 (1999).

Torchilin, V. P., Rammohan, R., Weissig, V., and Levchenko, T. S. TAT peptide on the surface of liposomes affords their efficient intracellular delivery even at low temperature and in the presence of metabolic inhibitors. *Proc. Natl. Acad. Sci. USA* **98,** 8786–8791 (2001).

Vaughan, A. T., Betti, C. J., and Villalobos, M. J. Surviving apoptosis. *Apoptosis* **7,** 173–177 (2002).

Vives, E., Brodin, P., and Lebleu, B. A truncated HIV-1 TAT protein basic domain rapidly translocates through the plasma membrane and accumulates in the cell nucleus. *J. Biol. Chem.* **272,** 16010–16017 (1997).

Vocero-Akbani, A. M., Heyden, N. V., et al. Killing HIV-infected cells by transduction with an HIV protease-activated caspase-3 protein. *Nature Med.* **5,** 29–33 (1999).

Wadia, J. S., and Dowdy, S. F. Protein transduction technology. *Curr. Opin. Biotechnol.* **13,** 52–56 (2002).

Wajant, H., Pfizenmaier, K., and Scheurich, P. TNF-related apoptosis inducing ligand (TRAIL) and its receptors in tumor surveillance and cancer therapy. *Apoptosis* **7,** 449–459 (2002).

Wang, C. Y., Cusack, J. C. Jr., Liu, R., and Baldwin, A. S. Jr. Control of inducible chemoresistance: Enhanced anti-tumor therapy through increased apoptosis by inhibition of NF-kappaB. *Nature Med.* **5,** 412–417 (1999a).

Wang, H. Y., Fu, T., Wang, G., et al. Induction of CD4(+) T cell-dependent antitumor immunity by TAT-mediated tumor antigen delivery into dendritic cells. *J. Clin. Invest.* **109,** 1463–1470 (2002).

Wang, J., Dai, H., Yousaf, N., et al. Grb10, a positive, stimulatory signaling adapter in platelet-derived growth factor BB-, insulin-like growth factor I-, and insulin-mediated mitogenesis. *Mol. Cell. Biol.* **19,** 6217–6228 (1999b).

Wu, W. H., Lee, W. L., Wu, Y. Y., et al. Expression of constitutively active phosphatidylinositol 3-kinase inhibits activation of caspase 3 and apoptosis of cardiac muscle cells. *J. Biol. Chem.* **275,** 40113–40119 (2000).

Wu, X., Bayle, J. H., Olson, D., and Levine, A. J. The p53-MDM-2 autoregulatory feedback loop. *Genes Dev.* **7,** 1126–1132 (1993).

Xia, H. B., Mao, Q. W., and Davidson, B. L. The HIV TAT protein transduction domain improves the biodistribution of beta-glucuronidase expressed from recombinant viral vectors. *Nature Biotechnol.* **19,** 640–644 (2001).

Yamamoto, Y., and Gaynor, R. B. Therapeutic potential of inhibition of the NF-kappaB pathway in the treatment of inflammation and cancer. *J. Clin. Invest.* **107,** 135–142 (2001).

Yang, Y. L., and Li, X. M. The IAP family: Endogenous caspase inhibitors with multiple biological activities. *Cell Res.* **10,** 169–177 (2000).

Zbar, B. Von Hippel-Lindau disease and sporadic renal cell carcinoma. *Cancer Surv.* **25,** 219–232 (1995).

Zimmermann, K. C., Bonzon, C., and Green, D. R. The machinery of programmed cell death. *Pharmacol. Ther.* **92,** 57–70 (2001).

Zornig, M., Hueber, A., Baum, W., and Evan, G. Apoptosis regulators and their role in tumorigenesis. *Biochim. Biophys. Acta* **1551,** F1–37 (2001).

CHAPTER 7

Drug Screening: Assay Development Issues

Steven S. Carroll, James Inglese, Shi-Shan Mao, and David B. Olsen

Molecular Cancer Therapeutics: Strategies for Drug Discovery and Development, edited by George C. Prendergast
ISBN 0-471-43202-4

In the pharmaceutical industry, the standard method for discovering small molecule drugs aimed at interfering with the activity of enzyme or receptor targets has followed a time-worn path. Potential targets are identified using a variety of epidemiological, biological, and/or biochemical data (as described earlier in this book), including animal model systems that involve gene knock-out or knock-down techniques, and population-based genetic screening and correlative approaches. Biochemical or cell-based assays are then developed to measure the target activity, adapted for compatibility with high-throughput screening (HTS) and used to test all compounds that are available. One confirms HTS hits, defined as lead compounds that produce some predefined degree of inhibition or greater in the assay, and eliminates artifacts with a variety of follow-up experiments designed to focus efforts on the most promising leads. Validated leads then serve as the structural templates for medicinal chemistry efforts to optimize potency, specificity, and in vivo efficacy. In many cases, random screening is one part of a multitargeted approach to identify lead inhibitors, which may include rational design based on previous knowledge of inhibitor structure or on modifications of the substrate or a co-factor.

The standard approach outlined above is not innovative, but recent technological innovations have increased the speed of the screening process. The approach is designed to be systematic and thorough, without introducing any bias by a priori elimination of any compounds from testing. Molecular modeling and in silico screening methods (Toledo-Sherman and Chen, 2002) have become increasingly sophisticated, but these methods have yet to supplant HTS as the industry standard for lead discovery. This chapter focuses on the practical issues related to the design of assay methods that are compatible with HTS and on the methods that are used to streamline the process of lead identification and to eliminate artifactual inhibition. The efficiency of screening has occasionally created a situation in which the identification of a compound with in vivo activity has allowed a target to be pharmacologically validated, as discussed in more detail later in this chapter. Thus screening methodologies can offer a route to pharmacologically validate targets that have not previously been validated by genetic or biochemical criteria.

7.1 HTS Versus UHTS and the Drive to Miniaturize

In the early 1990s, HTS emerged as the industry standard for discovering leads in compound archives. Most screens have been based on a 96-well plate format, with the majority configured to assay ligand binding, enzyme activity, and cell-based responses (Burbaum, 2000). The automation of 96-well plate assays became important with the increasing size of compound libraries; but a manual workstation approach, in which individual plates or stacks of plates are moved by hand between liquid dispensers, washers, and plate readers,

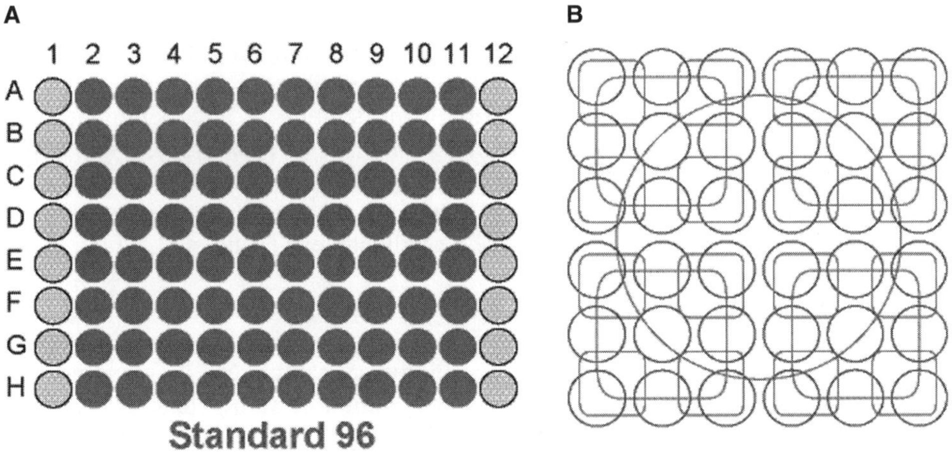

Figure 7.1 Microtiter plate designs. A, The standard 96-well plate format in which rows are designated by the letters A–H and columns by numbers 1–12. In this example the center 80 wells of the plate are occupied by test compounds while columns 1 and 12 are left empty or used for controls. This plate layout reflects the common practice of storing compound archives in the center 80 wells of deep-well storage plates. Reformatting an array of 80 compounds to higher density plates such as 384- or 1536-well plates will result in additional unused columns. In a 384-well plate, columns 1, 2, 23, and 24 will be empty, and in a 1536-well plate, columns 1–4 and 44–48 will be empty. B, Microtiter plate well sizes relative to that of a 96-well plate. The large circular center represents a single well from a 96-well plate. The large square represents a well from a 384-well plate (4 × density of 96-well), the smaller square a well from a 1536-well plate (16 × density of 96-well), and the small circle a well from a 3,456-well plate (36 × density of 96-well). Figure Courtesy of Kurt Berry.

continues to be widely used in industry. This situation is especially the case for smaller biotech companies not wishing or unable to invest in expensive robotic platforms.

Assay miniaturization, defined as the reduction of assay volumes to a few microliters or less, is driven primarily by economies of scale. The smaller the volume, the greater the well density per plate (Fig. 7.1). Therefore, for the same number of plates processed in a high throughput screen, correspondingly more samples are tested with a reduced consumption of reagents. This trend parallels the surge in test samples available for screening and in the improvement of fluid dispensation technology needed to accurately address and deliver submicroliter volumes to the smaller wells of these high well density plates (Dunn and Feygin, 2000).

Strategies for screening larger sample collections have been devised as compound libraries have grown in size, due to years of combined activity by the synthetic organic chemist or industry consolidation or as a result of parallel and combinatorial synthesis. Basically, two approaches were taken: screening single samples or pooled samples. In the former, a single compound per well is tested, whereas in pooling, 10 or more samples per well are tested simultaneously (Auld et al., 2003). The mixture approach has generally yielded to the single-compound approach, which is more direct and foolproof, although methods continue to be developed to exploit the potential power of pooling, for example, using techniques such as affinity selection (Lenz et al., 2000). In general, larger libraries tend to be populated by only a few milligrams of compound per sample or fractions of micrograms of compound when derived from combinatorial bead-based libraries (Chabala, 1995); so sample conservation is an important consideration in HTS deployment. In the limit case, such as occurs with certain kinds of very large combinatorial libraries (10,000–1,000,000 compounds) or those prepared on low-capacity solid supports (e.g., 90–200 μm tentagel beads), compound resynthesis is required to validate the activity of a candidate lead (Auld et al., 2002; Dunn et al., 2000; Tan et al., 1998).

Screening individual samples that number in the millions can require a significant consumption of reagents, making it prohibitive to screen targets for which only minute amounts of biological sample can be prepared. For example, in a screen designed to identify inhibitors of the MEK1- extracellular signal regulated kinase 2 (ERK2) pathway, a constitutively active MEK1 mutant enzyme was present at 11 nM, and its target was the wild-type inactive ERK2 kinase (p42-MAPK), a 43-kDa enzyme, at 60 nM. Screening 1×10^6 compounds in 96-well microtiter plate format, with a typical assay volume of 80 μL/well, would require 40 and 200 mg of the respective proteins. Using standard methods (Fig. 7.1), this effort would require 12,500 96-well plates, taking 3 months to complete if a rate of 200 plates/day were achieved. In contrast, assays run in a 3,456-well Nanoplate, which handles 1-μL per well, could be accomplished with 500 μg and 2.5 mg of the same proteins (Rodems et al., 2002) in 1 week. To put these reagent requirement costs in perspective, MEK1 protein that sells for \$100/0.1mg from Cell Signaling Technology, Inc. (www.cellsignal.com) would entail a cost of \$40,000 in MEK1 purchased for the assay, whereas in the miniaturized format the cost entailed would be only \$500. For this reason, such ultra-high-throughput screening (UHTS) is becoming the standard format for many laboratories.

The distinction between HTS and UHTS is not well defined but the technique has been suggested to enter the realm of "ultra" when the number of individual assays per day exceeds 100,000. Several ways to breach this number exist. One is to use efficient robotic systems that can process microtiter plates either very rapidly, the rate of which depends on the so-called cycle time, which is defined as the amount of time an assay plate resides on the robotic system. Cycle times depend on various factors, including the assay protocol itself and the time taken for system peripherals, such as a plate reader, to perform a given operation. Generally the slowest step in the process is the rate-limiting step, but this is not always the case (Cohen and Trinka, 2002; Rutherford and Stinger, 2001, www.lab-robotics.org). Of course, the number of microtiter plates that need to pass through the system is influenced by the number of assay wells per plate: fewer 3456-well plates will need to travel through a robotic system than 96-well plates to achieve a rate of 10^5 wells in 24 h, for example (Table 7.1). The liquid handlers, plate readers, and database needed to enable 3456-well plate screening are highly specialized and make up the Aurora UHTS System (Mere et al., 1999). Such systems are not commonplace in industry; thus they will not be discussed in further detail. However, these systems illustrate the some of the most advanced current technologies.

Table 7.1 Relationship of Microtiter Plate Density, Assay Volume, Reagent Use, and Apparent Throughput

Plate 5 or chip type (well no.)[a,b]	Plates/10^5 well (e.g., 100,000 equivalent)	Assay volume range (minimum, maximum)	Assays/month reagent (reagent efficiency ratio)[c]	Equipment Needed
Standard 96 well	1042	200–50 μL	8 (1)	Standard
Standard 384 well	261	50–35 μL	23.5 (2.94)	Standard
Small-volume 384 well	261	30–5 μL	57.1 (7.14)	Standard
864 well[d]	116	16–5 μL	95.2 (11.9)	Standard
Standard 1,536 well	66	10–2 μL	166.7 (20.83)	Standard
Low-profile 1,536 well[e]	66	2 μL	500 (62.5)	Standard
Virtual 1,536 well[f]	66	10–100 μL	200 (25)	Standard
Evotec 2,080 well Nanocarrier[g]	49	1.5–0.5 μL	1,000 (125)	Specialized
Aurora 3,456 well[h]	29	2–1 μL	666.7 (83.3)	Specialized
DuPont 9,600 well[i]	11	1–0.2 μL	1,666.7 (208.3)	Standard
BioTrove 10,000 holes[j]	10	0.05 μL	20,000 (2,500)	Specialized

[a] Sources of plates reviewed at www.the-scientist.com/yr1999/sept/profile1_990927.html (Sept 1999).
[b] For history of microtiter plate formats see www.microplate.org/content/history.htm.
[c] Reagent efficiency ratio is defined as (assays/milliliter reagent)/(assays/milliliter reagent for a standard 96-well plate) or (assays/milliliter reagent)/8.
[d] Comley, J. C. W., Binnie, A., Bonk, C., and Houston, J. G. A 384-well HTS for human factor VIIa: Comparison with 96- and 864-well formats. *J. Biomol. Screen.* **2,** 171–178 (1997).
[e] Trombley, A., Veilleux, J., Dunn, D., Orlowski, M., Zhang, M. and Boyd, D. 1536-well assay plate for HTS. *HTS Forum,* **9,** p. 1–3 (2000).
[f] Garyantes, T. Virtual wells for use in HTS assays. [WO99939829]. *PCT Intl. Appl.* (1999).
[g] Evotec OAI, Hamburg, Germany.
[h] Mere, L., Bennett, T., Coassin, P., et al. Miniaturized FRET assay and microfluidics: Key components for ultra-high-throughput screening. *Drug Discovery Today* **4,** 383–368 (1999).
[i] Oldenburg, K. R., Zhang, J., Chen, T., et al. Assay miniaturization for ultra-high throughput screening of combinatorial and discrete compound libraries: A 9600-well (0.2 ul) assay sytstem. *J. Biomol. Screening* **3,** 55–62 (1998).
[j] www.biotrove.com/spie/spie1.html.

In most laboratories, screening is most commonly accomplished in 96-, 384- or 1536-well plates, with 384- and 1536-well plates becoming the current industry standard (Garyantes, 2002; Wolcke and Ullman, 2001). Many liquid handling methods are available for 384-well plates and some can also be used for 1536-well plates by taking advantage of the indexing capabilities of the fixed-head based systems. For example, a popular liquid handling unit from CyBio, Inc. (Hamburg, Germany) uses a fixed 384-tip positive displacement head to aspirate from a reagent 384-well plate to a 384- or 1536-well assay plate. The Hummingbird (Cartesian Technologies, Irvine CA) is designed to be a noncontact (unlike the CyBio device), low-volume compound reformatting system. This system employs either 96- or 384-narrow-bore glass capillary tips to aspirate sample (50, 100 or 250 nL fixed volumes) using capillary action, which is followed by an air-pulse dispensation into 96/384 or 384/1536 assay plates. Both of these dispensers are useful in the delivery of sample compounds from archive plates to assay plates, but the volume flexibility (200 nL to 25 μL) of the CyBio system is ideal for delivering assay

Table 7.2 Dispense Volumes Required to Achieve Specific Test Compound Concentrations for Given Assay Volumes

Compound Source Concentration[a]	Aspiration (250 nL)[b]	Capillary action (250, 100, 50 nL)[c]	Pin tool (25 nL)[d]	Piezo-based (1–25 nL)[e]
500 μM	13 μM[f] 2.5%[g]	13, 5, 2.5 μM 2.5, 1, 0.5%	1.3 μM 0.25%	0.3–6.3 μM 0.05–125%
4 mM	100 μM 2.5%	100, 40, 20 μM 2.5, 1, 0.5%	10 μM 0.25%	2.4–50 μM 0.05–1.25%
Assay Volume	10 μL	10 μL	10 μL	2 μL

[a] Assuming samples are dissolved in 100% DMSO.
[b] See, e.g., CyBi-Well (www.cybio-ag.com'english/index.html).
[c] See www.cartesiantech.com.
[d] Slotted pin tool volumes can range from 5 nL to 1 uL; see Dunn, D. A., and Feygin, I. Challenges and solutions to ultra-high-throughput screening assay miniaturization: Submicroliter fluid handling. *Drug Discovery Today* **5,** S84–S91 (2000).
[e] E.g., Aurora's Piezo Dispensing Robot and Evotec's NanoDispensing Technology.
[f] Concentrations refer to [sample] in final assay volume.
[g] Percents refer to DMSO% in final assay volume.

reagents such as scintillation proximity assay (SPA) beads, mammalian cells, and buffers. Other systems such as the PreSys 4040 Integrated Dispensing System (Cartesian Technologies) uses from one to eight dispense heads that rapidly add reagents (50–500 nL volumes) to a variety of plate types. This system is based on a gated release under positive fluid pressure, which is established by a combination of a syringe and solenoid valve designed to handle assay reagents. Table 7.2 includes examples of compound concentrations that can be achieved with these dispensing systems. Regardless of which liquid handlers are used, the need to transfer compounds from either 96-, 384-, or 1536-well compound storage plates and the addition of assay reagents (e.g., cells, enzymes, membranes, buffers, antibodies, detection reagents, etc.) are requirements for the majority of screening activities. Needless to say, properly functioning, robust, accurate, and precise liquid handlers are critical for the success of any screening endeavor.

7.2 Assay Format

Efficiency tends to be maximized in the so-called mix-and-read format, also known as non-separation-based and homogenous assay formats. These latter terms, while often used to describe the mix-and-read format, are not necessarily the same. For example, while the SPA format is an example of one popular mix-and-read format, there is indeed a physical separation of radioligand–receptor complex that is bound to the polyvinyltoulidine (PVT) bead used in the format from the radioligand that remains in solution. Moreover, the assay mixture is certainly not homogenous. On the other hand, assays based on time-resolved fluorescence resonance energy transfer (TR-FRET) typically tend to be both non-separation and homogenous in nature.

A reason for the better efficiency in mix-and-read assays stems from the fact that separation steps in small volume plates are not currently accomplished with high speed, precision, or accuracy. The filter-binding assay has not generally been successful beyond the 96-well plate format, and even here automation is far from ideal. For ELISA assays that require wash steps, 96- and 384-well plate formats are currently the only ones for which reliable plate washers exist. There is little argument that the signal to background ratio can be substantially increased using separation-based assays compared to mix-and-read assays. However, the trade-off is that overall screening efficiency tends to suffer greatly, and highly miniaturized assays such as those conducted in 1536-well plates are primarily addition only.

SPA and fluorescence-based detection systems are two of the most popular formats for mix-and-read assays. A variety of SPA-bead coatings are available to increase the versatility of the format, including streptavidin to capture biotinylated products, wheat germ agglutinin to capture glycoprotein-containing membranes, and antibodies to capture antibody ligands. ^{3}H, ^{35}S or ^{33}P can be used as the radiolabel in SPA-formats, for example, the latter being relevant for kinase assays (Sorg, et al., 2002). One useful technique for increasing the signal to background ratio in assays using either polystyrene or PVT beads is to add a high concentration (4 M final concentration) of CsCl after quenching the reaction and allowing for product binding to the bead. The addition of CsCl causes the SPA beads to float to the top of the quenched reaction solution where they are closer to the photomultiplier tube (PMT) of scintillation counters that have a top-read format such the Topcount (Packard), resulting in more efficient counting (Ferrer et al., 2003). Disadvantages to SPA include the same issues that relate to radiometric detection methods in general, including costs associated with handling and disposal of radioactivity and the general need to devote equipment exclusively to assays involving radioactivity.

A variety of fluorescence-based assays that are appropriate for the mix-and-read HTS format have been used to monitor enzyme activity (Oldenburg et al., 2001). Screening actives can result in an increase or decrease in signal, depending on how the assay is configured. FRET assays, for example, can be based on substrates (or products) that separate or bring together two fluorophores by which the emission spectrum of the first fluorophore (the donor) overlaps the excitation spectrum of the second (the acceptor). Product detection then monitors the emission of the second fluorophore if the reaction brings the fluors together (true FRET), as may occur in a kinase assay, or the increasing emission of the first fluor if the reaction results in their physical separation (quenched or QFRET), such as generated by a protease activity with a dual-labeled peptide, by which the fluorescence of one label is quenched by a second. A large number of donor–acceptor pairs is available from commercial sources such as Molecular Probes (www.probes.com) or Trilink (trilinkbiotech.com). The extent of the spectral overlap will determine the efficiency of the energy transfer, which will factor into the signal to background ratio that can be achieved. The addition of a TR-FRET measurement using a long-lived fluorescence donor (e.g., europium cryptate) and organic fluorochrome acceptor (such as allophycocyanin) decreases interference due to background fluorescence of test compounds, since the half-life of background fluorescence is usually short (Bazin et al., 2002; Karvinen et al., 2002). Fluorescence-based methods can suffer from quenching of the signal due to compound absorbance, giving rise to false positives, which usually necessitate a follow-up assay with a different readout.

Detection systems or "readers" are the ultimate destination of the assay plate and interface with the informatics component of all screening and lead optimization operations. The reader is that point at which biology or biochemistry is converted to numerical data. Handling and transforming the raw reader counts are important areas of lead discovery, especially as it relates to large volumes of data and/or very high content data (i.e., such as that extracted from a high-throughput-scanning microscope or patch-clamp system) and the storage and visualization of such data. A discussion of data handling is beyond the scope of this chapter but can be found elsewhere (Taylor et al., 2000).

The choice of reader, in fact, depends on both the assay format and the plate format. For example, SPA formats are readily performed in 96- and 384-well plates employing standard plate-based scintillation counters; however, these standard PMT-based plate counters (e.g., Wallac Microbeta or Packard TopCount) cannot index 1536-well plates. Conversion of a SPA to 1536-well format requires a special charge-coupled device, CCD-based imager that is plate format independent (e.g., LEADSeeker) or a system capable of indexing a 1536-well plate with the necessary sensitivity (e.g., Imagetrak SPA/Lumi). The LEADSeeker is equipped for measurements of bead emission in the red region of the spectrum (610 nm) using specialized polystyrene or yttrium oxide (YO_x) beads, as opposed to the standard scintillant emission at around 420 nm. Color quenching due to yellow compounds that typically absorb 400 nm light and are prevalent in many compound collections is, therefore, minimized with the LEADSeeker instrument (Ramm, 1999; Zheng, et al, 2001).

The reduction in assay volume that accompanies assay miniaturization will generally result in a reduced total signal output for absorbance and radiometric assays (e.g., shorter path length or fewer cpm, respectively). Detection of fluorescence is affected less by scaling down a sample due to the fact that fluorophores can in principle be excited multiple times in the course of a measurement. A useful comparison between a prototypical kinase assay performed in a SPA and TR-FRET has been given by Park et al. (1999). In this study assay parameters were discussed that can significantly influence assay quality, such as secondary reagent optimization (e.g., antibodies used in TR-FRET assays) and the use of "quench correction" in SPA-based formats.

A study by Zheng et al. (2001) of a miniaturized RNA polymerase assay comparing a PMT detector (e.g., TopCount) versus cooled CCD imager (LEADSeeker) detection mode provides a practical illustration as to the limits of detection and signal-to-background (S/B) ranges that can be achieved when converting a radiometric assay from a standard SPA format to an imager-based format. This study found that approximately 1000 cpm total signal (as determined in a 15 μL 384-well SPA bead format) translated into a robust

LEADSeeker assay. In general, while both the 96 and 384-well assays using conventional SPA beads (PVT or yttrium silicate (YSi)) and PMT-based detector provide superior S/B, the CCD-based imager (using red-shifted PS or YOx beads) allowed scaling to a 6 μL 1536-well plate format, with reduced signal acquisition time of up to 15-fold and significantly less interference from colored samples.

7.3 Basic Issues of Assay Design

The goal of HTS of compound collections for identification of inhibitors of enzyme targets is to identify high-quality leads, as discussed below, that serve as the starting point for medicinal chemistry efforts to optimize the many pharmacological parameters required for in vivo efficacy. The overall chance of eventual success in the development of product clinical candidates and the speed with which that goal is achieved depend in part on the quality of the initial lead structure. As the numbers of compounds in screening collections grow into the millions, not only screening technologies have to keep pace but so must the ability to triage compounds for follow-up. Intelligent assay design helps streamline the process of compound follow-up by reducing the number of artifactual inhibitors, allowing efforts to be focussed on characterizing valid hits. This section focuses on practical considerations for the development of enzyme assays.

The kinds of assays used at the laboratory benchtop for mechanistic studies, small-scale compound evaluations, and basic target investigations are typically unsuitable for conversion to an HTS-compatible format. The characteristics of an HTS-compatible assay include robustness; sensitivity, in terms of both signal generation and sensitivity to inhibition; the ability to be adapted to small volumes and to automated reagent delivery and product detection; economical use of reagents; and reproducibility of signal in the presence and absence of inhibitors. Generally, some modification of an existing laboratory scale assay methodology will be necessary to achieve an acceptable HTS assay format. Prior knowledge of some of the general biochemistry of the target enzyme may be available to facilitate either the design of a new assay format compatible with HTS or adaptation of an existing assay to HTS. Some of the information regarding optimal enzyme assay conditions that will aid in the design of an acceptable HTS assay includes the pH activity profile, salt effects on activity, co-factor requirements and optimal concentrations; the conditions necessary to quench the assay if it is an end point determination, any requirements for dithiothreitol (DTT) or β-mercaptoethanol, enzyme (and product) stability with time and temperature, and the sensitivity of the enzyme to DMSO content.

The desire to use conditions that are optimal for enzyme activity is based primarily on the goal of conserving enzyme. As the number of assays required to screen a compound library increases into the millions the effort to supply the necessary amount of enzyme can become time-consuming and expensive. As previously mentioned, assay miniaturization alleviates much of the problem of reagent generation. However, an opposing consideration is the generation of a robust signal to give sufficient confidence to the results while maintaining the sensitivity of the assay to inhibition. Thus it is generally worthwhile to optimize the assay conditions to maximize the assay signal. The choices of salt and its concentration in the reaction are important considerations. Inclusion of the optimal concentration of KCl or NaCl will spare enzyme. Increasing salt concentration will also decrease the importance of ionic interactions between the inhibitory compounds and the enzyme target and emphasize the importance of hydrophobic interactions. Hydrophobic compounds are often advantageous because they generally have favorable oral bioavailability and cellular uptake properties. Reaction temperature is another variable to consider, with higher temperatures yielding generally higher reaction rates and, therefore, greater sensitivity; the trade-off is a reduction in the relative stability of the enzyme activity that can result in a reduced linearity of the reaction product with time. Higher reaction temperatures also increase the rate of evaporation, which becomes more of a factor as the assay volume is reduced. Typically, the effect of evaporation can be minimized by either using plate covers or a humidified cabinet during the reaction. The time required to achieve the desired reaction temperature should also be considered when determining the reaction time. Longer reaction times decrease the effect of the time required for the plate to reach the reaction temperature.

Stability of the enzyme to the DMSO content, the typical diluent for test compounds, must also be determined. A typical range of concentrations of DMSO would be from 1 to 10%, depending on several factors, including the pipettor to be used for compound delivery (which may set lower limits on the deliverable volume) and final compound concentration to be tested. DMSO solutions of test compounds can undergo oxidation during storage over long periods of time. Often it is useful to include a reducing agent in the reaction such as dithiothreitol to maintain the reduced state of the enzyme and compound. DTT, since it contains a potentially nucleophilic sulfur, can also react with electrophilic compounds such as Michael acceptors (see orgchem.chem.uconn.edu/namereact/michael.html), which that may otherwise produce inhibition by covalent modification of the enzyme, a generally undesirable property for a lead inhibitor. If a compound can covalently modify the target enzyme, it is likely to react covalently with other proteins, producing deleterious side effects when administered in vivo.

A thorough understanding of the stability of the enzyme activity "on the deck" (i.e., under conditions of storage of the stock reaction solution and the reaction assay itself) is critically important. If the enzyme activity is sensitive to freeze–thaw cycles it should be aliquoted in samples large enough to supply a day's worth of screening reactions to maximize day-to-day reproducibility. Various additives have been employed to stabilize enzyme activity, including polyethyleneglycol of various molecular weights (e.g., PEG8000) (Jordan et al., 1992), bovine serum albumin at $\sim$ 1 mg/mL, or low concentrations of detergents such as 0.2% *n*-octylglucoside or Triton X-100. Inclusion of detergents can also effect compound potency either by segregating hydrophobic compounds into micelles, if the detergent concentration is above its critical micelle concentration (CMC), or assisting in the solubilization of poorly soluble compounds. Most automated pipetting stations can reproducibly pipette small volumes of solutions containing low concentrations of detergents. Optimally, enzyme activity will be reasonably stable over a period of a day of assays in the stock solution to allow for convenient once-daily reagent preparation. Most screening laboratories have the capability to store a stock solution at 4°C and pipet directly from the stock when formulating reaction plates. However, even if the enzyme activity is stable under stock storage conditions, to guarantee the most meaningful comparison of assay results with control reactions, each plate of assays should be designed as an independent experiment, including positive and negative controls for enzyme activity as well as a positive control for inhibition with sufficient replicates to generate statistical significance.

For enzyme assays, steady-state reaction conditions must be maintained so that the reaction is most sensitive to inhibition and so that the equations used to evaluate inhibition are valid. The conditions that must be met are (1) that the enzyme concentration employed in the assay falls within the linear range of reaction rate as a function of enzyme concentration and (2) that the reaction progress curve is linear with reaction time. Generally the reaction is most sensitive to inhibition by competitive inhibitors when the substrate concentration is at or below its K_M value. The concept is illustrated in the following equation for competitive inhibitors,

$$IC_{50} = (1 + [S]/K_M) * K_i \tag{7.1}$$

If $[S] > K_M$, the inhibitor is competing with a high concentration of substrate and the IC_{50} value is increased. For purely noncompetitive inhibitors, the substrate concentration does not effect inhibition. An additional requirement is that the percentage of substrate converted to product during the reaction must remain $< 10\%$ to ensure that the reaction rate remains constant. If $> 10\%$ substrate is turned over when the substrate concentration is $\leq K_M$, the reaction progress curve will lose linearity. The requirement for relatively low substrate concentration can lead to a trade-off between the sensitivity of the reaction to inhibition versus the product signal that will decrease as the substrate concentration is lowered. It is of critical importance to the success of the screen to maintain sensitivity to inhibition, since conditions that produce a strong signal will be useless if inhibitors cannot be identified. On the other hand, generating reproducible screening data with low coefficients of variation is easier with a higher signal to background ratio, which makes it easier to identify weakly inhibiting compounds that may be useful lead structures.

When optimizing reaction parameters, it is generally more efficient to vary more than one parameter simultaneously, since, for example, changing the salt concentration can also change the reaction linearity with time. Varying multiple parameters can be accomplished using a matrix approach in 96- or 384 well reaction plates with one parameter varied along the rows of the plate and the other varied along the columns. An example is cited from the literature (Ferrer et al., 2002).

The compound concentration to be tested is another parameter to be determined. To some extent the final concentration tested may depend on the maturity of a therapeutic program that can indicate what level of inhibitory potency would be of interest. Generally, with robust biochemical assays compound concentrations in the range of 5–20 μM are employed. Our experience has been that this concentration range is sufficiently high to identify weak leads while not causing a large number of hits that completely inhibit the assay, making it difficult to prioritize follow-up assays. A wide distribution of potencies of inhibition helps stratify hits into different priority levels for follow-up assays. Compound solubility can also become an issue at higher concentrations. Insoluble or poorly soluble compounds can form aggregates that can produce artifactual inhibition (McGovern et al., 2001). In practice, it is helpful to construct test plates containing known inhibitors, if available, as well as a wide variety of structural classes of compounds. Repeat screening of these test plates at different final concentrations will aid in deciding the final test concentration that ensures detection of known compounds and an acceptable distribution of inhibition percentages. The assay design should also allow for an incubation of test compound and enzyme before the addition of the substrate to initiate the reaction. The incubation allows for the association of slow-binding inhibitors with the enzyme target that might otherwise be missed during the assay. A preincubation step is less important if the overall reaction time is long (e.g. > 30 min).

A final test of assay acceptability must be run under the exact conditions proposed for the screen to determine the signal to background ratio in the assay, the absolute signal, and the standard deviation of both the uninhibited and fully inhibited signal in the presence of a high concentration of a known control inhibitor. A useful parameter to determine the acceptability of the assay, termed Z, has been proposed (Zhang et al., 1999). The Z factor is defined as

$$Z = 1 - \{[(3\sigma_{c+}) + (3\sigma_{c-})]/|\mu_{c+} - \mu_{c-}|\} \tag{7.2}$$

where σ_{c+} and σ_{c-} represent the standard deviations of the signals in the presence and absence of a positive control, and μ_{c+} and μ_{c-} are the mean of the signal in the presence and absence of a positive control. Thus Z is a dimensionless parameter that takes into account both the confidence interval of the assay and background signals and the dynamic range between the two signals. Small standard deviations of the signals and/or a large difference between signal and background contribute to a high Z value, ideally approaching unity. Empirically, assays with $Z \geq 0.5$ are deemed acceptable for HTS, with an acceptable dynamic range for identification of inhibitors. It is also recommended that day-to-day variation in the Z factor of HTS assays should be assessed.

7.4 Follow-Up Studies of Screening Hits

As the numbers of compounds screened increases so generally does the number of hits that require follow-up efforts. Typically, hits are prioritized for follow-up based primarily on potency of inhibition, with an arbitrary cutoff that results in a manageable collection of hits that can be studied in some greater detail, usually numbering 1000–2000 compounds. These hits are then subjected to further characterization to identify attractive compounds that inhibit by binding to the target enzyme, rather than by some artifactual means such as substrate or enzyme depletion. If the follow-up studies of the top group of compounds do not yield viable lead structures, a second group of compounds with less potent inhibition can be examined. A general example illustrating a flowchart for follow-up assays to characterize hits from the primary screen is shown in Figure 7.2.

A critical consideration of the successful screening effort in drug discovery is to distinguish quickly between promising leads and the many useless false positives. False positives and artifactual inhibitors should be identified and removed from the screening database at an early stage before the commitment of resources and time. Several causes of false positive inhibition are listed in Table 7.3 (Rishton, 2003).

Specificity of inhibition can be a useful means of establishing real inhibition of the target by the compound in question. A useful first line of counterscreening hits can be carried out electronically. As compound collections are screened several times for inhibitors of different targets using different assay formats, promiscuous compounds can be readily identified by comparison with hits from previous assays. As outlined in Table 7.3, chemically reactive compounds are undesirable as leads and often will inhibit in multiple assays. In addition, compounds can be identified that inhibit specific assay formats. For example, biotin analogs, which are numerous in many compound collections, will interfere with biotin-streptavidin-based detection methods and can often be eliminated by comparison of hits with previously run biotin-streptavidin-based assays. Color quenching can either be identified as promiscuous inhibition or by visual inspection of the compound solution.

Specificity of inhibition is traditionally established in counterscreen assays in which hits are tested for inhibition of related enzymes or receptors. If counterscreen assays are run under similar assay formats to the primary screen and if the compound shows some significant degree of specificity for the target enzyme, many of the kinds of artifactual inhibition listed in Table 7.3 can be eliminated from consideration. Conversely, if the compound is scored as inhibitory in several different counterscreens, it becomes less attractive as a lead, although there are certainly cases in which higher specificity has been achieved due to medicinal chemistry efforts, particularly as higher potency was achieved.

Inhibitor titrations are necessary to establish potency accurately. The shape of the inhibition curve can also be informative. Well-behaved compounds producing inhibition by forming a 1:1 complex with the target enzyme should give rise to smooth titration curves that, when fit to the Hill equation (Eq. 7.3) or a functionally equivalent equation, give a Hill coefficient near unity, assuming that the enzyme target itself is monomeric with a single inhibitor binding site.

$$\text{Fraction inhibition} = [I]^n/\{[I]^n + \text{IC}_{50}^n\} \qquad (7.3)$$

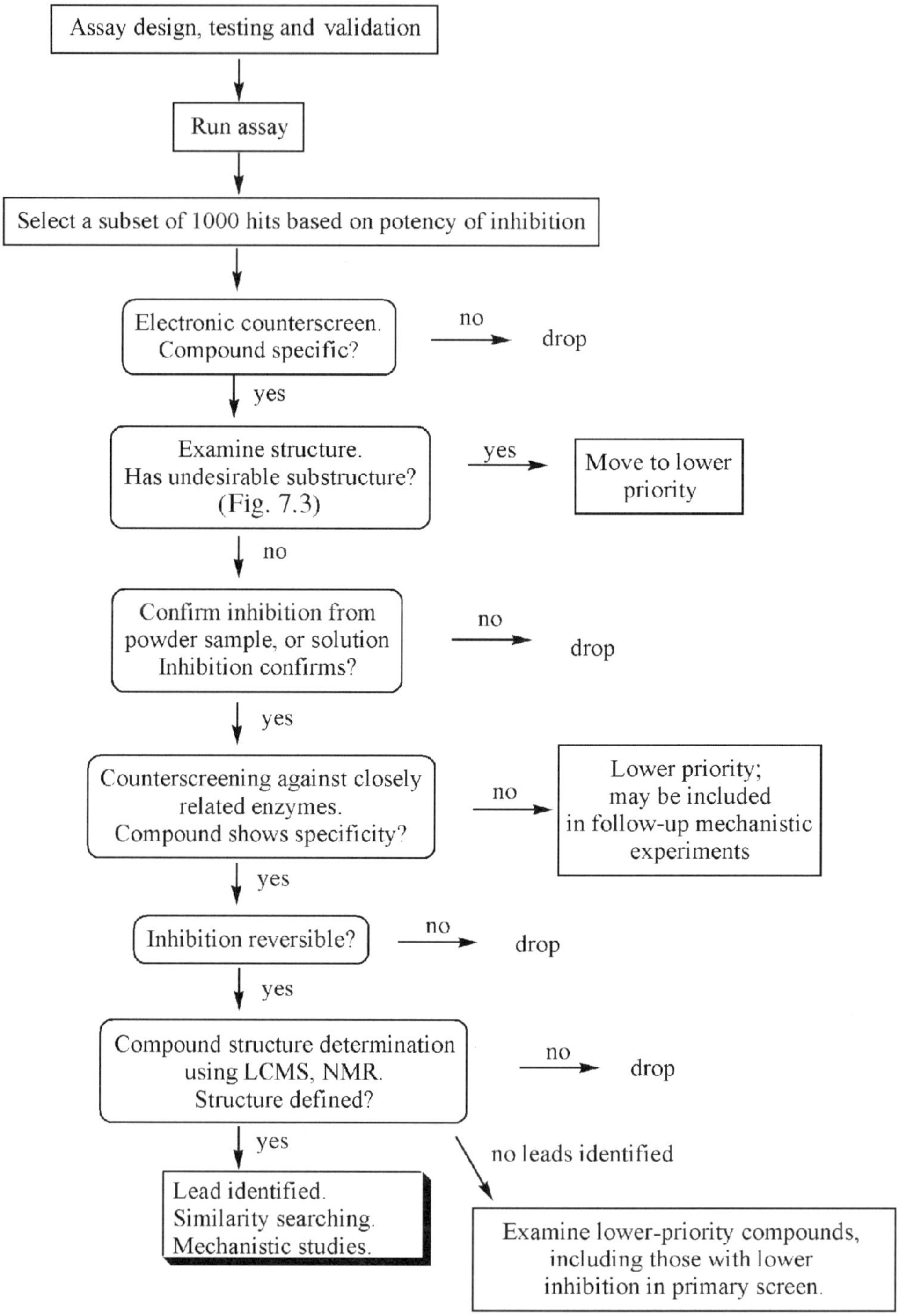

Figure 7.2 An example of a flowchart for follow-up studies of inhibitors identified during a primary HTS. A subset of the primary hits from HTS are subjected to a series of follow-up assays that are designed to eliminate false-positive compounds and to focus efforts on validated lead compounds for subsequent medicinal chemistry efforts.

Table 7.3 Diagnostic Tests for Artifactual Inhibition

Artifact	Diagnostic Tool	Prediction/Comments
Interference of product detection	Enzyme selectivity	Nonspecific inhibition
	Product analysis (e.g., HPLC) High background signals	Weaker or no inhibition
Precipitation/aggregate	Enzyme selectivity SAR	Nonspecific inhibition Little relationship
Depletion of inhibitor	Visible survey Electromicroscopy/light scattering Spectrophotometry HPLC analysis Inhibition analysis	Precipitation Aggregation High baseline Loss due to precipitation Saturation; lack of linear correlation at high $[I]$
	Ultracentrifugation	Loss due to sedimentation
Depletion of protein	Threshold of inhibition	Sharp rise of inhibition around IC_{50} High Hill coefficient
	Ultracentrifugation	Loss due to sedimentation
Depletion of substrate/ cofactor	Ultracentrifugation	Loss due to sedimentation
Interactions with substrate	Enzyme selectivity Substrate specificity Inhibition analysis	Nonspecific inhibition No inhibition with a different substrate Sigmoidal substrate saturation curve
Interference of micelles/ membrane	Selectivity	Nonspecific inhibition
	Inhibition analysis	High Hill coefficient

HPLC, high performance liquid chromatograph; SAR, structure-activity relationship.

Inhibitor titrations that produce a steep slope with $n > 1$ generally indicate artifactual inhibition, frequently due to compound solubility problems, as discussed below. The exception to this rule is when the target enzyme is oligomeric in its subunit structure, with some interaction between inhibitor binding sites. In this case, compounds can be expected to generate Hill coefficients > 1.

Compound collections can frequently contain series of structurally related compounds that may have been synthesized during a previous medicinal chemistry effort and then subsequently deposited into the collection. Occasionally, sets of structurally related compounds will exhibit significant inhibition in an assay, so that hits can be electronically clustered. Such relationships can be valuable, since the identification of common structural features by clustering software can provide information about the structure–activity relationships (SARs) of the hits and perhaps suggest the structure of a consensus lead that combines elements of different structures. A series of compounds that produces a range of IC_{50} values is useful to find, since these compounds offer an argument against the possibility that the inhibition produced by any particular hit is artifactual. It can be helpful to use hits for structure-based searches of the available chemical database (ACD), which can identify compounds available from commercial sources that are similar to lead compounds, since the acquisition and testing of such compounds may extend SAR studies and provide a route to potentially improve the potency of the original lead (chemfinder.cambridgesoft.com/result.asp).

It is common that some leads contain a chemically reactive functional group that produces the basis for inhibition. Reactive compounds such as aldehydes/ketones, epoxides, and acyl/alkyl halides are well known to react with proteins, in many cases forming irreversible covalent bonds (Rishton, 1997). Compounds containing Michael acceptors can produce enzyme inhibition simply by covalent modification of the enzyme. Inhibition by Michael acceptors can generally be abolished by addition of nucleophilic compounds, such as thiol-containing reagents like DTT. Metal ions are critical for many biological reactions; hence chelating agents will alter activity. Compounds with redox potential such as quinones could interfere with reactions involving reduction–oxidation. In general, these chemically reactive compounds will show poor specificity. A partial list of undesirable substructural motifs is shown in Figure 7.3.

As mentioned above, fluorescence-based assays are common technologies used in HTS and standard laboratory screening (Rogers, 1997). The widely used FRET assay generates a high percentage of false positives. One common cause is the interference of compounds on the detection of the fluorescence intensity. Fluorescence limitations, such as the inner filter effect and photobleaching, can also exhibit artifactual inhibition (Liu et al., 1999). Output fluorescent light can be absorbed by neighboring substrate or product molecules so that only a fraction of the fluorescence of the product reaches the detector. Attenuation due to absorption of the incident or emitted light are referred as the primary and secondary inner filter effects (Lakowicz, 1999). When the test compounds exhibit fluorescence excitation and emission properties that are close to those of the substrate, they will further enhance the effect of inner filter effects. The magnitude of the inner filter effect depends on the wavelength range, pathlength, and concentrations of the quenching components. The corrected fluorescence intensity (F_{corr}) is approximately calculated from the observed intensity (F_{obs}):

$$F_{corr} = F_{obs} \log^{-1}[(OD_{ex} + OD_{em})/2)] \qquad (7.4)$$

where OD_{ex} and OD_{em} are optical density at the excitation and the emission wavelengths, respectively. As a general rule, the preferred OD_{ex} of the running assay is < 0.1.

Although chromogenic substrates that depend on the measurement of absorbance are not subject to the inner filter effect, a high absorbance of the test compounds will cause a deviation from the linearity of the Beer's rule ($OD < 1$). Claims of the extended linearity of newer instruments (0–3 OD linearity) may be useful. Artifactual inhibition due to interference with fluorescence or absorbance will in general exhibit low selectivity on various enzymes and will exhibit little or no SAR. In practice, a follow-up product analysis that uses a secondary assay based on a different method for product detection, for example, such as high performance liquid chromatograph (HPLC), is worthwhile to verify lead inhibitors.

A common cause of artifactual inhibition is precipitation or aggregation of the test compounds, which typically leads to poor selectivity and SAR.

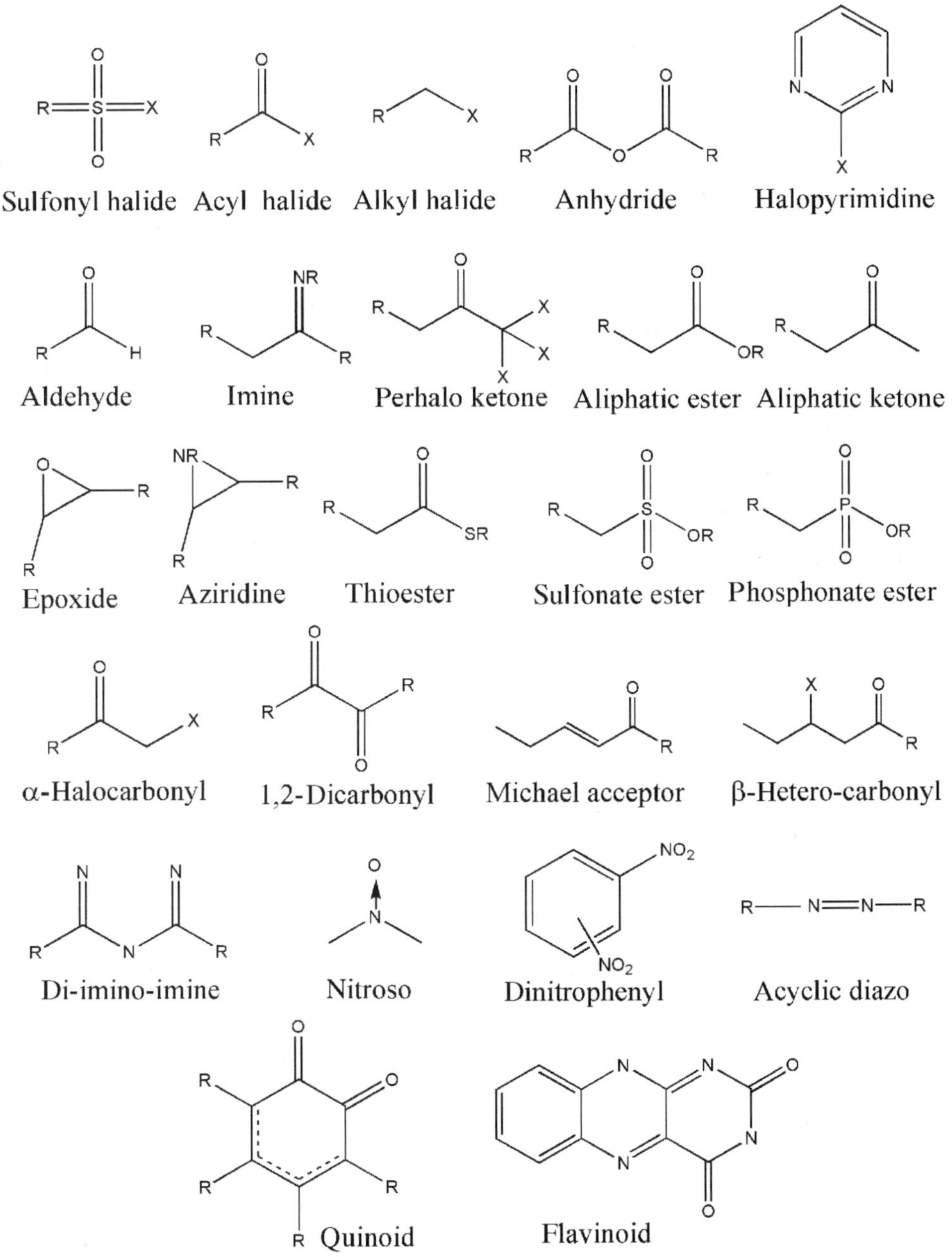

Figure 7.3 A partial list of substructures that are undesirable, either due to the expectation of poor pharmacokinetics, toxicity, or the likelihood that inhibition is artifactual due to covalent modification of the target. Data compiled from several sources including Rishton (1997).

Precipitation due to low solubility can be easily detected by a visible survey of compounds at higher concentrations. Electromicroscopy and light scattering have been used to identify false positives (McGovern et al., 2001). However, elevated baselines in UV spectral analysis can easily diagnose precipitation or aggregation that is potentially invisible to the naked eye. The Dixon plot is a common method of determining the dissociation constants of the enzyme-inhibitor complexes. In most cases, the linearity of V_0/V_i as a function of the inhibitor concentrations, $[I]$, is observed in the modified Dixon plot, as follows:

$$V_0/V_i = 1 + [I]/\mathrm{IC}_{50} \tag{7.5}$$

where V_0 and V_i are the reaction velocity in the absence and presence of the test compound, respectively. A maximum extent of inhibition will be observed when saturation of inhibitor at high concentrations occurs due to compound insolubility. Hence a plateau occurs in the Dixon plot. It is possible that the inhibition from some weak hits will be underestimated due to this saturation of inhibition. In rare occasions, a decrease in inhibition at high inhibitor concentration occurs, which may be due to a faster precipitation rate during the assay time period.

The precipitation of the compound can physically act on the enzyme and substrate. Proteins can co-precipitate with inhibitors, and the small particulates that result from aggregation–precipitation of compounds form surfaces on which protein may adhere. Loss of enzyme from the assay solution due to precipitation will lead to false inhibition. One sign of such a problem can be a dramatic, sudden increase in inhibition as the compound concentration is increased, with almost no intermediate inhibition. A similar phenomenon can occur with co-precipitation of substrates. Similarly, as many biologically active complexes require the presence of a cofactor(s), the loss of a cofactor(s) due to co-precipitation with an inhibitor will also produce false positives. Ultracentrifugation is useful to investigate precipitation of the test compounds. Loss of protein, cofactor, or substrate can be easily detected by activity loss or by HPLC analysis after centrifugation. In all cases, the selectivity of inhibition will be poor.

A false positive can also result from a compound that binds to the substrate in solution. In many cases, such as the binding of a compound to a primer–template DNA or RNA duplex in a polymerase, integrase, or helicase assay, this mechanism of inhibition by the compound is undesirable, since it does not directly inhibit the target. Such compounds can usually be identified in counterscreening assays, because their interaction with a substrate or cofactor will cause poor selectivity when the same substrate or cofactor is employed. Another signature of such compounds is that their titration in the assay will typically show a sigmoidal pattern (Segel, 1975).

Membrane-bound enzymes include some important pharmacological targets, and the presence of the membrane is essential to their activity in many biological systems. Detergent-like compounds that can disrupt the membrane and diminish enzymatic activity will, therefore, be identified as false-positive hits in assays that include membranes. Such compounds will exhibit a nonlinear increase in inhibition in titrations with high Hill coefficients. Because the compounds produce their effect by acting directly on the membrane, they will also exhibit little selectivity in counterscreen assays.

Further study of the mechanism of inhibition by lead compounds is, of course, highly desirable. As mentioned above, routine analysis of the product by an alternate analytical method that differs from the one used in the primary screen (such as HPLC) is useful, in addition to counterscreening to establish selectivity. Detailed steady-state kinetic studies to determine the mode of inhibition by leads not only provides insight into interactions with the targeted protein but can also identify compounds with undesirable mechanisms of action. For example, an inhibitor of an intracellular, ATP-dependent enzyme with a mode of inhibition that is competitive with ATP might be given a lower priority status, given the fact that in cellular settings the compound would be competing with millimolar intracellular concentrations of ATP. The reversibility of inhibition should be investigated, since irreversible inhibitors are widely considered undesirable for medicinal purposes. This issue can be readily explored by incubating a high concentration of the protein-inhibitor solution (10 $\times$ the IC_{50}) followed by 20-fold dilution before the enzyme assay is performed. The enzyme activity of the diluted reaction is determined and compared to the enzyme activity of a control reaction that contains the same final concentration of inhibitor (0.5 $\times$ IC_{50}) and enzyme. Recovery of the enzyme activity from the diluted reaction at the same level as the activity of the control reaction demonstrates a reversible interaction between inhibitor and enzyme and eliminates the possibility of reactive compounds that act irreversibly by forming covalent bonds with the enzyme or other components of the assay.

As interesting leads are identified through the series of experiments outlined above, it becomes necessary to confirm the chemical structure of the compound. As the compound collection ages, compounds degrade. Although this degradation has the effect of expanding the structural diversity that is sampled during screening, it also necessitates efforts to determine precisely what compound has produced the inhibition in the assay. Confirming the inhibition from other samples that may be available and verifying the compound structure by liquid chromatography–mass spectrometry (LCMS) and nuclear magnetic resonance (NMR) is always desirable. If no alternate source of sample is available, LCMS can also be determined from the original DMSO screening solution.

7.5 Additional Considerations for Cell-Based Assays

Some special comments are warranted regarding cell-based assays that have also been adapted to HTS formats. In this setting, compound cytotoxicity constitutes an additional concern for interpretation of results as compared to biochemical assays. Thus additional follow-up assays to measure cytotoxicity or a simultaneous readout of cytotoxicity must be included to determine the possibility of an off-target activity that scores as an apparent inhibition in the assay. One advantage to a cell-based assay is that active inhibitors are already cell-penetrant, presuming that cytotoxic and membrane-interfering actions can be ruled out, a feature that increases the level of interest in confirmed hits. On the other hand, cell-based assays set a stringent criterion in terms of cell penetration, and compounds that may be active against the target protein could be missed. While medicinal chemistry efforts can, in some cases, improve on the bioavailability and cellular uptake of lead compounds, hits that already have these characteristics are more attractive starting points. Such hits are then further assessed in cell permeability assays using the human intestinal cell line Caco-2 and LCMS-based detection of compound. Caco-2 assays have traditionally been used in industry as an indication of bioavailability, and efforts have been made to convert such assays to high-throughput formats (Hidalgo, 2001).

7.6 Target Validation

Many of the techniques used for identification of potential therapeutic targets are, in reality, indirect. Before the initiation of a screening program, evidence for the importance of the target to a disease state may come from genetic, biological, or biochemical approaches, as discussed elsewhere in this volume. While obviously important, these techniques do not exactly replicate the situation that will arise during chemotherapy. For example, target validation based on homologous gene disruption in mice may not directly translate to humans, due to species specificity, compensation in the knockout mouse by upregulation or downregulation of other proteins, or different mechanistic effects of genetic inhibition as compared to pharmacological inhibition of the target. Similarly, genetic validation through the use of dominant inhibitory mutants or siRNA may inaccurately replicate the precise inhibitory effect of a pharmacological agent, which may inhibit the critical activity of the target in an indirect or partial fashion. The latter scenario is illustrated in chemotherapy treatment, when the target present in the target–drug complex – potentially sufficient for full medical benefit – actually represents only a fraction of the total target present in the disease tissue. In short, a number of higher-order issues related to target validation are not addressed by most of the standard types of target validation assays performed in the laboratory. In the last analysis, the most reliable means to show that a target is valid is to do so using a reliable pharmacological lead, in other words, to show that a specific pharmacological agent that inhibits the target in the disease tissue at an appropriate level can elicit the desired medicinal benefit. Pharmacodynamic studies – that is, those performed to learn how the organism handles the compound – are indispensable to learn what level of inhibition must be achieved by in vivo dosing to achieve a therapeutic effect. Consequences of inhibitor dosing on the target or target effector mechanisms must also be considered, however. For example, dosing may increase the expression of the target or other mechanistically relevant proteins or may alter protein turnover. In addition, alternate pathways for substrate metabolism, or changes in the in vivo interaction between the target and other proteins in the presence of the inhibitor should also be considered. Thus, in the absence of direct clinical validation of a proposed target with a chemotherapeutic agent, screening and subsequent medicinal chemistry programs proceed with some degree of risk.

HTS has made it possible to rapidly identify specific inhibitors of a desired target, indeed so rapidly that the effort may outpace other efforts to genetically or biochemically validate the desired target enzyme or receptor. The possibility, therefore, exists that a specific inhibitor identified from screening may be used to validate the target in a pharmacological sense, before optimization of the compound for pharmacokinetic and toxicological parameters. If the compound has even suboptimal bioavailability and pharmacokinetic parameters, allowing it to reach the necessary site of action and maintain an effective concentration, it can be used in an available animal model to validate the importance of the target and its inhibition in the treatment of a disease state. The use of target mutants that are active but that can escape inhibition by the lead compound are also useful tools for promoting pharmacological validation of a target. For example, kinase mutants that can remain active in the presence of the kinase inhibitor – mutants that can often be selected by mutagenesis-selection schemes in cell culture models – can be used to rule out off-target effects of a compound, since their overexpression in the appropriate biological response model will render it drug resistant.

7.7 Summary

In closing, the early involvement of drug metabolism and animal efficacy models in the evaluation of compounds is encouraged, because these efforts not only can save time that might be wasted on poor structural classes of inhibitor but also can provide a way to confirm or rule out the validity of an hypothesized target for drug development. Since learning when to quit a project is always invaluable, exploiting HTS toward pharmacological validation may speed the drug discovery process by helping define when to drop one target in favor of screening another.

References

Auld, D. S., Dunn, D. A., Lehrach, J. M., et al. 1,536-well assay development and screening using whole cell displacement binding and laser scanning imaging. *Assay Drug Devel. Technol.* **1,** 167–174 (2003).

Bazin, H., Trinquet, E., and Mathis, G. Time resolved amplification of cryptate emission: A versatile technology to trace biomolecular interactions. *J. Biotech.* **82,** 233–250 (2002).

Burbaum, J. J. The evolution of miniaturized screening. *J. Biomol. Screen.* **5,** 5–6 (2000).

Cohen, S., and Trinka, R. F. Fully automated screening systems. *Methods Molec. Biol.* **190,** 213–228 (2002).

Chabala, J. C. Solid-phase combinatorial chemistry and novel tagging methods for identifying leads. *Curr. Opin. Biotechnol.* **6,** 632–639 (1995).

Dunn, D. A., and Feygin, I. Challenges and solutions to ultra-high throughput screening assay miniaturization: Submicroliter fluid handling. *Drug Discovery Today* **5,** S84–S91 (2000).

Ferrer, M., Hamilton, A. C., and Inglese, J. A PDZ domain-based detection system for enzymatic assays. *Anal. Biochem.* **301,** 207–216 (2002).

Ferrer, M., Kolodin, G. D., Zuck, P., et al. A fully automated [^{35}S]GTPγS scintillation proximity assay for high-throughput screening of G_i-linked G protein-coupled receptors. *Assay Drug Devel. Technol.* **1,** 261–273 (2003).

Garyantes, T. K. 1536-well assay plates: When do they make sense? *Drug Discovery Today* **7,** 489–490 (2002).

Hidalgo, I. J. Assessing the absorption of new pharmaceuticals. *Curr. Top. Med. Chem.* **1,** 385–401 (2001).

Jordan, S., Zugay, J., Darke, P., and Kuo, L. Activity and dimerization of human immunodeficiency virus protease as a function of solvent composition and enzyme concentration. *J. Biol. Chem.* **267,** 20028–20032 (1992).

Karvinen, J., Hurskainen, P., Gopalakrishnan, S., et al. Homogeneous time-resolved fluorescence quenching assay (LANCE) for caspase-3. *J. Biomol. Screen.* **7,** 223–232 (2002).

Lakowicz, J. R. *Principles of Fluorescence Spectroscopy.* New York, Kluwer Academic/Plenum (1999).

Lenz, G., Nash, H. M., and Jindal, S. Chemical ligands, genomics and drug discovery. *Drug Discovery Today* **5,** 145–156 (2000).

Liu, Y., Kati, W., Chen, C. M., et al. Use of a fluorescence plate reader for measuring kinetic parameters with inner filter effect correction. *Anal. Biochem.* **267,** 331–335 (1999).

McGovern, S., Caselli, E., Grigorieff, N., and Shoichet, B. A common mechanism underlying promiscuous inhibitors from virtual and high-throughput screening. *J. Med. Chem.* **45,** 1712–1722 (2001).

Mere, L., Bennett, T., Coassin, P., et al. Miniaturized FRET assay and microfluidics: Key components for ultra-high-throughput screening. *Drug Discovery Today* **4,** 383–368 (1999).

Oldenburg, K. R., Kariv, I., Zhang, J., et al. Assay miniaturization: Developing technologies and assay formats. *Drugs Pharm. Sci.* **114,** 525–562 (2001).

Park, Y., Cummings, R., Wu, L., et al. Homogeneous proximity tyrosine kinase assays: Scintillation proximity assay versus homogeneous time-resolved fluorescence. *Anal. Biochem.* **269,** 94–104 (1999).

Ramm, P. Imaging systems in assay screening. *Drug Discovery Today* **4,** 401–410 (1999).

Rishton, G. M. Nonleadlikeness and leadlikeness in biochemical screening. *Drug Discovery Today* **8,** 86–96 (2003).

Rishton, G. M. Reactive compounds and in vitro false positives in HTS. *Drug Discovery Today* **2,** 382–384 (1997).

Rogers, M. V. Light on high-throughput screening: fluorescence-based assay technologies. *Drug Discovery Today* **2,** 156–160 (1997).

Rodems, S. M., Hamman, B., Lin, C., et al. A FRET-based assay platform for ultra-high density screening of protein kinases and phosphatases. *Assay Drug Devel. Tech.* **1,** 9–19 (2002).

Rutherford, M. L., and Stinger, T. Recent trends in laboratory automation in the pharmaceutical industry. *Curr. Opin. Drug Discov. Develop.* **4,** 343–346 (2001).

Segel, I. H. *Enzyme Kinetics.* New York, Wiley & Sons (1975).

Sorg, G., Schubert, H. D., Buttner, F. H., and Heilker, F. R. Automated high throughput screening for serine kinase inhibitors using a LEADSeeker^{TM} scintillation proximity assay in the 1536-well format. *J. Biomol. Screen.* **7,** 11–19 (2002).

Tan, D. S., Foley, M. A., Shair, M. D., and Schreiber, S. L. Stereoselective synthesis of over two million compounds having structural features both reminiscent of natural products and compatible with miniaturized cell-based assays. *J. Am. Chem. Soc.* **120,** 8565–8566 (1998).

Taylor, P. B., Stewart, F. P., Dunnington, D. J., et al. Automated assay optimization with integrated statistics and smart robotics. *J. Biomol. Screen.* **5,** 213–226 (2000).

Toledo-Sherman, L. M., and Chen, D. High-throughput virtual screening for drug discovery in parallel. *Curr. Opin. Drug Discov. Devel* **5,** 414–421 (2002).

Wolcke, J., and Ullman, D. Miniaturized HTS technologies – μHTS. *Drug Discovery Today* **6,** 637–646 (2001).

Zhang, J. H., Chung, T., and Oldenburg, K. A simple statistical parameter for use in evaluation and validation of high throughput screening assays. *J. Biomol. Screen.* **4,** 67–73 (1999).

Zheng, W., Carroll, S., Inglese, J., et al. Miniaturization of a hepatitis C virus RNA polymerase assay using a–102°C cooled CCD camera-based imaging system. *Anal. Biochem.* **290,** 214–220 (2001).

CHAPTER 8

Gene Microarray Technologies for Cancer Drug Discovery and Development

Robert H. te Poele, Paul A. Clarke and Paul Workman

Molecular Cancer Therapeutics: Strategies for Drug Discovery and Development, edited by George C. Prendergast
ISBN 0-471-43202-4

8.1 Introduction

Drug discovery and development have benefited widely from the rapid implementation of new technologies (Workman, 2003a). Gene microarray technology is among the latest to affect cancer pharmacology. In this chapter we illustrate the power of the technology and describe its applications at various stages in the discovery and development of molecular cancer therapeutics, including target discovery, mechanism-of-action studies, toxicological profiling, identification of drug-resistant genes, and identification of pharmacodynamic markers that can be used to provide proof of concept and may predict drug response. Gene microarrays are central to the new field of pharmacogenomics and individualized cancer treatment. In addition, microarrays will be extremely important for clinical trials, particularly with respect to selecting patients and predicting outcome. These topics will also be discussed. The emphasis is on applications: this chapter is not intended to be a technical primer, although an overview of microarray methodology is provided with references to the technical literature. Because of the growing importance of single nucleotide polymorphisms (SNPs) and epigenetics in cancer research, we have devoted sections to describing microarray applications to these areas. Last, because microarrays bombard investigators with massive amounts of data that must be transformed into useful information, we also provide information about effective methods for data mining. Throughout the chapter, we have illustrated applications of microarrays for the development of the new generation of molecular cancer therapeutics. However, the methodologies described are equally applicable to studies on existing chemotherapeutic drugs and new agents of all types.

8.2 Cancer: Genes, Genomes, and Drug Targets

During the last decade, the molecular basis for cancer development and pathophysiology has come into much clearer focus. In particular, knowledge of the genetic changes that must occur in a normal cell to form a malignant cell has increased dramatically. Although our understanding of malignant progression is not fully complete, we have defined numerous genes and cell signaling pathways that are causally involved in the initiation and progression of cancer. Together, these studies have provided significant numbers of novel and interesting molecular targets for therapeutic intervention (Workman, 2001a and b; Workman and Kaye, 2002; Workman, 2003a).

Although there has been a lively discussion about the advantages of target-based cancer drug discovery, the general emerging consensus has been to move away from broadly cytotoxic agents to molecular therapeutics that target the genes and signaling processes that are causally involved in cancer (e.g. Workman and Kaye, 2002). Historically, compounds screened in animals and cell-based assays were selected because of their effective antiproliferative or cytotoxic properties, rather than their specific anticancer characteristics (see Chapter 2). These compounds were usually developed for clinical application without knowing the cellular target of the agent, and they often interfered with fundamental cellular processes that are equally essential to both cancer cells and normal cells. Because of the genetic foundation underlying their development, many investigators expect molecular-targeted therapeutics to be more effective, more cancer selective, and less nonspecifically toxic than current anticancer agents.

Strong evidence has been provided from the use of animal models of human cancer to support the potential therapeutic value of agents targeting particular cancer genes. In one example, a transgenic model was developed to address the basic question of whether an oncogene that is essential for the initiation of a specific tumor is still necessary to support the malignant phenotype of that tumor at a more advanced stage, when multiple genetic abnormalities have accumulated. Conditional transgenic mice were engineered to overexpress the c-*myc* oncogene, resulting in the formation of malignant osteosarcomas. Transient removal of c-*myc* overexpression caused the sarcomas to differentiate into mature osteocytes forming normal bone. Moreover, restoration of c-*myc* expression caused apoptosis of the osteocytes instead of the expected reversion to malignant proliferation (Jain et al., 2002).

There are a number of other examples that demonstrate how cancer cells become dependent on the ongoing activation of particular oncogenes. Furthermore, in model systems, when the expression of such oncogenes is turned off or attenuated, apoptosis commonly occurs. For example, transgenic mice overexpressing the human *H-ras* or *bcr-abl* oncogenes developed melanoma or leukemia, respectively, and apoptosis and tumor regression occurred on oncogene shut off (Chin et al., 1999; Pelengaris et al., 2002). Similarly, in cases of human cancer cell lines that constitutively overexpressed the *erb-B2/her-2/neu* or cyclin D1 oncogenes, attenuation of gene expression with antisense oligonucleotides blocked the ability of the cells to form tumors in immunocompromised nude mice, whereas cell lines that did not overexpress these oncogenes were unaffected by the treatment (Colomer et al., 1994; Weinstein, 2002). Experiments of this type have stimulated the development of the concept of *oncogene addiction* (Weinstein, 2002; see Chapter 5). According to this hypothesis, the multiple redundant signaling pathways in normal cells are lost in cancer cells through selection for critical oncogenic pathways, enhanced by genomic instability. Importantly, these experiments also support the strategy of therapeutic interventions to target oncogene function. By acting on the specific pathways on which cancer cells have become dependent, the new generation of molecular therapeutics is predicted to preferentially affect malignant cells with limited harm to normal cells. Of course, to develop such treatments in a rational way, it is essential to understand which pathways are activated in individual tumors.

One illustration of how oncogene addiction might be selectively exploited is provided by studies of the rapamycin analog CCI-779. This agent is an inhibitor of mTOR, which is a downstream target of the PTEN–PI3K–AKT pathway, a key cell survival pathway. Phosphatase and tensin homolog (PTEN) is the phosphatase that negatively regulates signaling by phosphatidylinositol 3-kinase (PI3K). In vitro and in vivo studies with PTEN+/+ and isogenic PTEN−/− cancer cells (both human and mouse) showed that proliferation of PTEN −/− cells, which have increased PI3K signaling, was preferentially blocked by treatment with CCI-779, and the accelerated growth of PTEN−/− xenograft models was reversed by administration of the drug (Neshat et al., 2001).

Alongside experimental support from animal models of cancer, the molecular dissection of human oncogenesis and malignant progression has also been extremely important to the development of new molecular therapeutics. It is now widely accepted that human cancer is caused by progressive acquisition of genetic and epigenetic abnormalities in susceptible cells (Ponder, 2001; Balmain et al., 2003). These abnormalities typically involve somatic mutations but inherited mutations can also play an important role. Multiple mutations (*hits*) are required before a fully malignant cancer develops, underscoring the concept that tumorigenesis is a multistep process. The genetic alterations in the development of colorectal cancers were among the first to be characterized (Kinzler and Vogelstein, 1996), but the mutations and pathways involved in colorectal cancers also apply to many other cancers. The mutation, deregulation, or attenuation of cancer genes results in a wide range of changes in cellular structure and function, all contributing in various ways to the classic hallmark traits of the malignant phenotype (Table 8.1) (Hanahan and Weinberg, 2000). Thus if multiple pathways combine to drive the particular cancer, a cocktail of inhibitors may be required to block the malignant phenotype (Workman, 2003a).

With the development of targeted molecular therapeutics, the particular genes and pathways that dictate sensitivity and resistance must be identified, to be able to predict which drug or drugs will be effective in different subgroups of individual patients. The selection of drug treatment will depend on which genetic abnormalities and hijacked pathways are driving the particular cancers. In many cases, only a genetically defined subgroup of tumors that depend on the continued activity of the drug target and its cognate pathway will be responsive to a particular molecular therapeutic drug.

It is equally important to identify biomarkers that can signal whether the drug actually modulates the intended molecular target (i.e. pharmacodynamic markers) and the biochemical pathways and biological processes in which it operates (Workman, 2003b). Indeed, it has been proposed that such molecular biomarkers are essential to allow the construction of a pharmacological "audit trail" that links the status and expression of the molecular target and the

Table 8.1 Hallmark Characteristics of the Malignant Phenotype[a]

Self-sufficiency in proliferative growth signals
Insensitivity to growth inhibitory signals
Evasion of apoptosis
Acquisition of limitless replicative potential
Induction of angiogenesis
Induction of invasion and metastasis

[a] Adapted from Hanahan and Weinberg (2000).

pharmacokinetic and pharmacodynamic effects of the drug (e.g., target and pathway modulation) to the clinical outcome of treatment (Workman, 2003c). Demonstrating proof of concept for a new molecular therapeutic is extremely important both in the preclinical discovery phase and in early phase clinical trials. The overall challenge is to develop these new molecular therapeutics as expediently and effectively as possible (Workman, 2001a and b; Workman and Kaye, 2002).

8.3 Gene Microarrays: Opportunities and Challenges

The publication of the draft human genome sequence (Lander et al., 2001; Venter et al., 2001) and the advent of the genomic era have fundamentally changed the drug development process. Not only does the human genome sequence itself contain a vast amount of information, for example for use in gene and target discovery, but it has also led to the enablement of techniques such as high-throughput DNA sequencing (Mullikin and McMurragy, 1999) and genome-wide expression profiling using DNA microarrays (Clarke et al., 2001).

DNA microarrays are based on the concept of nucleic acid blotting, in which DNA or RNA is immobilized on a solid support and an mRNA species or DNA sequence is quantified by hybridizing a gene-specific probe. Microarrays actually represent a reversal of this methodology. Thousands of unlabeled DNA probes are immobilized on a solid support and hybridized with one or more labeled single-stranded cDNA representations of a cellular mRNA pool. There are two main methodologies of manufacturing arrays. In one method, oligonucleotides are synthesized on the array in situ, using photolithographic or other techniques as pioneered by Affymetrix, Inc. In the second method, nucleic acids (PCR products, plasmids, or oligonucleotides) are robotically deposited onto a solid support, as pioneered in the laboratories of Brown and Botstein at Stanford University. The various steps involved in a cDNA microarray experiment are shown in Figure 8.1. The advantage of microarray technology over classical blotting methods is that samples can be screened genome-wide for changes in gene expression.

Measuring changes in the mRNA pool genome-wide can reveal a wealth of information about the cellular state, as the gene expression profile of a cell determines its phenotype, function, and response to the environment. Most cellular processes cause a modulation in gene expression. Transcription factors are frequently the downstream targets of signal transduction pathways, conveying the messages of internal and external stimuli. Signal transduction pathways that elicit a transcriptional response mediate the biological response to many drugs.

The successful application of the new and powerful genome-wide gene microarray screening techniques, including use in routine diagnosis and the monitoring of therapeutic and adverse responses to drugs, relies on several

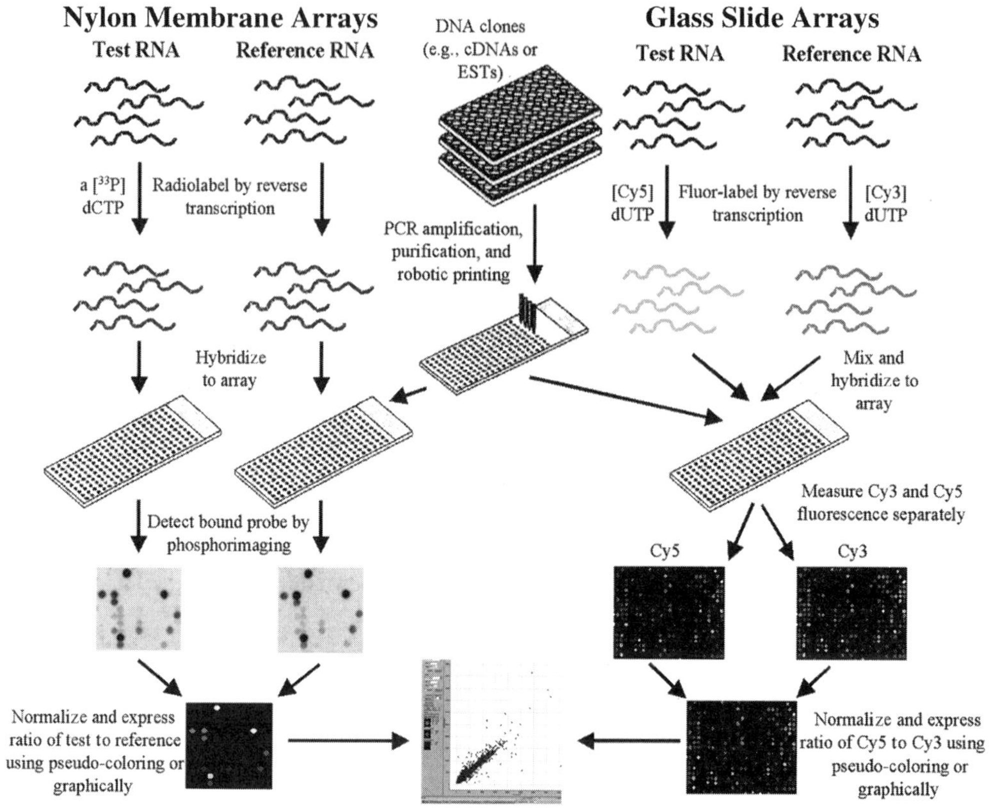

Figure 8.1 Various steps in a DNA microarray experiment. Plasmid clones are propagated in bacteria, and the cloned inserts are amplified by PCR and then purified. The purified PCR products are then robotically printed onto glass or nylon solid supports. Modifications of this approach include the use of oligonucleotides instead of PCR products and the in situ synthesis of oligonucleotides directly onto the glass support using photolithographic or other techniques. Separate nylon-based arrays are hybridized with 33P-radiolabeled cDNA prepared from the test and reference sample, whereas glass slide arrays are hybridized simultaneously with Cy5 and Cy3 fluorescently labeled test and reference samples, respectively. Following stringency washes, hybridization to nylon arrays is detected by phosphorimaging. Hybridization to glass slides is detected by excitation of the two fluophores at the relevant wavelength, and the fluorescent emission is collected with a charge-coupled device. The test and reference images are overlaid using specialist software and can be displayed in a number of ways, including as a scatter plot of the ratio of test:reference gene expression. Alternatively, a false color overlay can be generated, where green denotes a decrease in expression, red an increase and yellow no change in expression between test and reference. The brightness of the spot in false color overlay represents the magnitude of the change in expression between test and reference. Modified from Clarke et al. (2001).

criteria (reviewed by Petricoin et al., 2002). First, it is essential to have accurate amplification and location of sequence-verified probe molecules on microarray chips. Second, the probes have to be selected carefully. This is becoming less of an issue with the rapid improvement of microarray fabrication and the concomitant increase in the density of the probes. However, selection of sequences that distinguish between homologous protein family members may require subcloning of distinguishing sequences. Furthermore, since 40–60% of human genes are expressed as splice variants (Modrek and Lee, 2002) and a large part of the functional complexity of the human genome can be attributed to alternative splicing, probes that distinguish among different splice variants of the same gene need to be considered. At the moment only a small number of human genes have been verified for splice variants (reviewed by Modrek and Lee, 2002). Third and most important, the hybridization data must be reproducible and quality controlled, with confirmation by rigorous statistical analysis or by biological validation using alternative methods.

Most initial work with arrays concentrated specifically on measuring differences in mRNA expression between different cell or tissue samples. More recently, however, the use of arrays has broadened considerably, including to:

- Assess regions of genomic amplification and deletion by comparative genomic hybridization (CGH).
- Identify transcription factor binding sites.
- Map histone acetylation sites or DNA methylation sites (reviewed by Pollack et al., 2002).
- Monitor SNPs using SNP arrays (discussed in Section 8.6).
- Perform quantitative and functional proteomics by protein and protein function microarrays (reviewed by Kodadek, 2001, and MacBeath, 2002).
- Determine protein levels in tumor samples collected on tissue arrays (see Chapter 5).

In this chapter we illustrate how gene microarray technology is having a major impact on the efficiency of most stages of the cancer drug discovery process. Microarrays are providing new insights into the molecular pathology of human cancers and are helping to identify many new additional targets for drug discovery. Furthermore, by profiling the pharmacological effects of lead compounds on a genome-wide basis, microarrays are helping investigators to:

- Discover prognostic and pharmacodynamic markers of drug response.
- Define drug mechanisms of action.
- Identify 'on-target' and 'off-target' effects of drugs.
- Identify undesirable expression signatures of drug toxicity that may be resolved by medicinal chemical optimization.

Microarrays are also being used to help identify genes and expression patterns that are associated with drug sensitivity and resistance using in vitro models and also in retrospective analysis of clinical trials. In addition, they are being used to confirm and investigate the molecular modes of action of drugs in clinical trials, as well as preclinical studies, and to predict which patients are most likely to benefit from particular drugs, aiding individualized cancer treatment.

One of the powerful aspects of microarray analysis is that it generates vast amounts of data. At the same time, this is also one of the great challenges posed by the technology. How can one extract the most useful information, separating it from biological and experimental noise? A number of data-mining techniques are available and these are constantly improving; examples are presented throughout the text. An additional issue concerning the use of microarrays is that considerable effort is still required to improve the accessibility of data to third parties and to facilitate the comparison of data from different microarray platforms and laboratories. To get the most out of the data generated by microarray laboratories throughout the world, some kind of standardization is necessary. A worldwide initiative to address this problem is called 'Minimum Information About a Microarray Experiment' (MIAME). To quote from the objectives of this project (Brazma et al., 2001):

> MIAME aims to outline the minimum information required to unambiguously interpret microarray data and to subsequently allow independent verification of this data at a later stage if required. MIAME is not a dogma for microarray experiments to follow, but just a set of guidelines. This set of guidelines will then assist with the development of microarray repositories and data analysis tools. MIAME is been developed continuously in accordance with our understanding of microarray technology and its applications.

Microarrays can promote the twin goals of understanding the repertoire of genomic pathology that drives individual cancers and of exploiting this repertoire for diagnosis and therapy. With the annotation of the human genome that was essentially completed in 2003, the main challenge for biologists will be to define functions of the 30,000–40,000 human genes, licensing the era of functional genomics. To help address this challenge, Cancer Research UK and the Netherlands Cancer Institute have started an initiative to probe gene function using a global genome-wide strategy based on RNA interference (see Chapter 4). Microarray analysis could provide valuable clues to the function of uncharacterized genes that are knocked out in this manner. Furthermore, this type of global database could be interrogated with microarray data generated by an anticancer therapy, identifying candidate targets of the therapy from the ability of the agent to phenocopy the genetic fingerprint produced by specific gene knockout. The value of such an approach has been demonstrated in yeast. Using microarray analysis, the expression levels of 6,000 transcripts was determined under 300 experimental conditions, including 279 gene knockouts (Hughes et al., 2000). The resulting database was used to compare the effects of several compounds. Dyclonine, an anaesthetic of unknown mechanism of action, elicited expression changes that closely matched those causing disruption of ergosterol metabolism, suggesting that dyclonine induces anaesthesia by disrupting ergosterol metabolism. The feasibility and power of a genome-wide RNAi screen has been exemplified by recent work in *Caenorhabditis elegans* (Kamath et al., 2003).

The genomic understanding of cancer will in turn provides a basis for developing and using drug cocktails for individualized molecular therapeutics. By this strategy, information that is sufficient to develop specific new agents that act on particular genomically defined molecular targets in cancer cells has been obtained. Many agents targeting a specific genetic abnormality in cancer cells are in preclinical development or undergoing clinical trials (Baselga and Averbuch, 2000; Druker, 2002; Huang and Houghton, 2002; Johnstone, 2002; McClue et al., 2002; Neckers, 2002; Rosen, 2002; Senderwicz, 2000; Slamon and Pegram, 2001; Workman and Kaye, 2002) (Table 8.2). The success of this approach is exemplified by the regulatory approval of imatinib mesylate (Gleevec), trastuzumab (Herceptin) and gefitinib (Iressa) (Baselga and Averbuch, 2000; Slamon and Pegram, 2001; Druker, 2002).

Table 8.2 Genomic Targets for Cancer Drugs that are Approved or in Clinical or Preclinical Development

Target	Agent	Reference
EGF receptor	Iressa	Baselga and Averbuch, 2000
ERB-B2	Herceptin	Slamon and Pegram, 2001
BCR/ABL	Gleevec	Druker, 2002
HSP90	17AAG	Neckers, 2002
RAS–RAF–MEK–ERK	BAY-42-9006, PD184352 U0126	Herrera and Sebolt-Leopold, 2002
mTOR	CCI779, RAD001	Huang and Houghton, 2002
VEGF receptor	SU5416, Bevacizumab	Rosen, 2002
HDAC, HAT	Phenylbutyrate, Depsipeptide, MS-27-275, SAHA	Johnstone, 2002
CDK	Flavopiridol, UCN-01, CYC202	Senderowicz, 2000 McClue et al., 2002

CDK, cyclin-dependent kinase; HAT, histone acetyltransferase; HDAC, histone deacetylase; VEGF, vascular endothelial growth factor.

8.4 Array-Based Strategies to Identify Cancer Genes and Drug Targets

A number of methods have been employed to discover oncogenes and tumor-suppressor genes. Positional cloning was used to find many genes in regions of chromosomal gain or loss (e.g., *erb*-b2, *PTEN*) or in chromosomal translocations (e.g., *bcr* and *abl*). Many oncogenes, like *ras* and *myc,* were identified as the human homologs of viral-transforming genes. Linkage analysis of families with inherited predisposition to cancer led to the discovery of other genes such as *BRCA*-2 (Wooster et al., 1995). Studies in model organisms, such as yeast, *Drosophila, C. elegans,* and mouse, have also been important, as in the case, for example, of the identification of the mismatch-repair gene *hMLH*-1 as the human homolog of the bacterial *mutL* mismatch-repair gene (Papadopoulos et al., 1994).

It might be argued that the majority of the genes tractable to identification by such traditional methods, particularly genes that are amplified, deleted, mutated, or translocated, have now been discovered. Certainly the rate of discovery of new cancer genes by these approaches will decrease because the most tractable candidates – the "low hanging fruit" – has already been harvested (Futreal et al., 2001). The Human Genome Project, in additional to its enormous impact on biomedical research generally, is affecting specifically the discovery of additional cancer genes and potential drug targets, as well as the development of new molecular cancer therapeutics (Workman, 2001a and b). The initial working draft of the human genome (Lander et al., 2001; Venter et al., 2001) contained 93% of the human sequence and suggested the presence of 26,000–40,000 human genes. As a result of the output

Table 8.3 Useful Genome and Microarray Web Sites

Description	Web Site
Sanger Center	www.sanger.ac.uk
Cancer Genome Project	www.sanger.ac.uk/CGP
Human Genome Annotation	genome.ucsc.edu
	www.ensembl.org
UniGene Sequence Clustering	www.ncbi.nlm.nih.gov/UniGene
SNP Consortium	snp.cshl.org
Institute for Genomic Research	www.tigr.org/tdb
Whitehead Genome Center	www-genome.wi.mit.edu
European Bioinformatics Institute	www.ebi.ac.uk
European Bioinformatics Institute (Microarrays)	www.ebi.ac.uk/microarray
National Human Genome Research Institute	research.nhgri.nih.gov/microarray/main
Patrick Brown laboratory	cmgm.stanford.edu/pbrown
David Botstein laboratory	genomewww.stanford.edu/group/botlab
U.S. National Cancer Institute (Bioinformatics)	discover.nci.nih.gov
Microarray Gene Expression Database Group	www.mged.org
Microarray Protocols and Software	www.microarrays.org
Affymetrix	www.affymetrix.com
Agilent Technologies	www.chemagilent.com
Illumina	www.illumina.com
Orchid Biosciences	www.orchid.com

from the Human Genome Project (Table 8.3), the concomitant development of high-throughput sequencing technology (Mullikin and McMurragy, 1999) and bioinformatics, together with the rapid advance of techniques dependent on the human genome sequence (such as DNA microarray technology), the remaining cancer genes are likely to be identified in the next several years (see also Chapter 4).

Automated sequencing of genomic libraries constructed from cancer genomes and comparison of these with the normal human genomic sequence is now feasible and represents the most comprehensive and systematic way of identifying the majority of the remaining point-mutated cancer genes (Wooster, 2001). The UK-based Cancer Genome Project at the Sanger Centre has started the enormous task of systematic genome-wide mutation screening of human cancers (Futreal et al., 2001). This initiative has recently identified the *BRAF* gene as an oncogene mutated in 66% of malignant melanomas and a lower but significant proportion of other human cancers, including colorectal cancer (Davies et al., 2002). A large-scale, systematic mutational analysis of tyrosine kinases (the *tyrosine kinome*) in human colorectal cancers has identified previously unknown and potentially activating mutations in several kinase genes and suggested that a minimum of 30% of colorectal cancers contain at least one mutation in the tyrosine kinome (Bardelli et al., 2003).

Several array-based strategies can be employed to identify potential cancer genes. Expression profiling or CGH analysis on microarrays of normal versus tumor tissue can identify genes associated with disease. CGH conducted with bacterial artificial chromosome (BAC) DNA on microarrays can be used to identify regions of loss or gain on human chromosomes (Cai et al., 2002). This strategy is much more rapid and powerful than classical cytogenetic techniques, with the resolution depending on the size and overlap of the BAC DNAs. The regions of chromosomal gain or loss that are identified can then be examined further to determine which particular genes are amplified by CGH on cDNA sequences of the genes located within the region of genomic amplification or deletion. Expression profiling on gene microarrays can be used simultaneously to verify whether amplification of these genes gives rise to their overexpression, providing insights into which of the amplified genes in the amplicon have functional importance (Pollack et al., 2002; Fritz et al., 2002). Even in the absence of genomic losses or gains, expression profiling by microarray analysis can be used to identify overexpressed oncogenes or the absence of message, in the case of tumor-suppressor genes. This requires a comparison of expression profiles of cancers and those of the corresponding normal tissue (Birkenkamp-Demtroder et al., 2002; Welsh et al., 2001). Genes inactivated by epigenetic alteration – typically DNA methylation – can be identified by modified DNA microarray technology (see Section 8.6). Now that sequentially ordered, high-density maps of SNPs are available together with SNP arrays, it is possible to genotype thousands of SNPs simultaneously. This provides the opportunity to identify inherited genetic profiles that are statistically associated with disease, where classic linkage analysis of more complex genetic diseases, such as cancer, has failed.

8.5 Gene Microarrays in Drug Development

8.5.1 TARGET VALIDATION AND SELECTION

Hundreds of genes are already known to play a role in malignancy, and it is likely that many more will be discovered in the next few years. These genes will not all have equal importance in the initiation and maintenance of malignancy: some will be more critical than others. Furthermore, it is likely that the number of potentially "druggable" genes will continue to exceed the capacity of any drug development organization. Thus procedures are required to validate potential drug targets and to prioritize them before the drug discovery process is started.

Once a cancer gene that represents a potential drug target has been identified, some kind of validation is necessary to justify the allocation of resources needed for a new drug project (Workman, 2001a; Workman and Kaye, 2002). Potential drug targets differ greatly in their appeal in regard to tractability or druggability. For this reason, one must establish criteria for target validation and selection, including an assessment of the technical risk as well as an estimation of potential medical value of the agents that will emerge from a drug discovery program.

Although scientific and technical considerations are important, for pharmaceutical companies there are a number of additional issues involved in prioritization of targets, and decision making involves a complex balance of scientific, medical, and strategic factors (Workman, 2001c). For example, market size and the ability to identify suitable genomically defined patients for a clinical study are important. Evidence supporting the hypothesis that the putative drug target has a role in the malignant process can be obtained by screening human tumor cell lines and tumor samples for mutations, deletions, amplifications, or altered expression. The frequency of the abnormality, for example, mutation or deregulated expression, can tell us to whether the target likely plays an important role in malignant progression and can indicate the number of patients that should benefit from therapy based on target status. The type of cancers likely to respond can also be estimated from such an analysis. Measurement of the sequence and expression of potential target genes is greatly facilitated by microarray technology. However, expression data alone, though important, cannot provide sufficient evidence of causal involvement in disease pathology. Demonstrating that modulation of the target or the pathway in which the cancer gene operates can reverse the malignant phenotype gives a strong indication that once appropriate drugs have been developed they could prove useful therapy in patients who exhibit deregulation of the target. RNA interference (described in Chapter 4) can be a valuable technique, especially in combination with DNA microarrays, for understanding the molecular signatures of potential drug targets. Up-to-date techniques used to validate and select potential targets for drug development have been described in detail (Workman, 2001a and b; see also Chapters 3–7 and 10). A good example of this is the discovery and validation of the histone methyltransferase EZH2 as a drug target in metastatic prostate cancer (Varambally et al., 2003).

8.5.2 MOLECULAR MECHANISM OF ACTION

Modern mechanism-based drug discovery programs aim in prelinical work to achieve the desired profile of properties that is required of a clinical candidate (Workman, 2001c). The goal is to establish the necessary potency, selectivity, and therapeutic activity along with other factors, such as route and schedule of administration. Toward this end, a series of assays that form a biological test cascade are designed (Aherne et al., 2002; Workman 2001c). As described in detail in other chapters of this book, the top of the test cascade is a primary target screen, usually a biochemical or cell-based assay that is suitable for high-throughput screening, followed by assays for biological activity, pharmacokinetic properties, and therapeutic activity in an animal model. Throughout the preclinical development phase, and indeed in the early clinical trials, it is important to collect direct evidence that the drug effect occurs via the desired mechanism of action. Furthermore, in the case of cell-based screens, mechanistic assays are required to help identify the precise cellular target. Such assays can also provide valuable pharmacodynamic endpoints for use in animal studies and subsequent clinical trials (Workman, 2003c).

Definitive confirmation of the mechanism of action of a compound is quite challenging. In one approach, compounds can be submitted to be profiled across the National Cancer Institute panel of 60 human cancer cell lines (dtp.nci.nih.gov). Activity against this panel can then be correlated with the pattern of activity of compounds with known mechanisms of action by use of the COMPARE algorithm. In addition, correlation can also be sought with data on the expression of various molecular targets in the cell panel (Weinstein et al., 1997). One more comprehensive way to identify or confirm the target(s) of test compounds is to correlate sensitivity to the test compound with the output from genome-wide expression profiling. Microarray profiling (Scherf et al., 2000) and proteomics (see Chapter 10) are powerful methods for elucidating the mechanism of action underlying a cellular response to drug treatment (Clarke et al., 2000; Panaretou et al., 2002).

Using gene microarrays, we and others (Dracopoli, 2003) have moved toward compiling a database of gene expression signatures for the newer molecular targeted therapeutics (signal transduction inhibitors) as well as classical anticancer agents. In addition, new compounds from our drug discovery projects are profiled for the gene expression changes that they induce. These profiles are then compared to the molecular signatures in the database using cluster analysis (see Section 8.9), to obtain clues to the mechanism of action of the new compound and to determine whether a given compound hits additional cellular targets.

In situations in which no inhibitors of a pathway are available for a particular drug target, a gene expression signature of the pathway can be determined in cells in which the desired target is modulated by molecular biological techniques (e.g., RNA interference). For example, the target can be expressed ectopically and subsequent gene expression changes monitored by microarray analysis. Inhibition of the target can reasonably be expected to reverse the changes seen with overexpression of the target or alternative components of the same pathway. Approaches to phenocopy pharmacological modulation of the target include transfection or transduction of dominant negative constructs or neutralizing antibodies (Chapter 7), antisense constructs, antisense oligonucleotides (Ross et al., 2001), hammerhead ribozymes, and the use of homologous knockouts (Chapter 11). Recent experience with RNA interference technology (Hannon, 2002) suggests that this approach is particularly promising for genetic validation of targets, including in whole animals (Chapter 4).

There are a growing number of examples in the literature of using gene expression microarrays to profile the molecular signature of a drug response (reviewed by Clarke et al., 2001). In one of the first reports, our laboratory used cDNA microarrays to investigate the genes that showed altered expression in response to treatment with the HSP90 molecular chaperone inhibitor 17AAG, a promising agent that is completing Phase I clinical trials (Banerji et al., 2002; Clarke et al., 2000). In the initial analysis, genes showing altered expression at the mRNA level in human colon cancer cell lines included those encoding HSP90β, HSP70, keratin 8, keratin 18, and caveolin-1. Induction of HSP90 in a cell line that had recovered from HSP90 inhibition and reduction of HSP90 expression in a cell line that was particularly

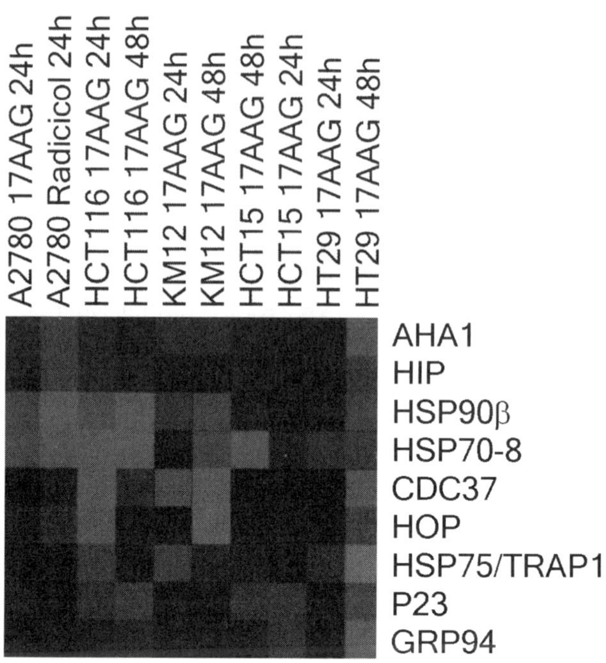

Figure 8.2 Grey scale representation of hierarchical clustering of HSP and co-chaperone gene expression data obtained with the HSP90 inhibitors 17AAG and radicicol. The horizontal axis shows the human cancer cell lines (A2780 ovarian and HCT116, KM12, HCT15, and HT29 colon) and exposure times for the two drugs. The vertical axis lists the genes. The HCT116 and A2780 samples cluster together, both show increased HSP90β expression (light gray, normally bright red in false color overlay), whereas the HT29 samples cluster away from these samples; the HT29 samples show no increased HSP90β expression following treatment (dark gray). Interestingly, the A2780 and HCT116 cells, which show induction of the target, recover more rapidly than the HT29 cells that do not induce HSP90β in response to treatment. Also of note is that several co-chaperones are co-induced by 17AAG and radicicol in some cell lines, including the recently identified AHA1. Modified from Panaretou et al. (2002).

sensitive to HSP90 inhibition suggested a potential resistance or recovery mechanism (Fig. 8.2). Increased expression of HSP70 was significant because of the potential for it to exert an antiapoptotic effect. Alterations in caveolin-1 and the keratin genes may relate to inhibitory effects of 17AAG on the PTEN–PI3K–AKT and RAS–RAF–MEK–ERK signal transduction pathways, each of which promotes cell survival. Apart from casein kinase, genes encoding client proteins for chaperoning by HSP90, including c-RAF-1, cyclin-dependent kinase 4 (CDK4) and AKT, were not affected at the mRNA level, despite the fact that the corresponding proteins were eliminated from the cell via the ubiquitin proteasome pathway. Whereas some genes were altered in all the cell lines investigated, most alterations only occurred in a cell-line-dependent manner.

Further studies that compared the transcriptome (cellular mRNAs) and the proteome (cellular proteins) after 17AAG-induced inhibition of HSP90 confirmed that depletion of client proteins did not occur at the transcriptional level; however, induction of HSP90β, HSP70, and a number co-chaperones was detected at both the mRNA and protein level (Fig. 8.2) (Maloney et al., unpublished). In the complimentary proteomic analysis, the new gene product

AHAl was found to be upregulated in tumor cells after 17AAG treatment (Panaretou et al., 2002). AHA1 is a previously uncharacterized co-chaperone that activates the ATPase activity of HSP90. Re-analysis of the gene expression microarray data showed that AHA1 gene expression was also upregulated at the level of mRNA. This may be therapeutically important, because increased expression of an activating co-chaperone together with increased expression of HSP90 itself could be seen as an attempt by the treated cell to overcome the inhibition of the intrinsic HSP90 activity. In summary, using a combination of gene expression microarrays, proteomics, western blotting and ELISA methodology, it was possible to identify a molecular signature or pharmacological fingerprint for HSP90 inhibition in human tumor cell lines, tumor xenografts, and peripheral blood lymphocytes (Maloney and Workman, 2002; Banerji et al., 2002). The specificity of the molecular signature is illustrated by the fact that it is shared with other active HSP90 inhibitors of similar or distinct chemical classes but is not shared with inactive analogs or cytotoxic agents like paclitaxel (Maloney et al., unpublished). We are currently exploring this molecular signature of HSP90 inhibition in clinical trials in cancer patients, and preliminary evidence demonstrates HSP90 inhibition in peripheral blood lymphocytes and tumor biopsies from 17AAG-treated patients (Banerji et al., 2002). Our experience with HSP90 inhibitors illustrates the value of global profiling of drug-induced changes in the transcriptome and proteome, including information on molecular mechanism, genes involved in drug sensitivity and resistance, and pharmacodynamic biomarkers that are useful for clinical trials and possibly for follow-up drug discovery programs.

Other studies support this experience. Using cDNA microarrays containing 1694 genes implicated in cancer, one group defined the transcriptional response of HCT116 colon carcinoma cells synchronized in S phase by aphidicolin (APH) treatment to two different concentrations of the topoisomerase I inhibitor camptothecin (CPT) (Zhou et al., 2002). At the lower concentration of 20 nM, a reversible G2 arrest was observed, whereas at the higher concentration of 1000 nM an irreversible arrest in the G2 phase of the cell cycle occurred. A total of 33 genes showed significantly altered expression after treatment; these genes could be divided into roughly three groups. Northern analysis validated 5 genes from different groups and showed an overall correlation coefficient of 0.86 with the microarray data. The first group of genes was upregulated in the APH-treated controls and those exposed to low-dose CPT, although the upregulation in the CPT-treated cells occurred at a later time point, matching the difference in timing of mitosis (e.g., genes encoding cyclin B1, aurora/STK15). Interestingly, 4 of the 6 genes in the first group were directly involved in the regulation of mitosis. The genes in the second group were downregulated when cells treated with low-dose CPT were recovering from the G2 delay. The third group of genes were upregulated only in response to high-dose CPT during S phase delay and/or G2 arrest. The genes in this group were DNA damage-inducible genes and were also associated with cell cycle arrest and apoptosis (e.g., those encoding DDB2, cyclin-dependent kinase inhibitor $p21^{WAF1}$, Fas). Of the 8 genes in this group, 5 are transcriptionally activated by the p53 DNA damage response pathway. Together, these observations suggest that there is a fundamental difference in the transcriptional response to mild DNA damage resulting in a reversible G2 arrest, as compared to the permanent G2 arrest following extensive DNA damage.

Microarray analysis can address the problem of intrinsic and acquired resistance to anticancer drugs, which seriously limits the efficacy of cancer treatment. A systematic microarray profiling methodology for measuring intrinsic drug resistance is discussed in Section 8.9. Molecular mechanisms of acquired drug resistance and downstream mediators of drug action can be examined in vitro by continuous exposure of cell lines to drugs until a subclone becomes resistant to the drug and can be selected from drug-sensitive parental cells. The resistant clone can then be compared to the parental line by gene expression microarray analysis.

Such an approach was used to investigate the molecular mechanisms of acquired resistance to CPT. The prostate cancer cell line DU145 and the selected subline RC0.1, resistant to 0.1 μM 9-nitro-camptothecin, were compared by cDNA microarray analysis (Reinhold et al., 2003). Using the statistical method of a stratum-adjusted Kruskal-Wallis test, expression changes over 1.5-fold were judged to be significantly altered. Of the 181 significant changes that were defined by this criterion, genes in several functional groups were found to be significantly overrepresented, including those involved in the MHC, nuclear factor κB (NFκB) signaling, and apoptosis. Resistance to a variety of cellular stresses seen with the RC0.1 subline suggested that the observed gene changes conferred a generalized apoptosis resistance phenotype. Several of the expression changes could explain the reduced apoptotic response in the RC0.1 subline, such as the downregulation of the pro-apoptotic *bad* and *caspase-6* genes and the reduced expression of genes involved in PI3K signaling, which may ultimately lead to reduced activity of the BAD protein. However, several of the changes observed, especially those in the NFκB and transforming-growth factor β pathway, were contrary to expectation, since they would generate pro-apoptotic signals.

This led the authors to suggest a two-step mechanism for the development of drug resistance in RC0.1 cells. The first step involves gene expression changes that directly conferred resistance to apoptosis. This step would allow changes that would normally favor selection if they did not also induce apoptosis to occur when the apoptotic pathway was effectively blocked downstream. For example, the dual-acting genes *E2F1* and c-*MYC,* which can drive both apoptosis and proliferation, each showed increased expression in the resistant subline. When the apoptotic pathway is blocked, increased proliferation would be a selective advantage. This model is analogous to one proposed for the development of malignant cancers (Hickman, 2002). Molecular carcinogenesis is inhibited whenever increased cell proliferation is linked to apoptosis. Once cells become resistant to apoptosis, increased proliferation can occur through dual-acting genes like *E2F1* and c-*MYC* without cell death as a consequence.

In a separate study, the use of gene expression microarrays revealed that the proposed CDK inhibitor flavopiridol had a different mechanism of action from two other CDK inhibitors studied; namely, roscovitine and 9-nitropaullone (Lam et al., 2001). Flavopiridol appeared to inhibit gene expression in a rather general fashion, causing changes similar to those produced by the transcription inhibitors actinomycin D and 5,6-dichloro-1-β-D-ribofuranosyl-benzimidazole. Investigating mRNA turnover following flavopiridol exposure revealed that different functional classes of genes had distinct distributions of turnover rates. Several apoptosis genes and key cell cycle regulators had very short half-lives, suggesting that flavopiridol may be particularly effective in cancers dependent on genes with high turnover rates, such as c-*myc*.

Another study used oligonucleotide arrays to compare the levels of total and polysome-bound RNA in response to rapamycin treatment in T cells (Grolleau et al., 2002). In this study, 159 transcripts remained polysome bound, mostly transcription factors, kinases, phosphatases, and members of the *ras* superfamily of small GTPases, suggesting that rapamycin does not affect translation of these genes. However, translation of 136 genes was repressed by at least 90%, including genes known to be modulated by rapamycin, such as translation initiation factors 4A and 5A. Interestingly, translation of 7 genes encoding subunits of the proteasome were completely inhibited, and translation of prothymosin α was also repressed. Both the proteasome and prothymosin α are involved in the immune, proliferative, and cytotoxic response in T cells, and this explains some of mechanisms involved in the biological activity of rapamycin.

Oligonucleotide microarrays also have been used to define a set of common genes regulated by histone deacetylase (HDAC) inhibitors (Glaser et al., 2003). In this study, the investigators compared gene expression profiles of two breast cancer cell lines and a bladder carcinoma cell line treated with suberoylanilide hydroxamic acid (SAHA), trichostatin A (TSA), both being hydroxamic acid derivatives, and the novel inhibitor MS-27-275. Concentrations of all three agents were selected to cause a similar, maximum level of induction of the cell cycle kinase $p21^{WAF1}$ after a 24-h exposure and a robust hyperacetylation of histone H4. The gene expression profiles of the three HDAC inhibitors were generally similar and distinct from those produced with the structurally related but inactive analogs of SAHA and MS-27-275, indicating that the changes observed with the HDAC inhibitors were mechanism related and not due simply to a common chemical backbone. The correlation of the expression changes induced by SAHA or TSA was higher than the correlation between either of these and the MS-27-275 compound, consistent with their different effects on cells. Because treatment with HDAC inhibitors was predicted to increase gene expression, it was surprising that as many genes were downregulated as upregulated after HDAC inhibitor treatment. All three inhibitors consistently altered the expression of only 13 genes across all three cell lines, of which most genes were involved in the control of the cell cycle, apoptosis, and DNA synthesis. These results show that mechanism-based gene expression changes can be identified, but also demonstrate the cell-line dependence of expression changes, highlighting the limitations of analyzing on by a single cell line.

A different array-based approach, *interventional profiling,* was used to test a compound library of FDA-approved drugs that would have the potential to provide protection in a hydrogen peroxide–induced oxidant injury model using neuroblastoma cells (Sarang et al., 2002). Twenty-six neuroprotective compounds were identified; of this group megestrol, meclizine, verapamil, methazolamide, sulindac, and retinol were examined by oligonucleotide microarray analysis. Five genes were either uniformly upregulated or downregulated with these compounds; namely, tissue inhibitor of matrix metalloproteinase 1, the *ret* proto-oncogene, clusterin, growth-associated protein 43, and the neuropeptide galanin. Treatment with the galanin peptide alone conferred resistance to oxidative stress. Interventional profiling represents a general and powerful strategy for identifying new bioactive agents for any biological process, as well as identifying key downstream genes and pathways that are involved.

Our experience is that time course experiments can be very important for determining primary and secondary effects of a drug. Exposures for longer periods can result in various cellular outcomes, including apoptosis, cell cycle arrest, and differentiation, which are secondary to and separable from the primary response to the drug at the initial biochemical level. Although identifying later endpoints of drug action can be valuable in evaluating new compounds, gene expression patterns that signify such cellular states will not be informative with respect to the molecular mechanism leading to such a particular biological outcome. In contrast, it is likely that initial drug-related responses to the modulation of the primary molecular target will be more mechanistically informative to understanding how a particular biological outcome is elicited. Thus, mechanistically informative changes are more likely to occur at earlier time points after drug treatment, whereas later changes will relate to secondary cellular effects. In our laboratory, we find it useful to look at the way gene expression profiles alter with both time and drug concentration. Furthermore, profiling structurally unrelated compounds and inactive analogs of the lead compound can aid in distinguishing mechanism-based expression changes from those caused simply by the chemical backbone of the test molecules.

The above examples illustrate the utility of the gene expression microarray approach in studying mechanism of action, discovering genes involved in sensitivity and resistance, and identifying pharmacodynamic markers of drug action for use in preclinical drug development and early clinical trials.

8.5.3 TOXICOLOGICAL PROFILING

Microarrays can provide a relatively simple and valuable alternative to the more detailed and expensive toxicology analysis that is performed on potential drugs, usually on attractive lead compounds at more advanced stages of the development. When a compound fails at later stages because of unacceptable toxicities, considerable resources will have already been spent. It would, therefore, be useful to develop methods capable of screening large numbers of compounds for potential toxicity. The increasing emphasis in drug development is on finding agents targeting essential oncogenic pathways in human cancers. Treatment with such agents can result in cell death and regression of tumors. They may, however, have cytostatic rather than cytotoxic properties, allowing one to contain or manage rather than cure disease. If the new generation of molecular therapeutics prove to act more commonly as cytostatic agents, then cancer will be treated as a chronic disease. In this scenario, patients will have to be treated for prolonged periods of time, making it even more important to carefully profile and predict potential toxicities at early times in development.

It is likely that all toxic chemical exposures alter gene expression to some degree, since signal transduction pathways that regulate transcriptional responses are known to mediate the toxicological effects of many agents. Toxic effects are commonly manifested as inflammation, proliferation, apoptosis, and necrosis or cellular differentiation. Although each of these can be used as a toxicological end point in assessing the toxicity of new agents in normal tissue, all of these cellular processes also result in particular expression changes that can be monitored by gene expression microarrays (Afshari et al., 1999; Nuwaysir et al., 1999). In this way, microarray analyses have the potential to be used as substitutes for biological assays to measure toxicity.

Construction of a database of gene expression profiles of toxic responses in cell lines, hepatic cells, and animals, using model compounds known from the literature to induce particular toxic reactions, could allow the identification of agents that are likely to give rise to various toxicological side effects, based on a specific molecular signature of toxicity. Certain types of toxic response, as in the case of inflammation, require the interaction of a number of different cell types and the contribution of paracrine signaling. These can be assessed only by using animal experiments. However, even in the absence of such complex interactions between tissue types, the measurement of gene expression in, say, hepatic cells in tissue culture should still detect transcriptional changes that indicate paracrine signaling and an immune response.

The use of microarrays will not be foolproof in identifying whether a compound will have particular toxic features in humans. However, it should give clear indications as to the induction of undesirable gene expression changes that are statistically correlated with adverse effects. Agents that are free of those effects can be given a higher priority over those that are not. A major benefit of this approach is the relatively low cost and high throughput of microarrays as compared to the extensive use of animals. Hence they can be employed to screen many compounds emerging from the chemical optimization process. Undesirable gene expression signatures can be flagged at an early stage and dealt with appropriately. Many pharmaceutical companies are now using expression profiling to weed out compounds that may cause unacceptable or undesirable side effects, such as hepatotoxicity. Unfortunately, such information will generally be of high proprietary value and hence will not at available in the public domain.

A pilot study was conducted to determine the feasibility of using DNA microarray analysis to classify known classes of toxins (Thomas et al., 2001). In this study, inbred mouse strains were exposed to acute doses of toxins and alterations in global gene expression in whole liver were monitored using cDNA microarrays. Two-dimensional hierarchical clustering of all genes that changed more than twofold grouped most of the compounds according to toxicogical class. Some exceptions were noted. This may not be very surprising, because, although compounds within a particular class will tend to share common mechanisms of toxicity, individual compounds may cause unique expression changes and indeed may not act via a single unique mechanism. A probabilistic approach based on Bayesian statistics was applied to obtain a molecular predictor of toxicity. This method ranks the genes in order of their estimated predictive value and adds individual genes to the predictor until the highest accuracy of prediction is reached. The accuracy of the prediction was measured using the "leave one out" method. The results showed that a signature predictor based on the expression of 12 transcripts had $\sim$100% accuracy in assigning the correct mechanism of toxicity across all compounds. The predictor contained some genes that were already known to be altered in response to the compounds, whereas other genes were not previously implicated. Here it is important to note that using inbred mouse strains will not take into account the likely considerable influence that SNP will have on the variability of the response to foreign molecules in outbred populations, including humans.

In another interesting experiment, the effects on the liver of a thienopyridine compound A-277249, an inhibitor of NFκB-mediated expression of cell adhesion proteins, were studied by microarray analysis (Waring et al., 2002). A-277249 caused hepatic hypertrophy and increased serum levels of metabolic enzymes. Hierarchical clustering of the thienopyridine expression profile with that of 15 known hepatotoxins demonstrated that the molecular signature was most similar to that of two activators of the aryl hydrocarbon nuclear receptor (AhR). A number of genes that are regulated by the AhR were found to show increased expression in response to A-277249 treatment, including the cytochrome P450 enzyme isoform CYP1A1. Thus these data suggested that the hepatotoxicity of A-277249 is at least partly caused by its effects on the AhR and nicely illustrates the potential usefulness of microarrays to assess mechanisms of toxicity.

Another illustration of how microarrays can be used to assess toxic side effects is illustrated by studies of inhibitors of the PTEN–PI3K–AKT pathway, which is implicated in tumor cell malignancy but also in insulin signaling, raising the potential for diabetic complications. A large body of evidence demonstrates that various molecular abnormalities in the PI3K pathway play a key role in malignant progression (Vivanco and Sawyers, 2002). The *PTEN* tumor suppressor gene encodes a lipid phosphatase that catalyses hydrolysis of the inositol lipid product of the PI3K reaction, thereby counteracting its potentially oncogenic effects. Notably, loss of PTEN function ranks second in the frequency of human cancer gene mutations only to the most commonly mutated *p53* gene (Ali et al., 1999). Oncogenic activation of the PI3K–AKT pathway, by loss of PTEN or increased expression of AKT, contributes to the malignant phenotype and cancer progression by stimulating proliferation, survival, migration, invasion, angiogenesis, and metastasis. It may also be involved in drug resistance. Because of these effects, inhibition of the PI3K pathway is seen as an attractive potential locus for therapeutic intervention. However, since the PI3K pathway is also involved in insulin signaling, there is a risk that drugs blocking this pathway may disturb the balance of insulin metabolism, resulting in diabetes.

We have shown that LY294002, a widely used broad-spectrum prototype inhibitor of the PI3K family, does indeed affect components of the insulin-signaling pathway (te Poele et al., 2002). However, the PI3K family comprises a large number of isoforms, and there are indications that certain isoforms are involved in insulin signaling, whereas other isoforms are responsible for driving tumor progression (Roche et al., 1998). Hence there is considerable interest in targeting specific PI3K isoforms. Profiling such inhibitors using gene expression microarrays may help to identify compounds that have anticancer effects but that do not block insulin signaling. Our laboratory is currently taking this approach to profile inhibitors that may have differential effects on the specific PI3K isoforms. Another possible approach to avoid affecting insulin metabolism is to identify druggable targets that are downstream of PI3K, where the pathway diverges to separate components that mediate survival, migration, proliferation, and insulin signaling. We have used gene expression microarrays to follow the time course of transcriptome changes in response to treatment with the well-known PI3K inhibitor LY294002 and used this information to identify genes that are affected downstream of PI3K inhibition. In this manner, we have identified several mitotic genes that appear to be co-regulated by the PI3K pathway, two of which were implicated previously in malignancy.

8.5.4 PHARMACOKINETICS AND DRUG METABOLISM

Pharmacokinetic behavior is commonly a rate-limiting step in taking compounds identified in cell-based assays into efficacy testing in animal models. Prediction of pharmacokinetic behavior in the whole animal can be difficult, and for this reason the issue is discussed in more depth in Chapters 12, 13, and 14. Pharmacokinetic prioritization screens and the use of cassette dosing can increase the speed and efficiency of the transition from in vitro to whole animal models (Rodrigues, 1997). Pharmacokinetic behavior is often controlled by metabolism of the compound, and knowledge of metabolic routes and rates of metabolism can be helpful in selecting compounds. Networks of nuclear receptors stimulated by xenobiotics control the transcriptional regulation of the genes for a large variety of drug-metabolizing enzymes. These nuclear receptors and the AhR described above mediate transcription as heterodimers or homodimers, together with tissue-specific transcription factors and transcription factors that respond to various cellular stresses and cytokines (Karpen, 2002). Drug-stimulated activation of these regulatory mechanisms can be readily monitored by gene microarray analysis.

A particular property to avoid in a clinical candidate is metabolism by polymorphic enzymes (see also Section 8.6), such as the cytochrome P450 CYP3A4, which can lead to extensive variability in metabolism and pharmacokinetics. Inhibition of cytochrome P450 enzymes that are responsible for drug metabolism should also be avoided, since such effects may lead to unwanted drug–drug interactions (for example increasing the exposure and toxicity of the molecular therapeutic itself or of other drugs in patients that receive multiple therapeutics). Measurement of the metabolism and potential inhibition of recombinant P450 enzymes can be extremely useful at early stages of drug development. Thus an assessment of the induction or repression of drug metabolism genes, particularly those encoding cytochrome P450 enzymes, can also provide valuable information. Gene microarray analyses are only beginning to be used in this manner, but they are likely to make significant contributions in this area in the future.

8.6 SNP Arrays to Identify Disease Genes and Predict Phenotypic Toxicity (Pharmacogenomics)

Pharmacogenomics is the term used to describe studies of the genetic basis for variable drug responses in different individuals due to inherited phenotypes, with emphasis on genome-wide analysis. Unpredictable toxicity in clinical trials or after regulatory approval is the principal reason that new therapeutic agents fail. For many years, pharmacogenetic studies have relied on the measurement of the status of individual drug-metabolizing enzymes to understand and predict the efficacy and toxicity of drugs in individuals. Inherited differences in DNA sequence contribute to phenotypic variation, influencing a given individual's risk of disease and also his or her reaction to the environment, for example, adverse or therapeutic response to drug treatment (Roses, 2002). Most sequence variation in humans can be attributed to SNPs. As their name implies, SNPs are specific nucleotides in the human genome sequence in which different individuals have different DNA bases. These single base variations, or polymorphisms, can lead to changes in protein-coding or regulatory sequences that can contribute to disease or adverse effects of drugs. For example, the antitumor activity and possibly the toxicologic properties of the HSP90 molecular chaperone inhibitor 17AAG in humans may be influenced by a polymorphism in the enzyme NAD(P)H: quinone oxidoreductase 1 (NQO1) (Kelland et al., 1999). With the publication of SNP maps and the rapid advance of high-throughput genotyping, statistical analysis, and bioinformatics, it is now possible to characterize both drug metabolism and disease genes and the response of individual patients to drugs and to determine whether toxicity or efficacy is associated with a particular phenotype (Roses, 2002, and references within).

In February 2001 a map of human genome sequence variation was published containing 1.42 million SNPs (Sachidanandam et al., 2001). This publication was the culmination of the efforts of the SNP consortium (snp.cshl.org/) and the analysis of clone overlaps by the International Human Genome Sequencing Consortium (Lander et al., 2001). Genome-wide linkage analysis and positional cloning have been used to identify hundreds of human disease genes (www.ncbi.nlm.nih.gov/entrez/query.fcgi?db=OMIM). However most of these diseases are rare conditions in which the mutation of a single gene is sufficient to cause the pathology. For most common diseases, genome-wide linkage analysis has had little success, consistent with a more complex genetic pattern. With several individual loci contributing modestly to disease genetics or drug reactions, more powerful high-resolution techniques are required to identify disease susceptibility genes. Sequentially ordered, high-density SNP maps could provide such a technique and allow identification of inherited profiles that are statistically associated with disease or drug response.

SNPs are distributed throughout the human genome with an average density of one SNP every 1.9 Kb (Sachidanandam et al., 2001). Global SNP analysis would, therefore, require genotyping millions of SNPs. However, SNP variants that are closely linked do not occur independently from each other, a phenomenon known as linkage disequilibrium (LD) between neighboring SNPs. Adjacent alleles are thus associated in a nonrandom manner, reflecting genetic haplotypes descended from single ancestral chromosomes (Reich et al., 2001). These haplotypes or LD blocks typically span 40 Kb but can extend >800 Kb (Dawson et al., 2002). The publication of an LD map of chromosome 22 (Dawson et al., 2002) has shown that developing genome-wide LD maps is feasible. The next step is to use the great abundance of SNPs and their clustering in LD blocks in association-type studies to identify disease genes and genes associated with drug efficacy or toxicity.

The fact that the allelic variants that contribute to complex disease are often fixed in haplotype blocks, creating disease haplotypes, means that all of the SNPs in a block will show association with disease. It may, therefore, be possible to type one SNP per block to identify the location of a disease or drug response gene, greatly reducing the amount of SNPs required to genotype individuals. If the haplotype blocks in a region are small, around 40 Kb as they commonly are, disease association implies a nearby susceptibility gene. To ensure that susceptibility genes are easy to identify once association between an LD block and disease has been established, the intervals between SNPs must be between 40 and 100 Kb (Cheung and Spielman, 2002). The approach of capturing SNPs in LD blocks is more sensitive than classical linkage analysis which relies on mapping recombinants in families. The frequency of recombination in humans results in a set of 350 markers throughout the genome at approximately 10 Mb intervals. Thus mapping SNPs in LD blocks requires genotyping between 30,000 and 100,000 SNPs per individual, and emphasises the need for high-throughput genotyping assays.

Several methods are under development, including those based on microarrays (Roses, 2002). Microarrays able to genotype a few thousand SNPs are now commercially available (Affymetrix, Illumina, Orchid; Table 8.3). However, for microarray-based genotyping to become the method of choice for the genetic analysis of complex disease and drug response, future SNP arrays will have to be genome-wide and use SNPs that capture variation in haplotype blocks. Nevertheless, the first experiments proving the power of association studies using high-density SNP maps have been published. For example, this approach was used to confirm that the apolipoprotein E allele is the susceptibility gene variant that is responsible for common, late-onset Alzheimer disease (Lai et al., 1998) and to identify the tumor necrosis factor α (TNFα) and HLA-B57 polymorphisms as susceptibility genes for hypersensitivity to abacavir, a reverse transcriptase inhibitor used to treat human immunodeficiency virus (Hetherington et al., 2002). Applications of SNP technology to the development and use of anticancer agents should be anticipated in the near future.

8.7 Epigenetics

In cancer genetics, most emphasis has been placed on identifying genes that are mutated, amplified, or deleted in human cancers. Recently it has become clear, however, that epigenetic alterations, especially DNA methylation, are frequently involved in the deregulation of both tumor supressors and oncogenes and are an important driving force in tumourigenesis (Baylin and Herman, 2000; Brown and Strathdee, 2002; Jones and Laird, 1999). DNA methylation is a covalent modification of cytosine residues in CpG dinucleotides, catalyzed by a family of DNA methyltransferases, and inherited in somatic cell division. Methylation of CpG rich sequences, known as CpG islands, in the promoter regions of genes often results in the transcriptional silencing of those genes. Whereas CpG islands in normal cells are usually unmethylated, they are frequently methylated in tumors. Epigenetic inactivation of genes involved in growth control, such as tumor-suppressor genes, cell cycle genes, DNA repair genes, and genes involved in invasion and metastasis, has been reported in numerous cancers. For example, *RB, p14ARF, APC,* and *BRCA1* are genes that are frequently inactivated epigenetically in human cancer (Esteller and Herman, 2002).

Reversing epigenetic inactivation of genes often results in the suppression of tumor growth or sensitisation to other anticancer drugs, for example by increased expression of mismatch-repair genes (Brown and Strathdee, 2002). Gene transcription is regulated by the local chromatin structure, which in turn is regulated by a complex interplay of DNA methylation and histone modification, including methylation and acetylation.

The pattern of DNA methylation that is established during development and differentiation is maintained by DNA cytosine-5-methyltransferase (DNMT1), which acts on the hemimethylated DNA of semiconservative DNA replication. The more recently identified enzymes DNMT3a and DNMT3b show no preference toward hemimethylated DNA and are thought to function as de novo methyltransferases. Mechanistically, there are several ways in which DNA methylation can contribute to transcriptional repression. Early hypotheses suggested that DNA methylation suppressed gene transcription by preventing the binding of transcription factors. It has been shown that methylation within the promoter-binding site of several transcription factors influences transcription (Tate and Bird, 1993). However, it has become evident that DNA methylation is part of a more general mechanism of transcriptional regulation (Bird and Wolffe, 1999; Tyler and Kadonaga, 1999). When DNA becomes methylated, a group of proteins containing methyl-binding domains (MBDs) are able to specifically bind methylated CpG sites (Hendrich and Bird, 1998). MBD proteins have been shown to directly repress transcription (Nan et al., 1997; Ng et al., 1999; Ng et al., 2000) as well as to recruite histone deacetylases (HDACs) and chromatin-remodeling activities as part of large repressive protein complexes (Jones et al., 1998; Ng et al., 1999). These data suggest the direct involvement of DNA methylation in the regulation of histone acetylation and higher-order chromatin structure. HDACs deacetylate lysine residues in the amino terminus of histone H3, ultimately leading to assembly of a chromatin structure that represses transcription (Bannister et al., 2001; Lachner et al., 2001). DNMT1, DNMT3a, and DNMT3b all contain transcriptional repressor domains and can recruit HDACs and other co-repressors in a manner similar to the MBD proteins, contributing to transcriptional repression. Another way that DNA methylation is believed to contribute to transcriptional silencing is by recruiting histone methylases (HMTs), which methylate the lysine 9 residue of histone H3, which in turn leads to the recruitment of the chromatin silencer HP1 (Bannister et al., 2001; Lachner et al., 2001).

Most research to date has focused on the effects of DNA methylation on higher-order chromatin structure. However, there is evidence that chromatin structure can likewise influence DNA methylation. For example, treatment with the HDAC inhibitor TSA leads to reduced DNA methylation in the fungus *Neurospora crassa* (Selker, 1998). More recently, a DNA methylation-deficient mutant gene in *Neurospora* was isolated from a screen of spores growing in the presence of the DNA methylation inhibitor 5-aza-2′-cytidine. Interestingly, the mutant gene encoded an HMT (Tamaru and Selker, 2001), suggesting a strong interplay between chromatin structure and DNA methylation. This work may provide insight into the molecular basis for the synergistic activation of epigenetically silenced genes by pharmacological treatments that target both DNA methylation and chromatin.

As mentioned before, epigenetic gene silencing plays an important role in the malignant process, and discovering genes silenced in such a way may provide valuable information about tumor biology and identify possible drug targets. Epigenetic mechanisms of gene inactivation include DNA methylation and histone modification, including histone methylation and acetylation. Early studies to identify genes with aberrant methylation of CpG islands in human cancer have used a candidate gene approach, selecting the CpG islands to be analyzed. Although this methodology has already identified numerous methylation-silenced genes, to establish the full extent of DNA methylation in human cancer it is necessary to conduct a genome-wide, unbiased study. Some recently established methods are now available to start genome-wide screening of CpG islands.

The first approach involves the use of microarrays to determine the upregulated transcripts after demethylation, which can be achieved by either genetic inactivation (Jackson-Grusby et al., 2001) or pharmacological inhibition of DNMT1 by 5-aza-2′-deoxycytidine (decitabine). The drawbacks of such an approach are that it requires dividing cells which make it difficult to apply to clinical tumor biopsies, and the fact that it is difficult to discriminate between the direct and indirect effects of demethylation. Time courses of decitabine effects may aid in differentiating between early direct effects of demethylation and later downstream effects.

In a second approach, a technique called *restriction landmark genomic scanning* (RLGS) was developed, based on end-labeling methylation

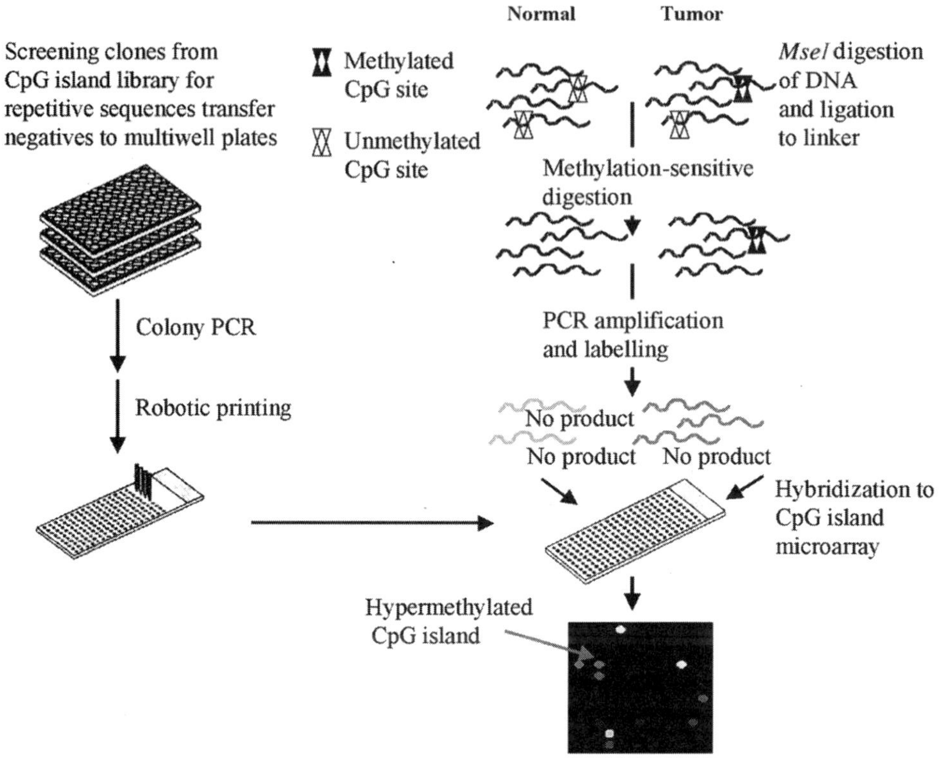

Figure 8.3 Schematic of differential methylation hybridization Genomic DNA to be compared is restricted to completion with *MseI. MseI* cuts DNA into small fragments but leaves CpG islands largely intact. The digests are purified and ligated to linkers. Repetitive DNA sequences are depleted from the ligated DNA using a *Cot-1* subtraction hybridization protocol. DNA is digested with a methylation-sensitive endonuclease. PCR reactions are performed with linker oligonucleotides. DNA fragments not containing restriction sites or methylated restriction sites are uncut and amplified, whereas fragments containing unmethylated sites are cut and are not amplified. The amplified products are labeled and co-hybridized on the CpG island microarray. In a false color overlay of the Cy3 and Cy5 signals a hypermethylated site will show increased signal relative to normal and be represented by a red spot (arrow hypermethylated CpG island). Clones from CGI genomic library are prescreened with labeled *Cot-1* DNA. Clones negative or weakly positive for the *Cot-1* hybridization signals are picked and transferred to 96-well plates. Inserts containing the appropriate restriction sites, verified by digestion and gel electrophoresis, are amplified by PCR and spotted onto solid support.

sensitive restriction sites and resolving the labeled products using two-dimensional gel electrophoresis (Costello et al., 2000). Analysis of the methylation status of 1184 unselected CpG islands in 98 primary human tumors identified patterns of CpG island methylation that were shared within each tumor type, together with patterns and genes that displayed distinct tumor type specificity (Costello et al., 2000).

The third approach, differential methylation hybridization (DMH), is based on a combination of methylation-sensitive DNA digestion and DNA microarrays. Using microarray technology allows thousands of CpG islands to be screened simultaneously (Fig. 8.3). Screening by DMH of paired primary breast tumor and normal samples revealed extensive hypermethylation in the majority of breast tumors relative to their normal controls, whereas other tumors had little or no detectable changes. Hypermethylation was associated with poorly differentiated tumors compared to moderately or well-differentiated tumors (Yan et al., 2000). Hierarchical clustering of the methylation patterns allowed segregation of the tumors and identified a methylation pattern that corresponded to the hormone-receptor status of the tumor (Yan et al., 2001). Using the same method to classify ovarian tumors the researchers identified two groups of patients with distinct methylation profiles. Progression-free survival after chemotherapy was significantly shorter for patients in the group with extensive methylation. In addition, a select group of CpG island loci were identified that could potentially be used as epigenetic markers for predicting treatment outcome in ovarian cancer patients (Wei et al., 2002).

Reversing epigenetic silencing has a distinct advantage over reversing mutational inactivation of genes, as the latter strategy requires the more technically challenging gene therapy approach to deliver the wild-type gene to tumor cells. The increased methylation patterns of CpG islands, as noted, rarely occur in normal cells. Therapies targeting methylation of genes, either directly or indirectly, could, therefore, provide selectivity for cancer versus normal tissue. Most therapies to date have focused on reversing methylation as a means to reactivate the expression of genes that negatively regulate growth or those that modify sensitivity to existing therapies (Brown and Strathdee, 2002; Costell and Vertino, 2002). Experiments in which methylation-silenced genes were overexpressed show that this can lead to tumor growth suppression and sensitization to existing therapies.

Currently, there are few effective inhibitors of DNMTs available. 5-aza-2′-cytidine and decitabine are frequently used in vitro to reactivate methylation-silenced genes. Decitabine has been used in clinical trials to treat hematopoietic malignancies (Pinto and Zagonel, 1993). Because of the toxicity of decitibine, the use of this agent has been limited, although protracted low-dose schedules and the use of decitabine in combination can partly overcome this problem (Lubbert et al., 2001; Plumb et al., 2000).

HDAC inhibitors have shown promising activity in preclinical studies. Butyrates, hydroxamic acids (including SAHA, *m*-carboxycinnamic acid bis-hydroxamide (CBHA), TSA, and pyroxamide), and the fungal tetrapeptides (such as depsipeptide) have each inhibited growth of a variety of solid and hematopoietic malignancies in vitro (Johnstone, 2002). Results in xenograft models have also been encouraging. CBHA inhibited growth of neuroblastoma tumor xenografts in a dose-dependent fashion, with treatment at a dose of 200 mg/kg causing complete suppression of tumor growth. The efficacy of lower doses was enhanced by co-treatment with *all-trans* retinoic acid (Coffey et al., 2001). SAHA suppressed the growth of human prostate tumor xenografts at dose of 25, 50, and 100 mg/kg/day. Treatment with 50 mg/kg/day resulted in a 97% reduction in the mean final tumor volume compared to controls without any detectable toxicity. Growth suppression of the xenografts was at least in part due to target inhibition as hyperacetylated histones accumulated within 6 h of SAHA administration. Given the success of HDAC inhibitors in preclinical studies, several are now being evaluated in clinical trials (Johnstone, 2002), most of which are still in progress.

A preliminary report of a Phase I clinical trial of depsipeptide in T-cell lymphoma has shown that therapeutic doses can be achieved with minimal side effects. Three patients had a partial response and one patient had a complete response (Piekarz et al., 2001). In a dose escalation study of depsipeptide in refractory solid tumors the dose-limiting toxicity was identified as fatigue, nausea and vomiting, thrombocytopenia, and cardiac arrhythmia. Biologically active plasma concentrations were achieved, some patients had stable disease, and one partial response was observed (Sandor et al., 2002). In two Phase I clinical trials of phenylbutyrate in refractory solid tumors, biologically active doses were achieved with minimal clinical and metabolic side effects. The most common toxicities observed were dyspepsia, fatigue, somnolence, hyperuricemia, and hypocalcaemia (Carducci et al., 2001; Gilbert et al., 2001). No partial responses were observed, but in one of the trials, 7 out of 28 patients had stable disease for >6 months while on the drug (Gilbert et al., 2001). In a recently completed Phase I clinical study of SAHA, the drug was shown to be well tolerated, accumulation of acetylated histones was demonstrated in peripheral blood lymphocytes, and four objective responses were observed (Kelly et al., 2003).

Microarray studies should be particularly useful during preclinical and clinical development of agents that modify the epigenetics of cancers, given that the altered gene expression is an intended downstream objective of the therapy (Glaser et al., 2003).

8.8 Clinical Trials: Patient Selection and Predicting Outcome

Gene microarrays can aid drug development during clinical trials, a critical stage in the process. Studies of molecular mechanism of action alongside the more established toxicity and pharmacokinetic evaluations needs to be a strong component in the early clinical testing of agents that act on new molecular targets. Cytotoxic drugs generally have a relatively nonspecific mechanism of action and, therefore, may have a relatively broad spectrum of activity in human cancers. In contrast, with drugs designed to act on a particular oncogenic target or pathway, which may have more narrow utility, considerable emphasis has to be placed on patient selection. Even in the case of targeting pathways that are activated in a large percentage of patients and across a variety of different cancers, there will be a considerable number of patients with tumors in which these pathways are not activated and do not contribute to tumor progression. Treating such patients with the molecularly targeted drug would not be expected to have much therapeutic benefit; on the contrary, such drugs may cause toxicity as well as prevent treatment with other drugs that are more likely to be active. Exclusion of patients that are unlikely to respond, but who might benefit from alternative treatment, generates obvious and considerable medical and pharmacoeconomic benefits.

Gene expression profiling is already beginning to rival classical pathology and immunohistochemistry in predicting clinical outcome in human cancers (van de Vijver et al., 2002; van't Veer et al., 2002). Although these early studies are very encouraging, the clinical utility of gene expression profiling has to be established in larger prospective studies. The next step will be to use this technique to profile tumors before treatment, to determine which oncogenic pathways are active within individual cancers, and thus to identify the most effective and logical treatment strategy.

As an example, platelet-derived growth factor receptor α (PDGFRα) and the downstream RAS-RAF-MEK-ERK pathway were found to be implicated in the progression from nonmetastatic to metastatic medulloblastoma using a class prediction algorithm (see Section 8.8) to analyze DNA microarray expression profiling data from 23 primary metastatic and nonmetastatic medulloblastomas (MacDonald et al., 2001). Such data could be used at diagnosis, when pretreatment biopsies could be profiled to assess the status of the RAS-RAF-MEK-ERK pathway. Patients with tumors that show activation of this pathway might be particularly suited for treatment with agents that inhibit the pathway. Using in vitro assays, MacDonald et al. showed that PDGFα stimulated migration and activated downstream mediators of the RAS-RAF-MEK-ERK pathway. Neutralizing antibodies to PDGFRα and the small molecule agent UO126, a specific inhibitor of MAP2K1 and MAP2K2, (MEK1 and MEK2) both inhibited unstimulated migration and also blocked PDGFα-stimulated migration.

Whereas the initial molecular classification of human cancers by gene expression microarray analysis was used to identify subtypes mainly by hierarchical clustering, more powerful data mining techniques (see Section 8.9) have been developed that are better suited to predict patient outcome based on tumor gene expression profile. Two recent studies have used gene expression profiling to successfully predict outcome in diffuse large B-cell lymphoma (DLBCL) following chemotherapy (Rosenwald et al., 2002; Shipp et al., 2002). Both studies found molecularly distinct subgroups as described previously (Alizadeh et al., 2000), according to a putative cell of origin; namely, germinal-center B-cell-like or activated B-cell-like. In addition, Rosenwald et al. found a third subtype that did not express the genes associated with the other subtypes. Alizadeh et al. used unsupervised hierarchical clustering to predict outcome in DLBCL and found that patients with germinal-center B-cell-like disease had a more favorable outcome than patients with activated B-cell-like disease. However, the overall response rate was not significantly correlated with any of the clusters, suggesting additional variation in the data that was not captured by hierarchical clustering.

The two other, more recent studies used supervized learning methods. In one study, a weighted voting algorithm and cross-validating testing (www.genome.wi.mit.edu/MPR) was used to identify patients with cured versus fatal or refractory disease (Shipp et al., 2002). Predictors containing 8–16 genes all resulted in statistically significant outcome predictions, separating the patients into two groups with a median survival at 5 years of 70% and 12%, respectively. The predictive gene expression pattern was independent of both the International Prognostic Index, and the putative cell of origin. The predictor genes encompassed those involved in B-cell receptor signaling, critical kinase cascades, and apoptosis. In the second study, a Cox proportional hazards model was used to identify genes that were significantly correlated with outcome (Rosenwald et al., 2002). These genes were assigned to functional groups, as proposed by Shaffer et al. (2001). The genes fell into four biological groups, of which the most variable genes were chosen for the outcome predictor. Within the outcome predictor a group of proliferation genes provided the best prediction of an adverse outcome, whereas the gene signatures associated with good outcome suggested that antigen presentation and the immune response might be critical determinants after chemotherapy.

The genetics and biological pathways in the pathogenesis of multiple myeloma (MM) are still largely unknown. As a consequence, the development of prognostic clinical markers and of effective treatment has been slow. The markers currently available only account for approximately 20% of the heterogeneity observed in MM. One study has profiled gene expression in bone marrow plasma cells (PCs) from newly diagnosed patients with MM, monoclonal gammopathy (MG), and healthy volunteers, together with MM cell lines (Zhan et al., 2002). Using hierarchical clustering, the normal and MM PC formed distinct clusters. Within the MM cluster, four subgroups could be identified (MM1–MM4). The profile of MM1 PCs most resembled that of normal and MG PCs, whereas that of MM4 was similar to the profile of the MM cell lines. The MM1–MM4 classification based on the expression data was found to correlate well with known clinical parameters by analysis of variance, indicating prognostically relevant clinical subgroups of MM. The MM4 PCs exhibited a profile indicating a more proliferative and autonomous phenotype; this profile was linked to poor prognosis. Using statistical methods (χ^2, Fisher exact test, and Wilcoxon rank sum test), the study also identified 120 genes that discriminated between normal and malignant PCs. Many of the genes were known to play an important part in the genesis of MM. The study also identified genes and classes of genes that were not previously implicated in the malignant progression of MM and may provide targets for drug development and markers of disease.

Expression profiling in colorectal cancer (CRC) revealed genes associated with Dukes classification as well as genes linked to disease progression (Birkenkamp-Demtroder et al., 2002). Most of the expression changes occurred during the progression from normal to early-stage CRC, whereas far fewer genes were altered during the progression through the different Dukes stages. This suggested that the bulk of the changes occur at initiation and relatively few gene changes are required for malignant progression. Using a functional classification that was suggested previously (Lander et al., 2001) revealed that most of the gene expression changes between tumors and normal tissue fell into the functional categories of metabolism (particularly mitochondrial metabolism), regulation of transcription and translation, cell growth and differentiation, cell cycle progression, cell adhesion, protein folding and degradation, transport, immune system, and nucleic acid metabolism. Notably, few genes involved in apoptosis and signal transduction had altered expression. The expression levels of genes classified within a functional category showed remarkable correlation. Most of the genes involved in nucleic acid metabolism, cell cycle regulation, translation, adhesion, or proteolysis were upregulated, whereas those involved in membrane and protein trafficking and lipid metabolism, together with most of the altered kinases and phosphorylases were downregulated. Interestingly, the upregulated or downregulated genes in CRC clustered to several distinct chromosomal locations, suggesting the possibility of some form of co-regulation through common transcription factors, through promoter methylation or opening of the DNA duplex, or as a result of chromosomal amplification or deletion. These types of data suggest that combinations of expression profiling, CGH and methylation analysis on microarrays may prove to be more powerful than the use of individual methods in classifying human cancers.

Renal cell carcinoma (RCC) is a clinically heterogeneous disease with prognosis based on tumor staging and grade. Patients presenting with metastatic disease have a poor prognosis and a life expectancy of about 12 months. Of patients without metastatic disease at diagnosis, 30% relapse after surgery, usually dying of this disease. Currently, there is little understanding of the molecular mechanisms underlying the differences in prognosis (Hughes et al., 2001). To study the pathways driving carcinogenesis in clear cell RCC (ccRCC) and identify molecular signatures of outcome, one group used cDNA microarrays to profile 29 ccRCC tumors and matched normal renal tissues (Takahashi et al., 2001). In this study, 9 of the patients died of the disease within 5 years and 19 were long-term survivors. Comparison of the ccRCC and matched controls identified genes with altered expression among the tumors and those that were in common in > 75% of the ccRCC. These genes may thus play a causal role in the molecular carcinogenesis of this malignancy and could perhaps be useful as molecular markers of disease and potential targets for drug discovery. For example, genes involved in angiogenesis were frequently altered, with vascular endothelial growth factor (VEGF) overexpressed in 96% of ccRCCs, whereas an inhibitor of angiogenesis, kininogen, was on average 27-fold downregulated. Another family of genes, those encoding metallothioneins, was also coordinately downregulated in ccRCC.

Hierarchical clustering and statistical analysis were used to analyze whether some of the heterogeneity in the gene expression observed in the ccRCC indicated the biology or the prognosis of particular subsets of tumors (Takahashi et al., 2001). Hierarchical clustering revealed that the 29 ccRCCs were divided in two main groups that were highly correlated with 5-year survival of the patients; only 1 patient was clustered inaccurately. The "leave-one-out" method was used to identify the genes that best discriminated the two groups of tumors, based on 5-year survival. The tumor that was left out was reclustered with the others, using the most discriminating genes; the tumor was classified according to whether it clustered with tumor samples with good or poor outcome. Using this technique, clinical outcome was correctly predicted in 96% of the cases.

Among breast cancer patients, the ability to more accurately predict prognosis would improve the selection of patients that might benefit from adjuvant therapy. Currently 70–80% of the patients receiving adjuvant therapy would survive without it. This was addressed in an important study of gene expression designed to identify an outcome predictor in breast cancer (van't Veer et al., 2002). 34 tumors were from patients that developed distant metastases within 5 years, 44 tumors were from patients who remained disease-free for a period of at least 5 years, 18 tumors were from *BRCA1* mutation carriers, and 2 tumors were from *BRCA2* mutation carriers. Labeled cRNA was derived from 5 μg total RNA and was co-hybridized with a reference cRNA from the pool of sporadic tumors (no germ-line *BRCA* mutations), on microarrays containing approximately 25,000 gene features deposited by ink-jet technology (Hughes et al., 2001). Unsupervised hierarchical clustering using the approximately 5,000 significantly altered genes resulted in two subsets of tumor profiles, associated with estrogen receptor (ER) status and lymphocytic infiltration. The expression data from 78 sporadic tumors were used to derive a molecular classifier. Of the 5,000 significantly altered genes, the correlation coefficient of each gene with disease outcome was calculated. A total of 231 genes were significantly associated with outcome and these were ranked in order of magnitude of the correlation coefficient. The molecular classifier was optimized by stepwise addition of 5 prognostic genes to the classifier until the best prediction was reached. Using a leave-one-out and cross-validation procedure, a 70-gene classifier was determined to be the best predictor for breast cancer outcome: Using this classifier it was possible to correctly predict actual outcome in 83% of the cases. In a validation set of 19 tumors, only two were classified incorrectly, outperforming currently available clinical and histopathological prognostic factors. The genes in the classifier reflected the molecular pathways likely to be involved in metastases. Genes involved in cell cycle regulation, invasion and metastasis, angiogenesis, and signal transduction pathways were significantly upregulated in the tumors with a poor outcome prognosis, offering potentially useful targets for drug development. The power of the 70-gene prognosis profile was confirmed in a large follow up study (van de Vijver et al., 2002), where the mean overall ten year survival in patients with a poor prognosis profile was 54.6%, whereas those with a good prognosis signature had a mean overall survival of 94.5%.

The examples above illustrate how differences in expression profiles can define tumor subclasses that are molecularly distinct and that have arisen from different transforming genetic aberrations. Although they share the same tissue of origin, distinct molecular pathways drive the malignant phenotype in these subclasses, and so different agents will be needed to target the molecular pathways activated in different subclasses to inhibit tumor progression.

In addition to identifying the prognosis classifier, van't Veer et al., 2002 used 38 ER-negative tumors to distinguish between sporadic cases and familial cases associated with *BRCA1* germ-line mutations. The resulting 100-gene classifier was enriched in lymphocyte-specific genes but was otherwise based on the magnitude of the differences in the relative levels of gene expression. Using this classifier, 17 out of 18 *BRCA1* tumors and 19 out of 20 sporadic tumors were classified correctly. The misclassified sporadic tumor was shown to have methylation of the *BRCA1* promoter, suggesting possible epigenetic silencing of the *BRCA1* gene. The *BRCA1* tumor incorrectly assigned to the sporadic tumors carried a mutation (BRCA15622del62) that affected only the last 29 amino acids of the BRCA1 protein. The same mutation was also misclassified using CGH data (Wessels et al., 2002). In an independent study, a *BRCA1* mutation carrier with a 5382insC mutation, leading to the truncation of the BRCA1 protein at the extreme C terminus (1829ter), clustered with normal samples using principal component analysis, again suggesting a different phenotype from other *BRCA1* mutations (Kote-Jarai et al., 2003).

Identifying germ-line *BRCA1* and *BRCA2* mutations remains difficult but can be assisted by microarray analysis. Current screening methods fail to identify germ-line mutations in approximately 70% of the families with high-frequency early-onset breast cancer. One group successfully used microarray CHG data to construct a molecular classifier to differentiate *BRCA1* mutation carriers from noncarriers (Wessels et al., 2002). CGH data from 28 proven *BRCA1* germ-line mutation carriers and a control group of 42 breast tumors with unknown *BRCA1* and *BRCA2* status was used to set up a simple Bayesian classifier (Domingos and Pazzani, 1997). The leave-one-out method of cross-validation was used to optimize the classifier.

Several chromosomal arms showed a significant difference between the *BRCA1* group and the control group. However the Bayesian classifier based on higher-resolution data was more predictive of *BRCA1* status. CGH data were clustered in chromosomal bands based on a high degree of correlation between CGH data from consecutive chromosomal positions. A *BRCA1* classifier based on chromosomal bands 3.1, 3.5, and 5.2 predicted *BRCA1* status with the highest accuracy (84%). Applying the classifier on the training set resulted in 10 false positives and one false negative. The classifier scores the probability of a tumor CGH profile according to the similarity to the *BRCA1* profile: The higher the similarity, the higher the score and the greater the probability that the tumor is correctly classified as *BRCA1* mutation carrier. In a validation set of 6 *BRCA1* tumors and 19 control tumors, a prediction accuracy of 84% was achieved, all the *BRCA1* tumors were classified correctly, but 4 out of 19 control tumors were classified as *BRCA1* mutation carriers. The false negative in the training set, the proven *BRCA1* mutation carrier misclassified as sporadic tumor, involved a deletion of the last 62 base pairs (5622del62), which may indicate a sporadic tumor or a tumor with a different phenotype from *BRCA1* carriers. Of the false positives some had a very high probability score, indicating possible *BRCA1* mutations missed in the initial screening.

A different study investigated whether gene expression profiling could be used to distinguish between normal breast fibroblasts from *BRCA1* mutation carriers and noncarriers after radiation-induced DNA damage (Kote-Jarai et al., 2003). Nine prophylactic mastectomy samples and five control samples from reduction mammoplasty were collected, and short-term breast fibroblast cell cultures were established. Confluent cells were irradiated, and total RNA was extracted from the cells 1 h after irradiation. Reference RNA was pooled from three cancer cell lines. Significance analysis of microarrays (SAM; see Table 8.4) was used to identify differentially expressed genes. Using a threshold value of 18% false discovery rate (FDR; as estimated by repeated permutation), 113 genes were identified as significantly downregulated in *BRCA1* mutation carriers compared to noncarriers. Several of the genes were involved in cell-cycle regulation, DNA repair, and transcriptional regulation. Hierarchical clustering and PCA analysis (Genesis; see Table 8.4) was performed with the subset of 113 genes and showed that the mutation carriers clearly clustered together, with the exception of one mutation-positive sample. The nonconforming sample was a carrier of the 5382insC mutation, which leads to truncation of the BRCA1 protein at the extreme C terminus (1829ter) and is a similar mutation to those misclassified in other studies (van't Veer et al., 2002; Wessels et al., 2002).

Table 8.4 Gene Microarray Analysis Software

Software	Web Site
Cluster[a]	rana.lbl.gov
GeneCluster[a]	www-genome.wi.mit.edu
Expression Profiler[a]	ep.ebi.ac.uk
ClustArray[a]	www.cbs.dtu.dk/services/DNAarray
R package[a]	www.r-project.org
SAM[a]	www-stat.stanford.edu/~tibs/SAM/index.html
Genesis	genome.tugraz.at/Software/GenesisCenter.html
Affymetrix Data Mining Tool	www.affymetrix.com
Affymetrix NetAffx	www.netaffx.com
Biomax Gene Expression Analysis Suite	www.bibomax.de
GeneData Expressionist	www.genedata.com
Informax Xpression NTI	www.informaxinc.com
Lion Bioscience ArraySCOUT	www.lion-bioscience.com
Rosetta Resolver	www.rosettabio.com
Silicon Genetics GeneSpring	www.sigenetics.com
Spotfire	www.spotfire.com
Axon Acuity	www.axon.com/GN_Acuity.html
SVM^{light}[a]	svmlight.joachims.org
Genesis[a]	genome.tugraz.at/Software
Clementine	www.spss.com
J-express	www.molmine.com

[a]Free download.

The techniques identifying *BRCA1* mutation carriers outlined above could be used to prescreen high-risk patients before complete sequencing of the *BRCA1* gene. In this instance, a high false-positive rate is preferable to a high false-negative rate, as very few *BRCA1* mutation carriers will be missed. Identifying *BRAC1* mutation carriers is important, because these individuals can be monitored more closely.

Chromosomal imbalances such as deletions and amplifications are common rearrangements in human cancer. Specific chromosomal aberrations are associated with certain types of cancer or with the stage of a particular tumor, implicating a gene or genes in a region of chromosomal imbalance in the initiation or progression of the malignant phenotype (Lengauer et al., 1998). With classical CGH, large chromosomal abnormalities can be identified. The technique, however, is not amenable to high-throughput and the resolution is low, necessitating the laborious identification of a possible oncogene or tumor suppressor-gene from numerous candidate genes. Microarray-based CGH overcomes both of these shortcomings.

In addition to identifying oncogenes and tumor suppressors, at genomic regions of amplification or deletion, microarray-based CGH is now increasingly used to classify human cancers, as shown above in this section. Microarray based CGH and expression profiling have been used together to classify 16 dedifferentiated liposarcomas (DLs) and pleomorphic liposarcomas (PLs) (Fritz et al., 2002). CGH on BAC DNA microarrays revealed amplifications of several known oncogenes, a subset of which showed increased expression on cDNA microarrays. A number of genes were differentially expressed in DL and PL. Comparing expression patterns of DLs and PLs to normal adipose tissue showed reduced expression of the tumor-suppressor genes *harakiri* and *KISS-1* and highly overexpressed genes in both PL and DL, including *RAL, SMARC1, DLK* and *CDK4*. Clustering algorithms (decision tree algorithm C5.0, Clementine; principal component analysis, J-express 2.01 d; and support vector machine, SVM^{light}; see Table 8.4), were able to successfully classify DLs and PLs based on the CGH data. In contrast, the classification based on the expression data misassigned some of the PLs as DLs. These data suggest that chromosomal imbalances are, in at least some instances, better for classifying human cancers. However, the tumor-expression profiles encompassed data for only 1600 cDNAs, and the classifier based on the expression data might well have been more powerful had expression levels of more genes been analyzed.

In addition to classifying human cancers more correctly and predicting outcome of patients, the microarray experiments outlined above in this section show the power of such experiments in identifying disease genes, i.e., genes that are most associated with a particular type of cancer or subclass of cancer. Some of these genes may well represent good targets for drug discovery and, as in the case of expression profiling, may influence the choice of treatment.

8.9 Exploring Possibilities to Predict Sensitivity to Treatment

As emphasized earlier in this chapter, when new agents are designed to act on specific molecular targets and pathways, it is imperative to define tumors that have become reliant on the activation of the target and its cognate pathways for maintaining the malignant phenotype. With the advance of microarray gene-expression profiling, it is now possible to derive molecular signatures that are associated with activation of certain oncogenic pathways and the sensitivity and resistance to anticancer agents. There are several ways in which information about sensitivity and resistance to new agents can be gathered. Model systems using cancer cell lines can be used to identify signatures associated with drug activity or resistance. Alternatively, clinical tumor biopsies can be used retrospectively for global gene expression profiling. These profiles can then be correlated with clinical outcome to identify genes that are associated with drug sensitivity (an example with Gleevec or imatinib mesylate is given later in this section). The profiles can then subsequently be used at diagnosis to categorize patient tumors as likely to respond or unlikely to respond to a particular drug. The disadvantages of using clinical material to identify sensitivity and resistance genes include the difficulty of obtaining samples from of enough patients to make useful predictions and the time frame of clinical follow-up. In contrast, although there are numerous drawbacks to using human cancer cell lines, there are no restrictions on sample supply, and hypotheses can be generated relatively quickly.

The approach we have taken is to use global gene expression profiling to establish a dataset of constitutive cell line expression patterns in a panel of 40 human cancer cell lines, and then to use data mining tools to associate expression signatures with drug sensitivity and resistance. The panel comprises the four major human cancers – breast, lung, colon, and prostate – together with brain, melanoma, and ovarian cancer. Cell lines are grown until mid-log phase and harvested for mRNA. Individual cell line mRNAs are labeled and co-hybridized to a labeled reference pool of the individual mRNAs (Fig. 8.1). A human whole genome array made up of 30,000 nonredundant cDNAs is used. Genes that have significant variance in the cell line panel are selected for association with drug sensitivity or resistance.

Sensitivity data for drugs under development at the Cancer Research UK Centre for Cancer Therapeutics are collected using the MTT assay to determine IC50 values after 96-h drug exposures. The IC50 data are used to separate the cell lines into a sensitive and a resistant population. Nearest-neighbor analysis and support vector machine analysis with leave-one-out cross-validation is used to build a molecular predictor based on the classification "sensitive" or "resistant." These analysis tools will select genes that are most significantly correlated with drug sensitivity or resistance and predict whether a sample will be sensitive or resistant according to the expression of the genes in the classifier. The same methodology using oligonucleotide arrays containing only 6800 genes has successfully been used to predict sensitivity to 88 of 232 compounds investigated (Staunton et al., 2001). Correlation of sensitivity of compounds in the National Cancer Institute (NCI) panel with constitutive gene expression profiles has already proved useful in determining how differences in expression of particular genes relates to the mechanism of drug sensitivity and resistance, for example to the clinical agents 5-fluorouracil and L-asparaginase (Scherf et al., 2000). One of the attractive features of this methodology is that the constitutive expression patterns can be used again and again to train new molecular classifiers based on the sensitivity data for new agents.

In the clinical setting, the idea would be to use the molecular classifier to predict sensitivity to a particular drug of a newly diagnosed tumor, according to the expression profile of pretreatment biopsy material. Obtaining biopsy material remains difficult because of the invasiveness of the procedure, and it may not possible to obtain enough material to carry out microarray analysis. There are, however, recent reports describing successful microarray studies using core needle biopsies (Ellis et al., 2002). Using amplification methodology (Wang et al., 2000; Xiang et al., 2003) it is now even possible to carry out gene expression microarray analysis using degraded RNA from archival paraffin-embedded material (Lewis et al., 2001) as well as small snap-frozen tumour biopsies. However, a predictor is usually based on a handful of the most significant differentially expressed genes, which opens up the opportunity to assess the expression levels of these predictive genes by the much more sensitive reverse-transcriptase (RT) PCR-based techniques.

In developing such predictive methods, it should be emphasized that there are numerous examples in the literature in which the situation in the clinic is different from that in cell line models. We have observed this in the context of studies measuring gene expression changes in response to a drug either in treated cancer patients or in cancer cell tissue culture. We have carried out an expression profiling study of rectal cancer patients in which changes in tumor gene expression during the period of treatment with 5-fluorouracil (5FU) were identified using gene expression microarrays. Gene expression data were obtained successfully from all of the patients in the study. Changes in expression were identified that were consistent with the treatment inhibiting c-*myc*-regulated gene expression (Clarke et al., 2003). In this study, a single patient treated with raltitrexed, a specific thymidylate synthetase (TS) inhibitor, had a similar expression pattern to the 16 patients treated with 5FU. In contrast, a patient treated with irinotecan, a topoisomerase inhibitor, gave a different expression response. Biochemical evidence for inhibition of TS by 5FU was detected in the biopsies that were profiled, suggesting that the pattern of expression in response to 5FU treatment was associated with TS inhibition. Interestingly, the expression pattern detected in patients could not be reproduced easily in a number of cell lines treated in vitro with 5FU or FdUrd, a metabolite of 5FU that results in TS inhibition. This disparity may not be so surprising, given the numerous differences between cells growing on a plastic dish and those growing in the complex environment of a solid tumor in a cancer patient in situ. The results highlight the need to validate patterns detected in vitro either using in vivo human tumor xenograft models or in early clinical studies of novel compounds. The study also illustrates the value of profiling changes in gene expression by microarray in clinical samples before and after drug treatment.

Probably the most promising new molecular therapeutics agent is the c-ABL kinase inhibitor imatinib mesylate (Gleevec). High response rates with imatinib mesylate have been achieved for the treatment of chronic myelogenous leukemia (CML) (Kantarjian et al., 2002). CML is characterized by deregulated c-ABL kinase activity arising from a chromosomal translocation involving the *bcr* and c-*abl* genes, resulting in overexpression of the fusion protein and continuous activation of the ABL kinase function. The same translocation of the *bcr* and *c-abl* genes also occurs in cases of acute lymphoblastic leukemia (ALL), although the response rate in this type of disease is lower than that in the case of CML. However, resistance to imatinib mesylate is a common clinical occurrence, suggesting preexisting mechanisms of resistance in CML cells or the development of resistance during treatment.

One group of workers used gene expression microarrays to profile bone marrow aspirates of 19 patients with BCR- and ABL-positive ALL to assess the possibility of predicting resistance to imatinib mesylate (Hoffman et al., 2002). A total of 17 samples were obtained before treatment and 8 samples were collected during treatment. Of the pretreatment samples, 10 were from patients classified as sensitive, 9 had a complete hematological remission, and 1 had a partial response. A total of 7 samples were from patients who did not respond to imatinib mesylate, including 5 patients without any hematological response. These samples were classified as primary resistant. Of the 8 bone marrow samples taken during treatment, 6 were from patients that initially responded but later relapsed and were designated secondary resistant. Nearest-neighbor class prediction was used to establish whether gene expression signatures could be used to classify the leukemic samples according to their response to imatinib mesylate. The authors identified 95 genes, the expression patterns of which could be used to predict sensitivity to imatinib mesylate. Of these, 50 genes were selected that distinguished between sensitive and primary resistant samples. A further 50 were selected that could correctly predict 9 out of 10 sensitive pretreatment samples from 8 that proved resistant. Another 25 genes were identified that were highly predictive for secondary resistance while on imatinib mesylate treatment. Analysis of these combined gene sets by hierarchical clustering resulted in two main clusters in which the sensitive samples were clearly separated from the resistant samples. There was no apparent functional association between the genes and the mechanisms of resistance, although the expression patterns correctly classified samples as sensitive or resistant.

However, of the 56 genes that characterized acquired resistance to imatinib mesylate, several could be organized into functional groups, some of which had previously been implicated in mechanisms of resistance or altered signaling in resistant cells. There was no increased expression of BCR-ABL protein or overexpression of the multidrug resistance gene 1 (*mdr1*) as reported in CML, although in 5 of the treated samples, a mutation in the ABL kinase domain was detected (Gambacorti-Passerini et al., 2003). Of note was overexpression of Bruton's tyrosine kinase (BTK), which may overcome inhibition of the ABL kinase function by phosphorylating downstream effectors of the pathway. Also of note was the greatly reduced expression of the proapoptotic BCL2 gene family member *BAK1* in the samples of patients with acquired resistance.

The results of this small study are encouraging. Not only was it possible to predict sensitivity and resistance of ALL bone marrow samples to imatinib mesylate, which if confirmed in prospective studies would be of great potential clinical benefit, but it was also feasible to identify mechanisms of resistance, some of which may be amenable to therapeutic intervention.

8.10 Data Mining from Gene Microarray Analyses

A major advance in gene expression profiling and related technologies is the genome-wide scale. However, this also creates a major challenge – that of bioinformatics and data mining. The easiest way to look at microarray data is simply to list the fold changes, rank them in order of magnitude, and then inspect the list visually. This approach can sometimes be useful in identifying the most obvious changes. However, discerning trends manually in such a manner quickly become impossible. Consider a relatively small experiment of 15 samples analyzed on a 30,000 gene array. This produces a data matrix containing 450,000 entries. The aim of data mining is to reduce the dimensionality of this matrix to allow visual inspection. Visualization is traditionally performed in two dimensions; accordingly, many of the methods allow reduction of a matrix of any size into just two dimensions. We briefly explain some of the methods and highlight the uses of the individual methods. For a more comprehensive introduction to the analysis of microarray data, see Knudsen (2002) and the Web sites listed in Table 8.4.

8.10.1 NORMALIZATION, FILTERING, AND STATISTICS

Samples to be analyzed must be scaled or normalized to ensure that the expression levels in the samples are directly comparable to those in the control (Quackenbush, 2002). The first possibility is to include so-called housekeeping genes, which are assumed to be constitutively expressed and relatively unchanged from experiment to experiment. These are compared in the test and in the control samples, and the values in the test sample are then multiplied by a factor so that the expression levels of the housekeeping genes are the same. Another method is to assume that the total amount of mRNA for each cell is constant or that the overall ratio of expression between test and control averaged over the ratios for every gene on the array is 1. The greater the number of genes on the array, the more likely it is that this assumption will hold true.

Before proceeding with any kind of analysis, it is useful to filter the data set. Genes that do not change in any of the samples will not contribute to the variation and discrimination among samples and should be excluded. When possible, replicates of the experiment should be included, which allows the elimination of false positives through significance testing of the genes that are upregulated or downregulated. If both the test and the control sample are repeated, a t-test can be used to determine whether a particular gene is significantly changed between sample and control. However, in most cases more then two conditions are tested; thus ANOVA should be used (Baldi and Long, 2001; see also 128.200.5.223/CyberT). Although most expression data are normally distributed and both the t-test and ANOVA can cope with small deviations from a normal distribution, in some cases the data will not be normally distributed and nonparametric testing should be used (Knudsen, 2002).

8.10.2 PRINCIPAL COMPONENT ANALYSIS

Principal component analysis (PCA) is useful for capturing as much variety in the expression data as possible in two dimensions. The principal components are constructed as the sums of the individual sample axes. The cloud that these genes will form is not spherical and will be extended in one direction according to expression in the samples, which is the first principal component. This component will not generally be made up of one of the sample axes but rather several samples will have projections on its axis. The second principal component captures the variation left in the data and is plotted perpendicular to the first axis. As an example, PCA analysis was used to successfully classify central nervous system embryonal tumor subtypes (Pomeroy et al., 2002).

8.10.3 HIERARCHICAL CLUSTERING

If the data are more complex, cluster analysis can be used. Hierarchical clustering treats each gene as a vector of *N* numbers, *N* being the number of samples. The algorithm then calculates the distance between two genes according to their respective expression in the different samples. This is done for all the genes, and a distance matrix is formed. Genes that are closest together in distance have similar expression in the samples and thus are grouped together. Clustering of the genes can be combined with grouping of the samples. The distance between the samples is calculated according to the expression of the individual genes. The samples are then grouped together according to the distance that separates them. This method can be used to identify genes, the expression of which is positively or negatively correlated to a group of samples. Numerous examples of the use of this approach are given throughout the earlier parts of this chapter.

8.10.4 K-MEANS CLUSTERING AND SELF-ORGANIZING MAPS

In K-means clustering, the distances between all the genes are not calculated. Rather the experimenter decides how many clusters are required. The K-means algorithm then randomly assigns each gene to one of the K clusters. Next, the distance between each gene in a cluster and the center of that cluster (centroid) is calculated. When a gene is actually closer in distance to the centroid of a different cluster, it is reassigned to that cluster. Following reassignment of all the genes to their closest cluster, the centroids are recalculated, and the distances of the genes can be reassessed. This process is repeated in an iterative fashion until the centroids remain unchanged.

Self organizing maps (SOMs) are similar to K-means clustering. However, instead of the centroids changing to accommodate the gene expression data as in K-means, they are confined to a two-dimensional grid specified by the user, such as 2 by 3 or 3 by 3. The algorithm then organizes itself to best fit the data to this grid (Kohonen, 1995). K-means clustering and SOMs are fast algorithms and are useful for initial identification of expression patterns, but they may not be powerful enough to distinguish subtle differences in expression among samples.

8.10.5 CLASSIFICATION

To classify, for example, cancer subtypes, nearest neighbor analysis, neural networks, or support vector machine analysis can be used. The simplest form is the nearest-neighbor analysis method, which can be used on a relatively small data set (stat.berkeley.edu/tech-reports/index.html). For each sample the k most similar samples are calculated. The sample is then classified according to the class that the majority of the k nearest neighbors belong to. For example if $k = 3$ was chosen, for each sample the three closest samples would be calculated; if two of them belonged to class A and one to class B the sample would be classified as class A (majority voting). Leave-one-out cross-validation can be used for training classifiers to prevent overtraining on the training set, especially when the amount of samples is limited, which can result in good prediction in the training set but a poor performance on test samples. Given a training set of N samples, the first sample in the training set is left out. The classifier is then trained on the $N - 1$ remaining samples. The resulting classifier is used to test the samples left out and scored (cross-validation). A correct classification is scored as 1 and an incorrect classification is scored as 0. Next, the left-out sample is reinserted and the second sample is left out. A new classifier is trained and tested on the second sample. This process is repeated until every sample has been left out once. The overall score of the classifier, the average of the individual scores, is a realistic indicator of the performance of the classifier in test samples.

If the number of samples is high, more advanced classification methods of neural networks and support vector machines can be used. One study used artificial neural networks (ANNs) to classify small round blue-cell tumors (Khan et al., 2001). The ANNs correctly classified all samples, including blind samples that were not included in the training set, and identified the genes most relevant to the classification. The support vector machine is particularly suitable to microarray data, because it is designed to work with vectors and can, therefore, encompass the multi-dimensionality of microarray data.

8.11 Summary

Breakthroughs in sequencing and genomics are providing us with a previously unimagined ability to comprehend biology and disease pathology. Gene microarray technology is a major spin-off beneficiary of the Human Genome Project and microarray-based methods are fast becoming indispensable in biomedical science. In this chapter, we described applications of gene microarray technology throughout the process of contemporary drug discovery and development. The benefits are already clear, from rapidly identifying targets, determining complex molecular mechanisms of drug action, investigating transcriptional structure–activity relationships, and identifying drug response genes and molecular biomarkers to facilitate patient selection and the prediction of treatment outcome. As we enter the era of increasingly personalized medical treatments, targeted to the genetics and molecular pathology of the individual patient, gene microarray technologies will continue to gain prominence. They will become cheaper and more user-friendly. Data mining and data handling techniques, currently bottlenecks for many if not most users, will dramatically improve. Because of their massively parallel capability, gene microarrays will contribute in a major way to the development and application of genome-based molecular therapeutics for the treatment of cancer and other diseases.

Acknowledgments

We are grateful to our colleagues in the Signal Transduction and Molecular Pharmacology Team and also many collaborators for valuable interactions and discussions. The authors' work (www.icr.ac.uk/cctherap) is supported primarily by Cancer Research UK. PW is a Life Fellow of Cancer Research UK.

References

Aherne, G. W., McDonald, E., and Workman, P. Finding the needle in the haystack: Why high-throughput screening is good for your health. *Breast Cancer Research* 4, 148–154 (2002).

Afshari, C. A., Nuwaysir, E. F., and Barrett, J. C. Application of complementary DNA microarray technology to carcinogen identification, toxicology, and drug safety evaluation. *Cancer Res.* **59,** 4759–4760 (1999).

Ali, I. U., Schriml, L. M., and Dean, M. Mutational spectra of PTEN/MMAC1 gene: A tumor suppressor with lipid phosphatase activity. *J. Natl. Cancer Inst.* **91,** 1922–1932 (1999).

Alizadeh, A. A., Eisen, M. B., Davis, R. E., et al. Distinct types of diffuse large B-cell lymphoma identified by gene expression profiling. *Nature* **403,** 503–511 (2000).

Baldi, P. and Long, A. D. A Bayesian framework for the analysis of microarray expression data: Regularized *t*-test and statistical inferences of gene changes. *Bioinformatics* **17,** 509–519 (2001).

Balmain, A., Gray, J., and Ponder B. The genetics and genomics of cancer. Nature Genetics 33 supplement, 238–244 (2003).

Banerji, U., O'Donnell, A., Scurr, M., et al. A pharmacokinetically (PK)-pharmacodynamically (PD) driven Phase I clinical trial of the HSP90 molecular chaperone inhibitor 17-allyamino17-demethoxygeldanamycin (17AAG). *Proc. Am. Assoc. Cancer Res.* **43,** 72–73 [Abstract 1352] (2002).

Bannister, A. J., Zegerman, P., Partridge, J. F., et al. Selective recognition of methylated lysine 9 on histone H3 by the HP1 chromo domain. *Nature* **410,** 120–124 (2001).

Bardelli, A., Parsons, D. W., Silliman, N., et al. Mutational analysis of the tyrosine kinome in colorectal cancers. *Science* **300,** 949 (2003).

Baselga, J., and Averbuch, S. D. ZD1839 ('Iressa') as an anticancer agent. *Drugs* **60,** 33–40 (2000).

Baylin, S. B., and Herman, J. G. DNA hypermethylation in tumorigenesis: Epigenetics joins genetics. *Trends Genet.* **16,** 168–174 (2000).

Bird, A. P., and Wolffe, A. P. Methylation-induced repression – Belts, braces, and chromatin. *Cell* **99,** 451–454 (1999).

Birkenkamp-Demtroder, K., Christensen, L. L., Olesen, S. H., et al. Gene expression in colorectal cancer. *Cancer Res.* **62,** 4352–4363 (2002).

Brazma, A., Hingamp, P., Quackenbush, J., et al. Minimum information about a microarray experiment (MIAME) – Toward standards for microarray data. *Nature Genet.* **29,** 365–371 (2001).

Brown, R., and Strathdee, G. Epigenomics and epigenetic therapy of cancer. *Trends Mol. Med.* **8,** S43–48 (2002).

Cai, W. W., Mao, J. H., Chow, C. W., et al. Genome-wide detection of chromosomal imbalances in tumors using BAC microarrays. *Nature Biotechnol.* **20,** 393–396 (2002).

Carducci, M. A., Gilbert, J., Bowling, M. K., et al. A Phase I clinical and pharmacological evaluation of sodium phenylbutyrate on an 120-h infusion schedule. *Clin. Cancer Res.* **7,** 3047–3055 (2001).

Cheung, V. G., and Spielman, R. S. The genetics of variation in gene expression. *Nature Genet.* **32 Suppl,** 522–525 (2002).

Chin, L., Tam, A., Pomerantz, J., et al. Essential role for oncogenic Ras in tumour maintenance. *Nature* **400,** 468–472 (1999).

Clarke, P. A., George, M. L., Easdale, S., et al. Molecular pharmacology of cancer therapy in human colorectal cancer by gene expression profiling. *Cancer Res.* 63, 6855–6863 (2003).

Clarke, P. A., Hostein, I., Banerji, U., et al. Gene expression profiling of human colon cancer cells following inhibition of signal transduction by 17-allylamino-17-demethoxygeldanamycin, an inhibitor of the hsp90 molecular chaperone. *Oncogene* **19,** 4125–4133 (2000).

Clarke, P. A., te Poele, R., Wooster, R., and Workman, P. Gene expression microarray analysis in cancer biology, pharmacology, and drug development: Progress and potential. *Biochem. Pharmacol.* **62,** 1311–1336 (2001).

Coffey, D. C., Kutko, M. C., Glick, R. D., et al. The histone deacetylase inhibitor, CBHA, inhibits growth of human neuroblastoma xenografts in vivo, alone and synergistically with all-trans retinoic acid. *Cancer Res.* **61,** 3591–3594 (2001).

Colomer, R., Lupu, R., Bacus, S. S., and Gelmann, E. P. erbB-2 antisense oligonucleotides inhibit the proliferation of breast carcinoma cells with erbB-2 oncogene amplification. *Br. J. Cancer* **70,** 819–825 (1994).

Costello, J. F., Fruhwald, M. C., Smiraglia, D. J., et al. Aberrant CpG-island methylation has non-random and tumour-type-specific patterns. *Nature Genet.* **24,** 132–138 (2000).

Costello, J. F., and Vertino, P. M. Methylation matters: A new spin on maspin. *Nature Genet.* **31,** 123–124 (2002).

Davies, H., Bignell, G. R., Cox, C., et al. Mutations of the *BRAF* gene in human cancer. *Nature* **417,** 949–954 (2002).

Dawson, E., Abecasis, G. R., Bumpstead, S., et al. A first-generation linkage disequilibrium map of human chromosome 22. *Nature* **418,** 544–548 (2002).

Domingos, P., and Pazzani, M. J. On the optimality of the simple Bayesian classifierunder zero-one loss. *Machine Learning* **29,** 103–130 (1997).

Dracopoli, N. C. Pharmacogenomic applications in clinical drug development. *Cancer Chemother. Pharmacol.* **52 Suppl 1,** 57–60 (2003).

Druker, B. J. STI571 (Gleevec) as a paradigm for cancer therapy. *Trends Mol. Med.* **8,** S14–18 (2002).

Ellis, M., Davis, N., Coop, A., et al. Development and validation of a method for using breast core needle biopsies for gene expression microarray analyses. *Clin. Cancer Res.* **8,** 1155–1166 (2002).

Esteller, M., and Herman, J. G. Cancer as an epigenetic disease: DNA methylation and chromatin alterations in human tumours. *J. Pathol.* **196,** 1–7 (2002).

Fritz, B., Schubert, F., Wrobel, G., et al. Microarray-based copy number and expression profiling in dedifferentiated and pleomorphic liposarcoma. *Cancer Res.* **62,** 2993–2998 (2002).

Futreal, P. A., et al. Cancer and genomics. *Nature* **409,** 850–852 (2001).

Gambacorti-Passerini, C. B., Gunby, R. H., Piazza, R., et al. Molecular mechanisms of resistance to imatinib in Philadelphia-chromosome-positive leukaemias. *Lancet Oncol.* **4,** 75–85 (2003).

Gilbert, J., Baker, S. D., Bowling, M. K., et al. A Phase I dose escalation and bioavailability study of oral sodium phenylbutyrate in patients with refractory solid tumor malignancies. *Clin. Cancer Res.* **7,** 2292–2300 (2001).

Glaser, K. B., Staver, M. J., Waring, J. F., et al. Expression profiling of multiple histone deacetylase (HDAC) inhibitors: Defining a common gene set produced by HDAC inhibition in T24 and MDA carcinoma cell lines. *Mol. Cancer Ther.* **2,** 151–163 (2003).

Grolleau, A., Bowman, J., Pradet-Balade, B., et al. Global and specific translational control by rapamycin in T cells uncovered by microarrays and proteomics. *J. Biol. Chem.* **277,** 22175–22184 (2002).

Hanahan, D., and Weinberg, R. A. The hallmarks of cancer. *Cell* **100,** 57–70 (2000).

Hannon, G. J. RNA interference. *Nature* **418,** 244–251 (2002).

Hendrich, B., and Bird, A. Identification and characterization of a family of mammalian methyl-CpG binding proteins. *Mol. Cell. Biol.* **18,** 6538–6547 (1998).

Herrera, R., and Sebolt-Leopold, J. S. Unraveling the complexities of the Raf/MAP kinase pathway for pharmacological intervention. *Trends in Molecular Medicine* 8 (Supplement) S27–S31 (2002).

Hetherington, S., Hughes, A. R., Mosteller, M., et al. Genetic variations in HLA-B region and hypersensitivity reactions to abacavir. *Lancet* **359,** 1121–1122 (2002).

Hickman, J. A. Apoptosis and tumourigenesis. *Curr. Opin. Genet. Dev.* **12,** 67–72 (2002).

Hofmann, W. K., de Vos, S., Elashoff, D., et al. Relation between resistance of Philadelphia-chromosome-positive acute lymphoblastic leukaemia to the tyrosine kinase inhibitor STI571 and gene-expression profiles: A gene-expression study. *Lancet* **359,** 481–486 (2002).

Huang, S., and Houghton, P. J. Inhibitors of mammalian target of rapamycin as novel antitumor agents: From bench to clinic. *Curr. Opin. Investig. Drugs* **3,** 295–304 (2002).

Hughes, T. R., Mao, M., Jones, A. R., et al. Expression profiling using microarrays fabricated by an ink-jet oligonucleotide synthesizer. *Nature Biotechnol.* **19,** 342–347 (2001).

Hughes, T. R., Marton, M. J., Jones, A. R., et al. Functional discovery via a compendium of expression profiles. *Cell* **102,** 109–126 (2000).

Jackson-Grusby, L., Beard, C., Possemato, R., et al. Loss of genomic methylation causes p53-dependent apoptosis and epigenetic deregulation. *Nature Genet.* **27,** 31–39 (2001).

Jain, M., Arvanitis, C., Chu, K., et al. Sustained loss of a neoplastic phenotype by brief inactivation of *MYC. Science* **297,** 102–104 (2002).

Johnstone, R. W. Histone-deacetylase inhibitors: Novel drugs for the treatment of cancer. *Nature Rev. Drug Discov.* **1,** 287–299 (2002).

Jones, P. A., and Laird, P. W. Cancer epigenetics comes of age. *Nature Genet.* **21,** 163–167 (1999).

Jones, P. L., Veenstra, G. J., Wade, P. A., et al. Methylated DNA and MeCP2 recruit histone deacetylase to repress transcription. *Nature Genet.* **19,** 187–191 (1998).

Kamath, R. S., Fraser, A. G., Dong, Y., et al. Systematic functional analysis of the Caenorhabditis elegans genome using RNAi. *Nature* **421,** 231–237 (2003).

Kantarjian, H., Sawyers, C., Hochhaus, A., et al. Hematologic and cytogenetic responses to imatinib mesylate in chronic myelogenous leukemia. *N. Engl. J. Med.* **346,** 645–652 (2002).

Karpen, S. J. Nuclear receptor regulation of hepatic function. *J. Hepatol.* **36,** 832–850 (2002).

Kelland, L. R., Sharp, S. Y., Rogers, P. M., et al. DT-diaphorase expression and tumor cell sensitivity to 17-allylamino, 17-demethoxygeldanamycin, an inhibitor of heat shock protein 90. *J. Natl. Cancer Inst.* **91,** 1940–1949 (1999).
Kelly, W. K., Richon, V. M., O'Connor, O., et al. Phase I clinical trial of histone deacetylase inhibitor: Suberoylanilide hydroxamic acid administered intravenously. *Clin. Cancer Res.* **9,** 3578–3588 (2003).
Lewis, F., Maughan, N. J., Smith V. et al. Unlocking the archive-gene expression in paraffin-embedded tissue *J. Pathol.* 195, 66–71 (2001).
Khan, J., Wei, J. S., Ringner, M., et al. Classification and diagnostic prediction of cancers using gene expression profiling and artificial neural networks. *Nature Med.* **7,** 673–679 (2001).
Kinzler, K. W. and Vogelstein, B. Lessons from hereditary colorectal cancer. *Cell* **87,** 159–170 (1996).
Knudsen, S. *A Biologist's Guide to Analysis of DNA Microarray Data*. New York: John Wiley & Sons, (2002).
Kodadek, T. Protein microarrays: Prospects and problems. *Chem. Biol.* **8,** 105–115 (2001).
Kohonen, T. *Self-Organizing Maps*. Berlin: Springer (1995).
Kote-Jarai, Z., Cattini, N., Copeland, M., et al. Gene expression profiling after radiation induced DNA damage is highly predictive of *BRCA1* mutation carrier status. *Am. J. Hum. Genet.* **71,** S424 (2003).
Lachner, M., O'Carroll, D., Rea, S., et al. Methylation of histone H3 lysine 9 creates a binding site for HP1 proteins. *Nature* **410,** 116–120 (2001).
Lai, E., Riley, J., Purvis, I., and Roses, A. A 4-Mb high-density single nucleotide polymorphism-based map around human APOE. *Genomics* **54,** 31–38 (1998).
Lam, L. T., Pickeral, O. K., Peng, A. C., et al. Genomic-scale measurement of mRNA turnover and the mechanisms of action of the anti-cancer drug flavopiridol. *Genome Biol.***2,** 41, 1–11 (2001).
Lander, E. S., Linton, L. M., Birren, B., et al. Initial sequencing and analysis of the human genome. *Nature* **409,** 860–921 (2001).
Lengauer, C., Kinzler, K. W., and Vogelstein, B. Genetic instabilities in human cancers. *Nature* **396,** 643–649 (1998).
Lubbert, M., Wijermans, P., Kunzmann, R., et al. Cytogenetic responses in high-risk myelodysplastic syndrome following low-dose treatment with the DNA methylation inhibitor 5-aza-2′-deoxycytidine. *Br. J. Haematol.* **114,** 349–357 (2001).
MacBeath, G. Protein microarrays and proteomics. *Nature Genet.* **32 Suppl,** 526–532 (2002).
MacDonald, T. J., Brown, K. M., LaFleur, B., et al. Expression profiling of medulloblastoma: PDGFRA and the RAS/MAPK pathway as therapeutic targets for metastatic disease. *Nature Genet.* **29,** 143–152 (2001).
Maloney, A., and Workman, P. HSP90 as a new therapeutic target for cancer therapy: The story unfolds. *Expert Opin. Biol. Ther.* 2, 3–24 (2002).
McClue, S. J., Blake, D., Clarke, R., et al. In vitro and in vivo antitumor properties of the cyclin dependent kinase inhibitor CYC202 (R-roscovitine). *Int. J. Cancer* **102,** 463–468 (2002).
Modrek, B., and Lee, C. A genomic view of alternative splicing. *Nature Genet.* **30,** 13–19 (2002).
Mullikin, J. C., and McMurragy, A. A. Techview: DNA sequencing. Sequencing the genome, fast. *Science* **283,** 1867–1869 (1999).
Nan, X., Campoy, F. J., and Bird, A. MeCP2 is a transcriptional repressor with abundant binding sites in genomic chromatin. *Cell* **88,** 471–481 (1997).
Neckers, L. Hsp90 inhibitors as novel cancer chemotherapeutic agents. *Trends Mol. Med.* **8,** S55–61 (2002).
Neshat, M. S., Mellinghoff, I. K., Tran, C., et al. Enhanced sensitivity of PTEN-deficient tumors to inhibition of FRAP/mTOR. *Proc. Nat. Acad. Sci. USA* **98,** 10314–10319 (2001).
Ng, H. H., Jeppesen, P., and Bird, A. Active repression of methylated genes by the chromosomal protein MBD1. *Mol. Cell. Biol.* **20,** 1394–1406 (2000).
Ng, H. H., Zhang, Y., Hendrich, B., et al. MBD2 is a transcriptional repressor belonging to the MeCP1 histone deacetylase complex. *Nature Genet.* **23,** 58–61 (1999).
Nuwaysir, E. F., Bittner, M., Trent, J., et al. Microarrays and toxicology: The advent of toxicogenomics. *Mol. Carcinog.* **24,** 153–159 (1999).
Panaretou, B., Siligardi, G., Meyer, P., et al. Activation of the ATPase activity of hsp90 by the stress-regulated cochaperone *Aha1*. *Mol. Cell.* **10,** 1307–1318 (2002).
Papadopoulos, N., Nicolaides, N. C., Wei, Y. F., et al. Mutation of a mutL homolog in hereditary colon cancer. *Science* **263,** 1625–1629 (1994).
Pelengaris, S., Khan, M., and Evan, G. I. Suppression of Myc-induced apoptosis in beta cells exposes multiple oncogenic properties of Myc and triggers carcinogenic progression. *Cell* **109,** 321–334 (2002).
Petricoin, E. F. III Hackett, J. L., Lesko, L. J., et al. Medical applications of microarray technologies: A regulatory science perspective. *Nature Genet.* **32 Suppl,** 474–479 (2002).
Piekarz, R. L., Robey, R., Sandor, V., et al. Inhibitor of histone deacetylation, depsipeptide

(FR901228), in the treatment of peripheral and cutaneous T-cell lymphoma: A case report. *Blood* **98,** 2865–2868 (2001).

Pinto, A., and Zagonel, V. 5-Aza-2′-deoxycytidine (decitabine) and 5-azacytidine in the treatment of acute myeloid leukemias and myelodysplastic syndromes: Past, present and future trends. *Leukemia* **7 Suppl 1,** 51–60 (1993).

Plumb, J. A., Strathdee, G., Sludden, J., et al. Reversal of drug resistance in human tumor xenografts by 2′-deoxy-5-azacytidine-induced demethylation of the hMLH1 gene promoter. *Cancer Res.* **60,** 6039–6044 (2000).

Pollack, J. R., Sorlie, T., Perou, C. M., et al. Microarray analysis reveals a major direct role of DNA copy number alteration in the transcriptional program of human breast tumors. *Proc. Natl. Acad. Sci. USA* **99,** 12963–12968 (2002).

Pomeroy, S. L., Tamayo, P., Gaasenbeek, M., et al. Prediction of central nervous system embryonal tumour outcome based on gene expression. *Nature* **415,** 436–442 (2002).

Ponder, B. A. Cancer genetics. *Nature* **411,** 336–341 (2001).

Quackenbush, J. Microarray data normalization and transformation. *Nature Genetics* 32, 496–501 (2002).

Reich, D. E., Cargill, M., Bolk, S., et al. Linkage disequilibrium in the human genome. *Nature* **411,** 199–204 (2001).

Reinhold, W. C., Kouros-Mehr, H., Kohn, K. W., et al. Apoptotic susceptibility of cancer cells selected for camptothecin resistance: Gene expression profiling, functional analysis, and molecular interaction mapping. *Cancer Res* **63,** 1000–1011 (2003).

Roche, S., Downward, J., Raynal, P., and Courtneidge, S. A. A function for phosphatidylinositol 3-kinase beta (p85alpha-p110beta) in fibroblasts during mitogenesis: requirement for insulin- and lysophosphatidic acid-mediated signal transduction. *Mol. Cell. Biol.* **18,** 7119–7129 (1998).

Rodrigues, A. D. Preclinical drug metabolism in the age of high-throughput screening: An industrial perspective. *Pharm. Res.***14,** 1504–1510 (1997).

Rosen, L. S. Clinical experience with angiogenesis signaling inhibitors: focus on vascular endothelial growth factor (VEGF) blockers. *Cancer Control* **9,** 36–44 (2002).

Rosenwald, A., Wright, G., Chan, W. C., et al. The use of molecular profiling to predict survival after chemotherapy for diffuse large-B-cell lymphoma. *N. Engl. J. Med.* **346,** 1937–1947 (2002).

Roses, A. D. Genome-based pharmacogenetics and the pharmaceutical industry. *Nature Rev. Drug Discov.* **1,** 541–549 (2002).

Ross, P. J., George, M., Cunningham, D., et al. Inhibition of Kirsten-*ras* expression in human colorectal cancer using rationally selected Kirsten-*ras* antisense oligonucleotides. *Mol. Cancer Ther.* **1,** 29–41 (2001).

Sachidanandam, R., Weissman, D., Schmidt, S. C., et al. A map of human genome sequence variation containing 1.42 million single nucleotide polymorphisms. *Nature* **409,** 928–933 (2001).

Sandor, V., Bakke, S., Robey, R. W., et al. Phase I trial of the histone deacetylase inhibitor, depsipeptide (FR901228, NSC 630176), in patients with refractory neoplasms. *Clin. Cancer Res.* **8,** 718–728 (2002).

Sarang, S. S., Yoshida, T., Cadet, R., et al. Discovery of molecular mechanisms of neuroprotection using cell-based bioassays and oligonucleotide arrays. *Physiol. Genomics* **11,** 45–52 (2002).

Scherf, U., Ross, D. T., Waltham, M., et al. A gene expression database for the molecular pharmacology of cancer. *Nature Genet.* **24,** 236–244 (2000).

Selker, E. U. Trichostatin A causes selective loss of DNA methylation in Neurospora. *Proc. Natl. Acad. Sci. USA* **95,** 9430–9435 (1998).

Senderowicz, A. M. Small molecule modulators of cyclin-dependent kinases for cancer therapy. *Oncogene* **19,** 6600–6606 (2000).

Shaffer, A. L., Rosenwald, A., Hurt, E. M., et al. Signatures of the immune response. *Immunity* **15,** 375–385 (2001).

Shipp, M. A., Ross, K. N., Tamayo, P., et al. Diffuse large B-cell lymphoma outcome prediction by gene-expression profiling and supervised machine learning. *Nature Med.* **8,** 68–74 (2002).

Slamon, D., and Pegram, M. Rationale for trastuzumab (Herceptin) in adjuvant breast cancer trials. *Semin. Oncol.* **28,** 13–19 (2001).

Staunton, J. E., Slonim, D. K., Coller, H. A., et al. Chemosensitivity prediction by transcriptional profiling. *Proc. Natl. Acad. Sci. USA* **98,** 10787–10792 (2001).

Takahashi, M., Rhodes, D. R., Furge, K. A., et al. Gene expression profiling of clear cell renal cell carcinoma: Gene identification and prognostic classification. *Proc. Natl. Acad. Sci. USA* **98,** 9754–9759 (2001).

Tamaru, H., and Selker, E. U. A histone H3 methyltransferase controls DNA methylation in Neurospora crassa. *Nature* **414,** 277–283 (2001).

Tate, P. H. and Bird, A. P. Effects of DNA methylation on DNA-binding proteins and gene expression. *Curr. Opin. Genet. Dev.* **3,** 226–231 (1993).

te Poele, R., Cattini, N., Maillard, K., et al. The gene expression profile of the phosphatidylinositol

3-kinase (PI3K) inhibitor LY294002 in human colon adenocarcinoma cells. *Proc. Am. Assoc. Cancer Res.* **43,** 271 [Abstract 1345] (2002).

Thomas, R. S., Rank, D. R., Penn, S. G., et al. Identification of toxicologically predictive gene sets using cDNA microarrays. *Mol. Pharmacol.* **60,** 1189–1194 (2001).

Tyler, J. K., and Kadonaga, J. T. The "dark side" of chromatin remodeling: Repressive effects on transcription. *Cell* **99,** 443–446 (1999).

van de Vijver, M. J., He, Y. D., van't Veer, L. J., et al. A gene-expression signature as a predictor of survival in breast cancer. *N. Engl. J. Med.* **347,** 1999–2009 (2002).

van 't Veer, L. J., Dai, H., van de Vijver, M. J., et al. Gene expression profiling predicts clinical outcome of breast cancer. *Nature* **415,** 530–536 (2002).

Varambally, S., Dhanasekran, S. M., and Zhou, M. The polycomb group protein is involved in progression of prostate cancer. *Nature* 419, 624–629 (2003).

Venter, J. C., Adams, M. D., Myers, E. W., et al. The sequence of the human genome. *Science* **291,** 1304–1351 (2001).

Vivanco, I., and Sawyers, C. L. The phosphatidylinositol 3-kinase AKT pathway in human cancer. *Nature Rev. Cancer* **2,** 489–501 (2002).

Wang, E., Miller, L. D., Ohnmacht, G. A., et al. High-fidelity mRNA amplification for gene profiling. *Nature Biotechnol.* 28, 457–459 (2000).

Waring, J. F., Gum, R., Morfitt, D., et al. Identifying toxic mechanisms using DNA microarrays: Evidence that an experimental inhibitor of cell adhesion molecule expression signals through the aryl hydrocarbon nuclear receptor. *Toxicology* **181–182,** 537–550 (2002).

Wei, S. H., Chen, C. M., Strathdee, G., et al. Methylation microarray analysis of late-stage ovarian carcinomas distinguishes progression-free survival in patients and identifies candidate epigenetic markers. *Clin. Cancer Res.* **8,** 2246–2252 (2002).

Weinstein, I. B. Cancer. Addiction to oncogenes – The Achilles heal of cancer. *Science* **297,** 63–64 (2002).

Weinstein, J. N., Myers, T. G., O'Connor, P. M., et al. An information-intensive approach to the molecular pharmacology of cancer. *Science* **275,** 343–349 (1997).

Welsh, J. B., Sapinoso, L. M., Su, A. I., et al. Analysis of gene expression identifies candidate markers and pharmacological targets in prostate cancer. *Cancer Res.* **61,** 5974–5978 (2001).

Wessels, L. F., van Welsem, T., Hart, A. A., et al. Molecular classification of breast carcinomas by comparative genomic hybridization: A specific somatic genetic profile for BRCA1 tumors. *Cancer Res.* **62,** 7110–7117 (2002).

Wooster, R. Richard Wooster on cancer and the Human Genome Project [interview by Ezzie Hutchinson]. *Lancet Oncol.* **2,** 176–178 (2001).

Wooster, R., Bignell, G., Lancaster, J., et al. Identification of the breast-cancer susceptibility gene *Brca2. Nature* **378,** 789–792 (1995).

Workman, P. Auditing the pharmacological accounts for Hsp90 molecular chaperone inhibitors: Unfolding the relationship between pharmacokinetics and pharmacodynamics. *Mol. Cancer Ther.* **2,** 131–138 (2003c).

Workman, P. Challenges of PK/PD measurements in modern drug development. *Eur. J. Cancer* **38,** 2189–2193 (2003b).

Workman, P. New drug targets for genomic cancer therapy: Successes, limitations, opportunities and future challenges. *Curr. Cancer Drug Targets* **1,** 33–47 (2001c).

Workman, P. The opportunities and challenges of personalized genome-based molecular therapies for cancer: Targets, technologies, and molecular chaperones. *Cancer Chemother. Pharmacol.* **52 Suppl 1,** 45–56 (2003a).

Workman, P. Overview: Changing times: developing cancer drugs in genomeland. *Curr. Opin. Investig. Drugs* **2,** 1128–1135 (2001b).

Workman, P. Scoring a bull's-eye against cancer genome targets. *Curr. Opin. Pharmacol.* **1,** 342–352 (2001a).

Workman, P., and Kaye, S. B. Translating basic cancer research into new cancer therapeutics. *Trends Mol. Med.* **8,** S1–9 (2002).

Xiang, C. C., Chen, M., Ma, L. A new strategy to amplify degraded RNA from small tissue samples for microarray studies. Nucleic Acids Res. 9, e53 (2003).

Yan, P. S., Chen, C. M., Shi, H., et al. Dissecting complex epigenetic alterations in breast cancer using CpG island microarrays. *Cancer Res.* **61,** 8375–8380 (2001).

Yan, P. S., Perry, M. R., Laux, D. E., et al. CpG island arrays: An application toward deciphering epigenetic signatures of breast cancer. *Clin. Cancer Res.* **6,** 1432–1438 (2000).

Zhan, F., Tian, E., Bumm, K., et al. Gene expression profiling of human plasma cell differentiation and classification of multiple myeloma based on similarities to distinct stages of late stage B-cell development. *Blood* **101,** 1128–1140 (2002).

Zhou, Y., Gwadry, F. G., Reinhold, W. C., et al. Transcriptional regulation of mitotic genes by camptothecin-induced DNA damage: Microarray analysis of dose- and time-dependent effects. *Cancer Res.* **62,** 1688–1695 (2002).

CHAPTER 9

Transgenic Mouse Models of Cancer

T. J. Bowen and A. Wynshaw-Boris

Molecular Cancer Therapeutics: Strategies for Drug Discovery and Development, edited by George C. Prendergast
ISBN 0-471-43202-4

Oncogenes were first discovered by studying oncogenic viruses and by transfection studies of tumor DNA into cell lines, and tumor-suppressor genes were first identified through loss of heterozygosity in human tumor studies. However, the biological foundations for understanding of the mechanism of action of oncogenes and tumor suppressors in vivo was first made possible by the use of genetically engineered organisms, particularly the mouse, by which the technology for mammalian germline genetic manipulation initially became available. Indeed, the mouse has become the workhorse model organism for many types of genetic studies, including studies aimed at assessing the activity of genes that control the initiation and progression of cancer.

The basic experimental design for studying oncogenes and tumor-suppressor genes differs considerably. For oncogenes, transgenic mice are typically produced that overexpress oncogenes in specific tissues under the guidance of tissue-specific promoters. The expectation is that such mice will develop tumors in the tissue in which the oncogene is overexpressed.To study tumor suppressors, however, it is necessary to inactivate the gene to study its function. Therefore, knock-out mice are produced with inactivation of tumor-suppressor function. If the tumor suppressor is required for embryonic development or viability of the organism, then special conditional knockouts can be produced that allow for gene inactivation in specific tissues at specific developmental times or adulthood.

The discovery of oncogenes and tumor suppressors revolutionized the way cancer research is approached. It was realized over two decades ago that when key genes in these categories either gained or lost function cancer could result. For many years, cell culture techniques represented the state of the art for isolating and studying cancer genes. Novel oncogenes were discovered by the study of tumor viruses and retroviruses and by transfecting cells in culture with genomic DNA from tumor cells. Similarly, tumor suppressors were characterized through the generation and characterization of cell hybrids between tumor and normal cells, in which the cell hybrid would display the normal rather than the tumor phenotype. These findings propelled the field of cancer research forward and helped develop the knowledge needed to more directly attack the issues of cancer development and molecularly targeted therapies. However, since these first studies were all done in vitro, they were unable to effectively address the pressing issues concerning how cancer functions and progresses in a living organism.

As it was necessary to develop a system in which these questions could be asked on the whole organism level, genetically altered organisms became the next important tool for studying cancer. The technology for germ-line genetic manipulation of a mammalian system was first created in the mouse, which was also used historically for studies of cancer development and treatment; therefore, genetically manipulated mouse models developed logically as the system of choice in cancer genetics. The ability to genetically alter the mouse genome through transgenic and knock-out techniques has significantly advanced our understanding of the complexities involved in carcinogenesis in the living animal, encompassing tumor cells but also the key role played by the stromal environment of the tumor. In recent years, it has become possible to selectively activate oncogenes and eliminate tumor-suppressor genes in different tissues or at different times in development and even to turn the genes on and off repeatedly during tumor development. These advances have enabled us to address questions about requirements in initiation and maintenance of disease. These are critical advancements in knock-out technology, because they allow an investigator to selectively activate or ablate genes that may have lethal effects if performed throughout the animal's life or at early stages of its development. New strategies are being developed to find novel tumor suppressors and oncogenes using transgenic and knock-out technologies, which continue to develop, and investigators continue not only to identify the molecular interactions involved in tumorigenesis in a more precise manner but also to more accurately model the random nature of mutations that lead to sporadic cancer.

In this chapter, the methodology for genetic manipulation of the mouse is outlined. Then, examples of the application of these techniques for the study of oncogenes and tumor suppressors are presented. These techniques have been used to validate cancer gene products as targets for therapeutic intervention. The mouse models that have been produced in many cases have also proved useful as models to assess novel therapeutic agents. Detailed protocols and brief descriptions of the history of the development of transgenic manipulation and gene targeting techniques are available in several excellent laboratory manuals (Hogan et al., 1994; Joyner, 1993; Wasserman and DePamphilis, 1993). These techniques have allowed for the stable transfer of normal, altered, or chimeric genes into the mouse germ line, thus providing powerful tools with which to study mechanisms underlying development, function, tumorigenesis, and gene expression within a physiologic context.

9.1 Development of Genetically Altered Mice

In the 1950s and 1960s, several scientists developed techniques that would become critical in the ability to grow embryos in vitro as well as to transfer uteri and oviducts. Whitten (Australian National University and the Jackson Laboratory) and Biggers and Brinster (University of Pennsylvania) were at the forefront of preimplantation embryo culture, and McLaren (Cambridge University) developed uterine and oviduct transfer techniques. These techniques would eventually lead to the development of teratocarcinoma cell lines that were pluripotent, leading to a variety of somatic tissue. These cells could be injected into early embryos and contribute to numerous cell lineages. Next came the derivation of the embryonic stem (ES) cell, which could contribute to somatic tissues as well as the germ line. Martin (University of California at San Francisco) and Evans (Cambridge University) independently demonstrated that ES cells derived from blastocysts were able to contribute to germ cells after reinjection into blastocysts. The development and integration of these tools produced the foundation that made genetic manipulation in the mouse possible.

Two different methods for generating mice with altered genomes were developed in the early 1980s. Pronuclear injection of cloned DNA into the male pronucleus or the fertilized egg was one technique. The other method involved targeted introduction of altered genes by homologous recombination in ES cells. Several investigators found that cloned genes that were injected into fertilized eggs could be stably passed through the germ line. During the same period, several investigators were also determining the conditions needed for homologous recombination to occur in cultured cells. These findings were successfully applied in ES cells, allowing for the production of genetically modified ES cells. Several groups then used these cells to generate mice with germ-line transmission of the modified alleles.

These techniques documented the ability to transfer genetic material, whether altered, normal, or chimeric, into a mouse germ line for the study of the effects that these genes may play in development, signal transduction, gene expression, and tumorigenesis in the context of a living mammalian organism. Several laboratory manuals offer detailed protocols and historical accounts of the development of transgenic mice (Hogan et al., 1994; Joyner, 1993; Wasserman and DePamphilis, 1993).

9.2 Method I. Homologous Recombination in Embyro Stem Cells

Homologous recombination describes the process by which DNA can be swapped in regions that contain the same sequences. This is the process by which crossing-over occurs during meiosis, leading to genetic variability in germ cells. This natural phenomenon has been harnessed and used by molecular biologists and geneticists for the purpose of inserting genetic material into desired locations of the genome. In contrast to random integration, which is mostly used for the expression of transgenes, homologous recombination has been used primarily to disrupt the expression of genes, although it is also used to "knock in" genes for the purposes of altering the regulation or structure of a gene product. The frequency of homologous recombination is low compared to nonhomologous recombination (about 1 in 1000 inserts is via homologous recombination). However, it is now fairly easy to select for ES cells that have undergone homologous recombination and eliminate those that have random integrations. Constructs designed for homologous recombination contain dominant selectable markers that allow one to identify cells in which only homologous insertion of DNA has occurred. The systems used permit a positive selection for homologous recombination and an active negative selection for the more common event of nonhomologous insertion (Fig. 9.1). Once ES cells have been selected that contain the appropriate insertion, they are injected into a host blastocyst that was flushed from the uterus of a 3.5-day pregnant female mouse. The injected blastocyst is transplanted into the uterus of a pseudo-pregnant female, where the embryo develops; the offspring are subsequently screened for the inserted allele. These mice are called chimeras and contain some cells from the host blastocyst and some from the injected ES cells that contain the altered gene. Since this means that not all of the cells in the chimera will contain the altered allele, the chimeric mice that have the mutant allele in their germ line must be selected so that the altered allele can be passed to future generations. The chimeras with

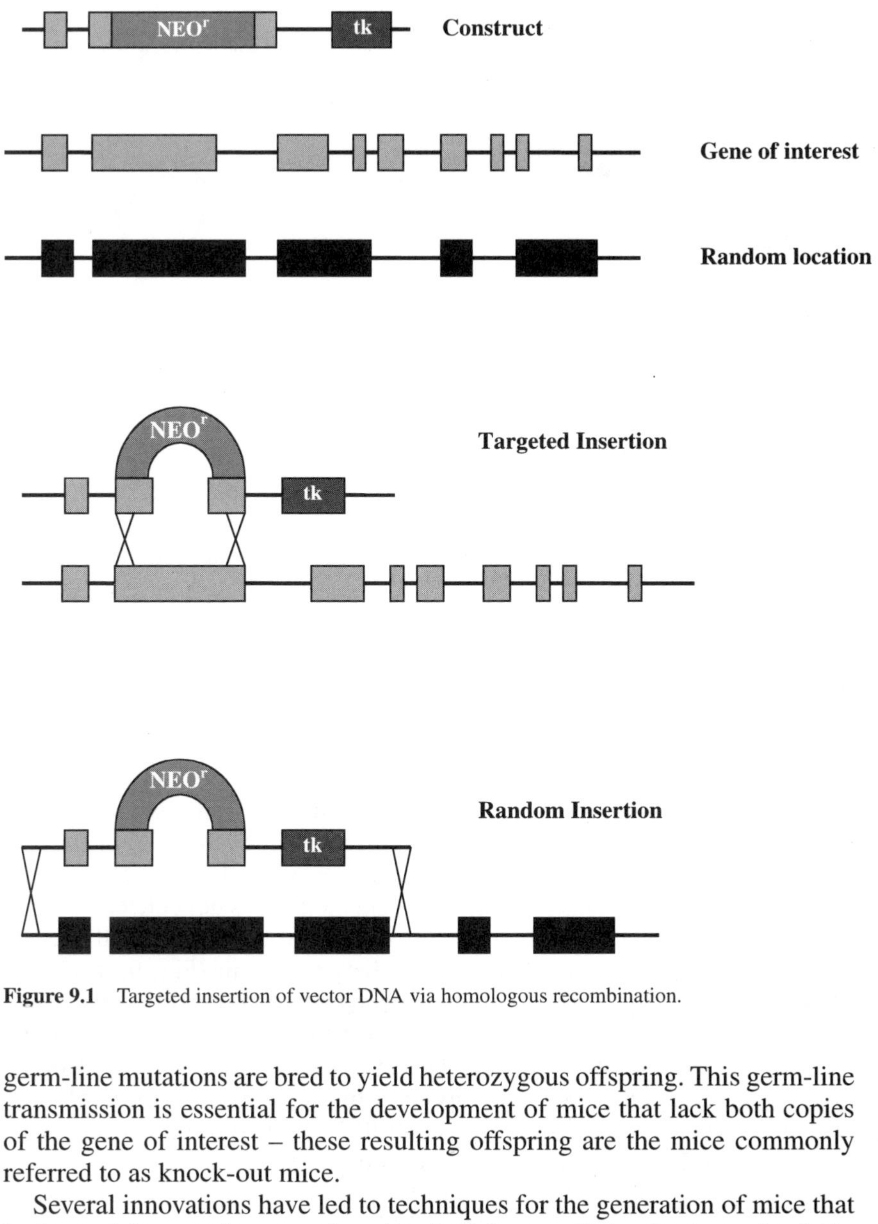

Figure 9.1 Targeted insertion of vector DNA via homologous recombination.

germ-line mutations are bred to yield heterozygous offspring. This germ-line transmission is essential for the development of mice that lack both copies of the gene of interest – these resulting offspring are the mice commonly referred to as knock-out mice.

Several innovations have led to techniques for the generation of mice that harbor subtle mutations, nonfunctional alleles, or alleles that are selectively abrogated in a tissue type or at a particular time in development, including at different points in the adult. For gene knock-out mice, a common approach is to insert by homologous recombination a selectable marker gene (generally neomycin phosphotransferase; neo) at a noncoding site between exons in the gene of interest. This tactic introduces stop codons into the reading frames of the gene downstream of the insertion by disrupting the splice pattern of the gene, thereby abrogating gene expression. However, an investigator must be careful in the design of the gene-targeting construct to avoid a scenario in which a stable truncated protein or a splice variant is produced that may retain activity. An alternative to this strategy is to replace the exons

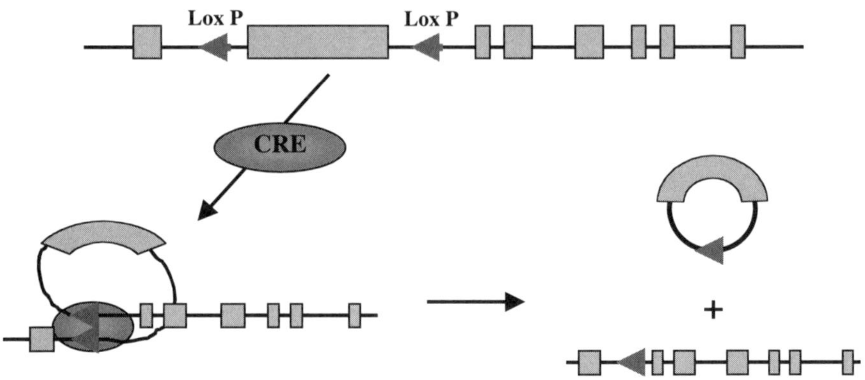

Figure 9.2 Conditional knockout.

containing critical protein domains with an inserted neo gene, albeit with the same caveats as above. If successful, one of these strategies produces a complete loss-of-function allele (referred to often as a nullizygous allele or null allele).

It is also possible to generate so-called conditional knockouts in which genes are inactivated by homologous recombination only in specific tissues or at specific times in development. The Cre/LoxP system is a system commonly used for this purpose. In this system, a gene-targeting construct is engineered with short bacterial sequences, called LoxP elements, that flank the part of the gene to be deleted. Typically, LoxP sites flank one or more exons that encode critical regions of the gene product, but they can also be used to target the gene promoter for deletion. Transgenic mice that harbor the so-called floxed allele are unaffected, because the noncoding LoxP sites are localized at benign regions (e.g., introns) and the gene is not deleted without co-expression of the Cre recombinase, a bacterial protein that binds the LoxP sites and mediates the excision of DNA between them (Fig. 9.2). Cre is introduced in vivo by interbreeding the floxed mouse strain with a transgenic mouse strain that expresses Cre from a general or tissue-specific promoter. By the use of a tetracycline-regulated system to control Cre activity (Corbel and Rossi, 2002), it is possible to impose further specificity on the Cre/LoxP system, allowing gene excision to be controlled by the time of an exogenous stimulus as well as by a tissue-specific promoter. It has also been shown recently that it is possible to create large deletions in the mouse germ line using Cre/loxP excision; deletions of up to 1,000,000 bases have been demonstrated to be possible (Zheng et al., 2000).

9.3 Method II. Pronuclear Injection

As mentioned above, transgenic mice can be produced by the direct injection of plasmid DNA vectors into pronuclei of fertilized mouse eggs, resulting in the random integration of the introduced DNA into the mouse genome. The gene inserted in the transgenic mouse that is generated contains all of the information necessary to efficiently express the genes of interest, including promoter/enhancer regions, open reading frames (ORFs), RNA splice sites, and polyadenylation sequences. Because the genes of interest must be expressed in the appropriate cell type, the promoter used is selected to direct the desired specificity of expression (both spatially and temporally). Several promoters have been developed with well-defined tissue specificity, for example, the mouse probascin promoter has been used to specifically direct gene expression in the prostate (Wu et al., 2001). Other promoters have been characterized that are specific for a certain time in development. The whey acidic protein (WAP) promoter has been used to specifically direct gene expression in the mammary gland when it develops at pregnancy (Andres et al., 1987). The use of these well-characterized and very specific promoter sequences has enabled transgenes (such as oncogenes) to be expressed in a controlled and predictable fashion.

Between 100 and 200 copies of the cloned gene are injected into the male pronucleus of the fertilized egg, with only a small fraction of this DNA randomly integrating into the genome at the one cell stage. Every cell of the transgenic mouse should contain the inserted genetic material if integration occurs at the one-cell stage. Several of these modified embryos are then injected into the oviduct of a pseudo-pregnant surrogate female, where they develop to delivery. Offspring are analyzed to determine if they contain the integrated gene; if so, they are used for husbandary as so-called founders. Founders generally have a randomly integrated transgene at a single site in the genome, but smaller genes (up to 50 kb fragments) are sometimes inserted as a concatomer of 1–50 copies of the transgene arranged in a head-to-tail fashion. This unpredictable duplicate insertion generates significant variability of transgene expression in different founders. Therefore, one must carefully assess transgene expression in different founders to select one that contains a trangene expressed at a suitable level for study. After a founder is selected with appropriate expression, the pattern of expression from the integrated transgene generally remains stable in offspring that are derived from the founder.

More recently, larger gene fragments have been generated that integrate and express in a more predicatable and uniform manner. These larger (up to 200 kb) inserts are generated from bacterial artifical chromosomes (BACs) or yeast artificial chromosomes (YACs), which can be engineered to hold much more DNA than bacterial plasmids. The additional DNA that can be accommodated in BACs and YACs can be exploited by flanking the gene with sequences that can help insulate it from the influences of neighboring genes, which may be proximal to the site of integration. In addition, the inclusion of additional flanking sequences in transgenes offers the potential to include more distal regulatory sequences for the gene of interest compared to smaller inserts. BACs have become the vector that is more preferentially used, because YACs have become notorious for instability leading to chimerism of DNA. Also, mouse BAC libraries have become commercially available, making it possible to easily obtain virtually any gene of interest in its normal genomic setting.

9.4 Oncogenes and Tumor Suppressors

9.4.1 ONCOGENES

Oncogenes encode cellular proteins that are critical to drive cell division, most commonly transcription factors, signal transduction proteins, and growth factor receptors that function in the wide diversity of mechanisms required to induce DNA replication and mitosis. During the development of human cancers, oncogenes become mutated and/or deregulated so that their ability to drive cell division becomes unregulated. In model systems, these alterations are necessary and sufficient to drive cells toward malignant transformation. The first oncogenes were discovered in the late 1970s and early 1980s as the transforming genes of retroviruses that could cause cancer in animals. Analysis of these genes led to the discovery of the normal cellular counterparts of these genes, sometimes called proto-oncogenes, which gained oncogenic potential when they were mutated in retroviruses or in human cancer cells. Current models of carcinogenesis suggest that oncogenic mutations that arise randomly in cells allow them to defeat regulatory mechanisms that prevent unrestrained cell division. In the most widely accepted model, benefits conferred by mutations in a single cell permit a clonal expansion that increases the likelihood of acquiring additional mutations that drive malignant progression. This stochastic process cannot be replicated easily in a transgenic mouse model. However, it has been found that transgenic expression of activated oncogenes that are derived from human cancers is sufficient to elicit cancers at an elevated frequency, or penetrance, in mouse models. The specificity and human relevance of these effects can be illustrated in models in which the oncogene is driven by a tissue-specific promoter, showing that cancer incidence in that tissue sometimes rises dramatically after the oncogene activation event. Since the first demonstrations of this capability, specific promoter sequences have been characterized to drive oncogene expression in a specific manner in a variety of tissues as a model of the development of premalignant or malignant lesions.

The first demonstration of how cancer incidence can be elevated in a transgenic mouse by expression of an oncogene was made in Leder's laboratory (Harvard University) (Stewart et al., 1984). The demonstration was based on specific expression of the c-*myc* oncogene in the adult mammary gland, using the promoter from the mouse mammary tumor virus (MMTV), a retrovirus that is activated by estrogen during pregnancy. The regulatory sequences in the long terminal repeat (LTR) of MMTV direct high-level expression of the downstream genes in the breast. In the normal situation, when MMTV integrates upstream of an oncogene, it increases the expression level of that gene in breast tissue, increasing the incidence of breast tumors. To generate the transgenic mouse, Leder's laboratory generated an expression plasmid in which the MMTV promoter was fused to the c-*myc* oncogene (which encodes a transcription factor). Transgenic mice that harbored this fusion gene were generated, many of which developed spontaneous adenocarcinomas in the mammary glands. The increased incidence of breast cancer in these transgenic mice was transmitted to all female progeny of these mice, which frequently developed breast tumors as well. These MMTV-c-*myc* mice were the first example of a transgenic mouse strain that was predestined to develop tumors due to expression of a human oncogene, in this case, c-*myc*. Since the first use of the MMTV promoter, other oncogenes placed under control of the MMTV-LTR promoter (e.g., *ras* or *her*-2/erb-2) have confirmed that deregulated oncogene expression is sufficient drive breast tumorigenesis in the mouse. Today, many examples of cancer-prone transgenic mice exist.

As a group, the transgenic mouse strains that express oncogenes and that exhibit an elevated susceptibility to cancer are often referred to informally as *oncomice*. In addition to their great scientific import, oncomice have also opened up new areas of intellectual property law that affect researchers: Patents that Harvard University sought and received for developing this technology mean that to use any transgenic oncomouse an investigator must sublicense the technology (from DuPont, which holds an exclusive license to the technology from Harvard), whether for academic or commercial research purposes.

9.4.2 TUMOR-SUPPRESSOR GENES

Tumor-suppressor genes oppose the action of oncogenes in cancer. In human cancers, tumor suppressor pathways that are controlled by key suppressor genes must be inactivated for unregulated cell proliferation and tumorigenesis. Tumor-suppressor genes can act in a variety of ways, but the canonical pathways thought to be most important are those that inhibit cell cycle progression, stimulate programmed cell death (apoptosis), or promote senescence. Different definitions of a tumor-suppressor gene exist: The most rigorous definition insisted on by some investigators requires evidence of mutational inactivation in cancer and a causal role of mutation in driving cancer initiation or progression. However, broader definitions accepted by others encompass evidence of epigenetic inactivation or even stromal downregulation that is causal and linked to a negative growth regulatory role in cancer. Transgenic mouse studies have helped define critical tumor-suppressor pathways, in cases in which it can be shown that tissue-specific inactivation of a gene product in the pathway is sufficient for initiation or progression of cancer in that tissue. Because genes that control cellular proliferation, apoptosis, and cell cycle play critical roles in every cell, as well as in the complexity that exists among the different pathways involved, the number of tumor suppressor genes encoded by the mammalian genome is likely to be vast. Here it should be noted that in oncomouse models, a loss of endogenous tumor-suppressor pathways occurs stochastically in the cells in which the oncogene is activated. The period needed for these secondary events to arise in the cells harboring the oncogenic lesion explains the delayed kinetics and penetrance that are characteristic of oncomice. Conversely, the delayed kinetics of cancer incidence in knock-out mice in which a key tumor suppressor has been inactivated is explained by the stochastic incidence of random oncogene activation.

This interplay between oncogene activation and tumor-suppressor inactivation has been formally demonstrated in transgenic mice engineered for both events, where tumors arise with significantly faster kinetics and penetrance (sometimes even in utero).

The best known example of a tumor suppressor protein is p53. p53 is well known mainly because its mutation is among the most common genetic lesions that occur in human cancers. p53 plays a central role in the regulation of cell cycle checkpoints, apoptosis, and response to cellular DNA damage after ionizing radiation, UV light, and carcinogenic chemicals. In the normal cellular state, p53 is expressed at low levels. When DNA damage occurs, however, the levels of p53 are quickly increased, and the protein activates several important downstream pathways that lead to growth arrest or apoptosis. p53 is a transcription factor that acts by stimulating the expression of genes that are important to cell cycle control and apoptosis, such as p21/*waf1* (p21) and *Bax* (Donehower, 1992; Jacks, 1994; Levine, 1997).

The prevalence of p53 lesions in cancer made p53 an obvious target for knockout to determine whether its loss could promote tumor formation in a mouse model. p53 null mice were found to be viable and to develop spontaneous cancers of different types by 4–6 months of age. Heterozygous mice also developed cancer, but at a longer latency (Jacks, 1994). The latter model replicates the situation in the human Li-Fraumeni syndrome, in which individuals carry heterozygous mutations in p53. Such individuals eventually acquire a second mutation in the normal p53 allele, causing the cell to be resistant to DNA damage-induced arrest, apoptosis, and senesence; thereby driving it toward malignant development. Tumors that arise in p53 heterozygous mice generally replicate this phenomenon by undergoing an event known as loss of heterozygosity (LOH) at the remaining wild-type p53 allele. At a particular genetic locus in human cancers, the appearance of LOH along with mutation of the second remaining allele is a hallmark signature of a tumor-suppressor gene. The ability of mouse models to replicate this signature argues that the cancers that arise in the models have biological relevance to the human situation.

9.5 Conditional Knockouts and Tumor Suppressors

The ability to ablate the functions of genes in a specific spatial and temporal fashion is of particular importance in the case of tumor-suppressor genes. As discussed above, loss of function in a tumor-suppressor gene can promote tumorigenesis, starting with the cell in which the gene is lost. Often, these genes are lost relatively more frequently in tissues that have a high proliferative index. Therefore, the generation of a mutant tumor-suppressor allele that is systemic may not be a good model for cancer. Often tumor suppressors are involved in cell fate determining events, such as cell cycle checkpoint controls, which arrest cell division if a chromosomal lesion occurs. If a gene that is involved in such a checkpoint is lost during development, cell fates may be altered and affect the overall development of the embryo, leading to embryonic lethality. An example of one such tumor suppressor is the *BRCA1* gene, which controls cell cycle checkpoint responses. *BRCA1* is mutated in many cases of familial breast cancer. As the first breast cancer susceptibility gene to be identified in the 1990s, *BRCA1* received a large amount of attention. Indeed, studies of *BRCA1* proved to be valuable in identifying DNA repair processes that are required to maintain genomic integrity and to stave off cancer, with knock-out mouse studies firmly establishing the causal role of *BRCA1* lesions in breast cancer development.

Knock-out studies of *BRCA1* illustrate a common theme for tumor-suppressor genes in development. Due to the obvious importance of this gene in tumorigenesis in the breast, several groups pursued the generation of *BRCA1* knock-out mice by removing the large exon 11 and replacing it with a neomycin cassette. This deletion led to an early embryonic lethal phenotype. Thus the role of *BRCA1* is not limited to adult breast tissue but also has a key role in cell fate during development (Gowen et al., 1996; Hakem et al., 1996; Liu et al., 1996; Ludwig et al., 1997). Since a null phenotype did not permit the study of the *BRCA1* gene in breast tumorigenesis, tissue-specific knock outs were generated, using the Cre-loxP system, described previously, in combination with a tissue-specific promoter. In this manner, tissue-specific *BRCA1* knock-out mice were generated that were deficient for the gene only in breast epithelial cells. For this strategy, exon 11 was flanked with loxP sites and removed by Cre-mediated excision, abolishing the production of normal *BRCA1* transcripts. The ensuing reduction in *BRCA1* levels resulted in abnormal ductal outgrowth during pregnancy and increased levels of apoptosis. This demonstrated a specific role of *BRCA1* in apoptosis and a specific aspect of mammary development. More important, however, the loss of *BRCA1* led to the formation of mammary tumors. These tumors had similar characteristics to those seen in human *BRCA1*-deficient tumors, such as karyotype abnormalities (translocations and aneuploidy), alterations of p53 transcription, and abnormal cell cycle control (Xu et al., 1999). These experiments demonstrated a vital role for the use of a conditional knock-out mouse model for cases in which the traditional, or straight, knockout cannot be used to explore an hypothesized role in cancer development.

9.6 Inducible Genes and Other Applications

Recent technology that affords temporal control of gene expression has been shown to be effective for cancer genetics studies. One strategy that has aided in the understanding of oncogene function is the induction of a transgene through addition of a drug to the diet. With the use of tissue-specific promoters that can be switched on or off by the drug, one is able to control the time at which the gene is activated or inactivated. One of the most common inducible systems uses a tetracycline-inducible system developed by Bujard. In this system, the vector includes a transgene placed downstream

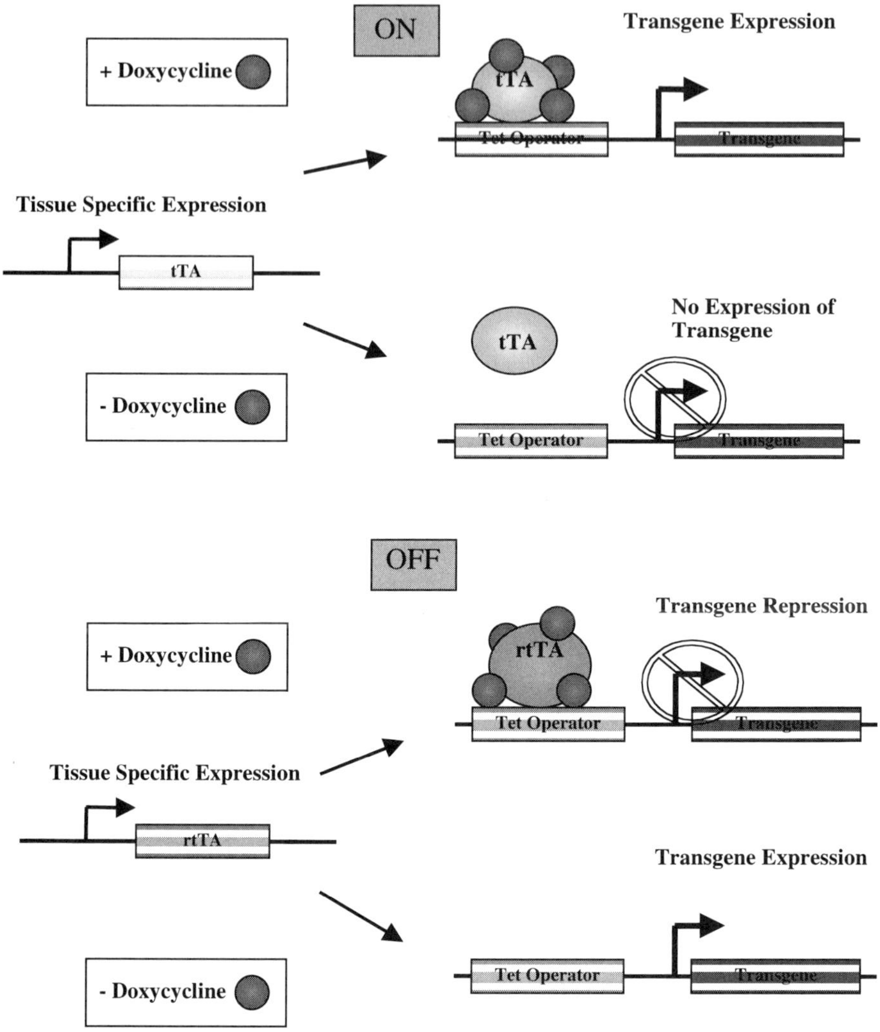

Figure 9.3 Inducible transgene (on/off system).

of a tissue-specific promoter that is engineered to respond to a tetracycline-responsive transcriptional repressor, termed tTA. This protein binds a Tet operator (TetO) DNA recognition element that is embedded in the proximal region of the promoter. In the presence of the tetracycline analog, doxycycline, tTA is switched off and the promoter activity is derepressed, allowing expression of the oncogene. Reciprocal vector systems in which doxycycline represses transgene expression also exist (Fig. 9.3). Thus, depending on the configuration of the system, one can either induce or repress transcription of the transgene.

This tetracycline-based gene regulatory system was used to conditionally express the SV40 T-antigen (T-Ag) in mouse mammary tissue, illustrating that a single oncogene can drive the production of cancer. By initiating

T-Ag expression and then eliminating its expression, it was shown that at early stages of tumorigenesis, hyperplasias were reversible when the gene expression was eliminated. However, after prolonged periods of T-Ag expression, the cancer was no longer reversible (Ewald et al., 1996). This latter result demonstrated that a single oncogene was sufficient to initiate cancer but that it was not required to maintain it, implicating secondary events in maintenance. It also suggested that mutations in additional critical genes were required to allow tumors to continue to progress and escape growth control. The controllable expression of genes in specific tissues at defined times in development is a flexible tool for addressing questions that relate to such temporal as well as tissue-specific aspects of cancer development.

Recently, Cre-mediated excision was demonstrated, in Bradley's laboratory, to be capable of producing genomic deletions as large as 22 centimorgans in transgenic mice, permitting new strategies to map tumor-suppressor genes. Using this method, Bradley's group identified regions of chromosomes that contain potential oncogenes leading to hyperplasias and tumors, while deletions of large regions of chromosome 11 have led to embryonic lethality. This technique has been used more recently in combination with *N*-ethyl-*N*-nitrosourea (ENU) mutagenesis to attempt to discover novel tumor-suppressor genes in Justice's laboratory. ENU is a mutagen that introduces intragenic mutations in spermatogonia at a high frequency. When a large region of a chromosome has been deleted using Cre-mediated excision, the ENU strategy will randomly mutate the complimentary alleles of the intact chromosome as well as all other chromosomes (Liu et al., 1998). This can lead to recessive phenotypes that would not be discovered in a standard mutation screen, because it would require both alleles of the same gene to be mutated to see the recessive phenotype. In summary, the increasingly numerous and flexible technologies for discovering and regulating cancer genes in transgenic mice will continue to increase the potential value of the models for studies of cancer pathophysiology and treatment, in ways that more closely approach human disease settings.

9.7 Limitations of Transgenic Mouse Models

As with any model for a disease, there are limitations that are inherently associated with the transgenic mouse models of human cancers. Although currently, these models offer the best systems for analyzing the intricate and complex pathways and mechanisms that lead to cancer, they are by no means ideal in replicating human disease as it relates to cancer development and treatment. At the current time, many investigators believe that the shortcomings of transgenic models derive from relatively subtle differences in the control, expression, and action of genes that are considered homologous in the mouse and human.

The mouse and human genomes are similar in content, but they also have many differences, particularly in terms of chromosomal organization and noncoding sequences. It is also evident that there are numerous differences in the genes between mouse and human that weaken the assumption that homologous genes in each organism will have the same function. Among the differences that are evident are gene duplications, loss of genes, and differences in gene regulation. Gene redundancy can be a problem in the prediction of the function of a gene in humans based on the action of the gene in the mouse. There also may arise differences in the pathways in which a particular gene product is involved, for example, as in the case of *ras* signaling in fission versus budding yeasts. In either human or mouse, it is possible that a gene may have a more specific function than in the other organism; the possibility that the homologous genes could have evolved more refined functions in one organism relative to the other could lead to somewhat different observable phenotypes when a knockout is generated. Thus it is possible that a knockout or transgenic mouse may not completely recapitulate a mutation in the homologous human gene.

Another dissimilarity could be seen in gene regulation. Genes could be identical in sequence and action, but if they are expressed behind different promoters, the resulting function could differ as a result of unique contexts of expression. Differential expression could involve the tissues the gene is expressed in, the time during development that the gene is expressed, or the quantity of the gene product that is expressed. Each of these differences could make a drastic difference in the way the protein functions in an organism, limiting the ability to translate findings in mouse models to human settings.

Other differences at the organism level likely affect the utility of mouse models in cancer studies. The shorter life span of the mouse, which is about 2 years, limits the study of the natural progression of tumorigenesis as it occurs in humans, in whom cancer may develop over the course of decades. Given that the major risk factor for cancer in humans is age, the use of younger cancer-prone transgenic mice may skew interpretation of cancer genetics results. While there is emerging interest in the interfaces between aging and cancer, for example as demonstrated in the positive effects of senescent stroma on tumor progression, this aspect may be problematic to model in transgenic models.

Strain differences are another important issue in the use of transgenic mouse models of cancer. Mice of different backgrounds differ greatly in their susceptibility to cancer, and the inbreeding of strains reduces the extrapolation of findings to humans, who have a much more variable genetic makeup. Modifier genes that are likely responsible for the great differences in the cancer susceptibility of different strains are only beginning to be identified, and these genes may ultimately prove to have a dominant influence in studies of oncogene and tumor-suppressor pathways in mice. The complexities of human cancer no doubt have important modifier components, and it should be possible to replicate these effects in transgenic mice, in which different modifier alleles have been knocked in for comparison. However, the likelihood that many modifier effects are multigenic in nature will likely restrict the possible types of studies that can be achieved, even if all the alleles can be identified. In summary, cancer-related differences in transgenic models apply specifically to broad assumptions about the function of a gene in tumorigenesis based on an inbred background. For any novel genes, it will be important to examine the effects on additional strains in which potential modifier effects can be assessed.

9.8 Summary

The development of the transgenic mouse as a model mammalian organism for the study of cancer has been one of the leading achievements of the past few decades in biomedical research. The ability to determine the role a specific gene plays in the progression of cancer has led to a deeper and much more profound understanding of the way that cells are able to become cancerous. Gene targeting, transgenics, and the ability to precisely manipulate the genome of a model organism can provide important biological insights into how genes function in vivo; in particular, this technology has shed new light on the basic genetic processes that underlie the development of tumorigenesis. The approaches discussed in this chapter illustrate the value of the basic transgenic mouse models for cancer genetics studies. Refinement and new approaches in transgenic mice will continue to offer new breakthroughs in the pursuit of better understanding cancer at the organismal and molecular levels.

References

Andres, A. C., Schonenberger, C. A., Groner, B., et al. Ha-ras oncogene expression directed by a milk protein gene promoter: Tissue specificity, hormonal regulation, and tumor induction in transgenic mice. *Proc. Natl. Acad. Sci. USA* **84,** 1299–1303 (1987).

Corbel, S.Y., and Rossi, F. M. Latest developments and in vivo use of the Tet system: Ex vivo and in vivo delivery of tetracycline-regulated genes. *Curr. Opin. Biotechnol.* **13,** 448–452 (2002).

Donehower, L. A., Harvey, M., Slagle, B. L., et al. Mice deficient for p53 are developmentally normal but susceptible to spontaneous tumours. *Nature* **356,** 215–221 (1992).

Ewald, D., Li, M., Efrat, S., et al. Time-sensitive reversal of hyperplasia in transgenic mice expressing SV40 T antigen. *Science* **273,** 1384–1386 (1996).

Gowen, L. C., Johnson, B. L., Latour, A. M., et al. *Brca1* deficiency results in early embryonic lethality characterized by neuroepithelial abnormalities. *Nature Genet.* **12,** 191–194 (1996).

Hakem, R., de la Pompa, J. L., Sirard, C., et al. The tumor suppressor gene *Brca1* is required for embryonic cellular proliferation in the mouse. *Cell* **85,** 1009–1023 (1996).

Hogan, B., Beddington, R., Costantini, F., and Lacy, E., Ed. (1994). Manipulating the Mouse Embryo. New York, Cold Spring Harbor Laboratory Press.

Jacks, T. Tumor spectrum analysis in p53-mutant mice. *Curr. Biol.* **4,** 1–7 (1994).

Jackson-Grusby L. Modeling cancer in mice. *Oncogene* **21,** 5504–5514 (2002).

Joyner, A. L., Ed. (1993). A Practical Approach, Gene Targeting. Oxford, Oxford University Press.

Levine, A. J. p53, the cellular gatekeeper for growth and division. *Cell* **88**, 323–331 (1997).

Liu, P., Zhang, H., McLellan, A., et al. Embryonic lethality and tumorigenesis caused by segmental aneuploidy on mouse chromosome 11. *Genetics* **150,** 1155–1168 (1998).

Liu, C. Y., Flesken-Nikitin, A., Li, S., et al. Inactivation of the mouse *Brca1* gene leads to failure in the morphogenesis of the egg cylinder in early postimplantation development. *Genes Dev.* **10,** 1835–1843 (1996).

Ludwig, T., Chapman, D. L., Papaioannou, V. E., and Efstratiadis, A. Targeted mutations of breast cancer susceptibility gene homologs in mice: Lethal phenotypes of *Brca1*, *Brca2*, *Brca1/Brca2*, *Brca1*/p53, and *Brca2*/p53 nullizygous embryos. *Genes Dev.* **11,** 1226–1241 (1997).

Stewart, T. A., Pattengale, P. K., and Leder, P. Spontaneous mammary adenocarcinomas in transgenic mice that carry and express MTV/*myc* fusion genes. *Cell* **38,** 627–637 (1984).

Wassarman, P. M., DePamphilis, M. L., Ed. (1993). Methods in Enzymology. Guide to Techniques in Mouse Development. San Diego, Academic Press, Inc.

Wu, X., Wu, J., Huang, J., et al. Generation of a prostate epithelial cell-specific Crè transgenic mouse model for tissue-specific gene ablation. *Mech. Dev.* **101,** 61–69 (2001).

Xu, X., Wagner, K.-U., Larson, D., et al. Conditional knock-out of *Brca1* in the mammary epithelial cells results in blunted ductal morphogenesis and tumor formation. *Nature Genet.* **22,** 37–43 (1999).

Zheng, B., Sage, M., Sheppeard, E. A., et al. Engineering mouse chromosomes with Cre-loxP: Range, efficiency, and somatic applications. *Mol. Cell. Biol.* **20,** 648–655 (2000).

CHAPTER 10

Transgenic Versus Xenograft Mouse Models of Cancer: Utility and Issues

Ming Liu, W. Robert Bishop, Yaolin Wang, and Paul Kirschmeier

Molecular Cancer Therapeutics: Strategies for Drug Discovery and Development, edited by George C. Prendergast
ISBN 0-471-43202-4

Preclinical tumor models play an essential role in the evaluation of efficacy and optimization of lead compounds in the discovery and development of anticancer drugs. A robust, dependable animal model of human disease is critical to the evaluation of potential anticancer candidates. Although other animal species (e.g., rat, hamster, rabbit, dog) have been used as animal models for cancer research, the mouse has been the most important contributor due to its accessibility, short generation time, ease of propagation, lower consumption of test compounds, and advances in mouse genetics. This chapter focuses on mouse models.

The discovery of cancer drugs through in vivo screening methods traditionally used syngeneic transplantable murine tumors. Since 1955, the U.S. National Cancer Institute (NCI) has provided screening support to cancer researchers worldwide (Grever et al., 1992). The earliest in vivo screens were the fast-growing murine leukemias, L1210 and P388, implanted intraperitoneally. These tumors were derived from leukemias originally induced chemically in the DBA/2 mouse by painting the skin with methylcholanthrene (Dawe and Potter, 1957; Law et al., 1949). Using survival as the end point, these tumors provided a rapid and reproducible means for identifying potential anticancer drugs (Teicher, 2002; see Chapter 2). From 1975 until 1985, the in vivo P388 mouse leukemia model was used almost exclusively as the initial or primary screen at the NCI. With few exceptions, agents that showed minimal or no activity in the P388 system were not selected for further evaluation in other tumor models. It became evident that there were marked similarities in the drugs emerging from the murine leukemia screen. The classes of agents found active in the mouse tumor models were limited, mainly comprising alkylating agents and DNA interacting drugs (Fiebig et al., 1999).

Subsequently, panels of syngeneic murine solid tumors and human tumor xenografts have largely replaced the murine leukemias used in early anticancer drug screens. Syngeneic rodent tumor models provide an experimental model for evaluating the anticancer effects of therapeutic agents in animals with an intact immune system. However, their relevance to human cancer may be overestimated or underestimated (Amadori et al., 1992). Successful xenografting of human tumors into nude mice was first reported in the late 1960s (Giovanella et al., 1972; Rygarrd and Povlsen, 1969). Nude mouse models are now extensively used in the development of potential anticancer drugs and studies of tumor biology. Moreover, mice with severe combined immunodeficiencies (e.g., SCID, beige, xid) have enlarged the spectrum of possible models and enabled engraftments of human tumors that were previously difficult to explant, such as those of the hematopoietic system. Thus xenograft/human explants have become the gold standard in cancer drug development, and their use is highly recommended by various regulatory agencies (Fiebig and Burger, 2002).

In the late 1980s and early 1990s, the focus of new drug development shifted to molecularly targeted/disease-directed treatment strategies (Sausville and Feigal, 1999). In 1985, the NCI initiated a new project assessing the feasibility of employing human tumor cell lines for large-scale drug screening (Boyd, 1989). Cell lines derived from seven cancer types (brain, colon, leukemia, lung, melanoma, ovarian, and renal) were acquired from a wide range of sources and subjected to battery of in vitro and in vivo evaluations. In 1993, the composition of the cell line screen, often referred to as *the NCI panel of 60 cell lines,* was modified to include various prostate and breast tumor lines. As part of the evaluation, in vivo tumor models derived from this panel were used to assess the antitumor efficacy of new compounds. Although only a subset of cell lines would be used for each agent under screening, it was anticipated that for any selected compound, any cell line might be required as a xenograft model (Plowman et al., 1997).

In addition to the progress in the human xenograft models, the shift from compound-oriented to disease-oriented drug discovery at the NCI also prompted a realization that there was a need to identify more target-defined models. Specifically designed and bred transgenic/knock-out mice have proven useful to satisfy this need. During the past 20 years, an impressive range of tools has become available to the mouse geneticist and tumor biologist. As described in more detail in Chapter 12, groundbreaking experiments in several laboratories established the first transgenic mouse tumor models by expressing viral (Brinster et al., 1984; Hanahan, 1985) or cellular (Adams et al., 1985; Stewart et al., 1984) oncogenes in specific tissues. Germ-line inactivation of the prototype tumor-suppressors gene *Rb* (Jacks et al., 1992) and *p53* (Donehower et al., 1992) using gene targeting technology in mouse embryonic stem (ES) cells provided additional tools to model the scope of mutations in human tumors (Jackson-Grusby, 2002).

Both xenograft and transgenic/knock-out models are being increasingly used in the discovery of anticancer agents. However, both animal models have been criticized for failing to predict the response of human patients to new agents (Gura, 1997; Kerbel, 1999; Rosenberg and Bortner, 1999). It is generally believed that requirements for successful preclinical animal-tumor models should include the following characteristics: reproduction of the biology of human cancer, objective and quantitative evaluation of cellular and molecular events associated with cancer progression, reliability, availability, and affordability. In this chapter, we compare the utility and the issues of these two types of models, particularly from the viewpoint of the pharmaceutical drug development, capturing the advantages and disadvantages of both models for cancer drug evaluation.

10.1 Xenograft Tumor Models in Drug Discovery

10.1.1 IMMUNODEFICIENT MICE

The growth of human tumors in a different species (e.g., mouse) requires immunodeficiency in the host animal to prevent rejection of the transplanted foreign tissues. There are many strains of immunodeficient mice containing single mutations (e.g., *nude, scid, beige, xid, rag-1 null, rag-2 null*) or

Table 10.1 Immunodeficiencies in Nude and SCID Mice[a]

Cell	Nude	SCID
B cells	Defective maturation Precursor cells normal	Pre-B and B cells absent
T cells	Low or absent (mature and functional cells)	Nonfunctional
NK cells	High	Normal
Macrophages	Normal	Normal
LAK cells	Normal	Normal
Serum immunoglobulin	IgG low IgM normal IgA low	IgG low IgM low IgA low
Lymphoid organs	Athymic	Small lymphoid organs
Chemotherapy sensitivity	More hardy	More sensitive

[a]Modified from Clarke and Dickson (1997).

combined mutations (e.g., *bg/nu, bg/nu/xid, nude/scid, nod/scid*) available for cancer research (Amadori et al., 1992; Clarke and Dickson, 1997). These mice have mutations leading to different degrees of immunodeficiency in natural killer (NK) cells, lymphokine-activated killer (LAK) cells, macrophages, B cells, T cells, and blood immunoglobulin production.

Nude mice and severe combined immune deficiency (SCID) mice are the strains of immunodeficient mice most commonly used as the recipient for human tumor xenografts, with the nude mouse being more heavily used. Both nude mice and SCID mice are easily accessible in large quantities from commercial sources (e.g., Charles River Laboratories, Jackson Laboratory, and Harlan Bioproducts). The SCID mouse is in general the more immunodeficient of the two strains, with reduced NK cell, nonfunctional T cell, and defective B cells relative to the nude mouse. Thus higher take rates and metastasis are achieved in SCID mice for many human tumor cell lines (Clark et al., 1981; Williams et al., 1993). Since the development of the human–SCID chimera mouse models (Mosier et al., 1988), SCID mice have been broadly used in studies of anticancer immunotherapy by engrafting human tumor with various combinations of human peripheral blood leukocytes (HPBLs), subsets of the HPBLs, and/or intact human stromal tissue adjacent to the tumor tissue (Bankert et al., 2002). An outline of some of the immunological characteristics of the nude and SCID mice is in Table 10.1.

The nude mouse, due to its milder temper and hairless feature, is generally easier to handle and easier to observe/quantify the growth of transplanted tumors. The SCID mouse is more sensitive to the toxic effects of irradiation and some cytotoxic agents (Croy et al., 2001; Schuler and Bosma, 1989). The SCID mouse is also more expensive than the nude mouse, which may be a practical reason that it is used less frequently than the nude mouse for drug screening. Other immunodeficient mice, especially mice with combined mutations, are frequently used to explore the immunological mechanisms underlying tumor progression and compound efficacy (Bankert et al., 2002; Nielsen, 2000; Zheng et al., 1996).

10.1.2 CULTURED TUMOR CELLS VERSUS TUMOR FRAGMENTS

Implantation of cultured tumor cells into immunodeficient mice is widely practiced. In general, cultured tumor cells have a much higher take rate when inoculated as suspension into nude mice than human solid tumors of the same histological type that are transplanted directly from the patient (Giovanella et al., 1991). A wide variety of human cancer cells can be procured from institutions such as American Type Culture Collection (ATCC; www.atcc.org) and European collection of cell cultures (ECACC; www.ecacc.org.uk). Usually, the cryopreserved cell lines are thawed and cultured in medium (e.g., Dulbecco's Modification of Eagle's Medium (DMEM) and Roosevelt Park Memorial Institute medium (RPMI) supplemented with heat-inactivated fetal bovine serum, and expanded until the population is sufficient. Cells are harvested and implanted subcutaneously into various regions (e.g., axillary, flank, or back) of the immunodeficient mouse. Different cell lines require a different cell number inoculum for optimal tumor growth. Between 1 and 5 million cells per mouse is a suitable range for the majority of cell lines. Implanted animals are commonly monitored twice weekly for tumor growth using caliper measurements to determine the length (L), width (W), and height (H) of the tumor. The treatment of the implanted tumors can be started either immediately after inoculation (nonstaged model) or postponed until the tumor reaches a certain size (staged model).

Tumor fragments have also been subpassaged in animals and used as a tool to evaluate anticancer therapeutics by the NCI (Plowman et al., 1997) and others (Fiebig and Berger, 2002). The initial solid tumors established in mice are maintained in the mouse by serial passage of 30- to 40-mg tumor fragments implanted subcutaneously. For each line of tumor, both range and mean values of tumor doubling time are provided to demonstrate the inherent variability of growth. Serial passage is not allowed to exceed a defined range, with replacement starting from the frozen stocks around the 10th generation (Plowman et al., 1997). Issues with the use of tumor fragments include reports indicating contamination with malignant mouse cells, changes in hormone sensitivity, changes in histology patterns, and changes in response to anticancer agents following serial passage in mice (Gao et al., 1999; Horvath et al., 1991; Rydell et al., 1991). Despite these issues, research groups using human tumor xenografts established in serial passage believe that such methodology has a higher correlation with clinical drug response. In addition, serial passage of tumor fragments from clinical specimens also allows for preselection of responsive tumor types for follow-up studies (Fiebig and Berger, 2002).

10.1.3 SUBCUTANEOUS VERSUS ORTHOTOPIC TRANSPLANTATION

The subcutaneous xenograft model is easy to monitor and quantify; however, it is ectopic (i.e., out of the native place). The inhibition of the growth of a tumor implanted in the subcutaneous tissue space after administration of a cytotoxic compound may be a reliable assay for antitumor activity in vivo, but this same assay may be inappropriate to identify agents against cellular target molecules that are expressed only when the tumor resides (orthotopically?) in visceral organs (Fidler, 2001; Killion et al., 1999). To obtain improved models over subcutaneously growing human tumors, there have been efforts (Hoffman, 1999) to develop techniques of surgical orthotopic implantation (SOI) to transplant histologically intact fragments of human cancers, including tumors taken directly from the patient, to the corresponding organ of immunodeficient rodents. SOI models include spontaneous bone metastasis models of prostate cancer, breast cancer, and lung cancer and spontaneous liver and lymph node metastatic models of colon cancer. Comparison of the SOI models with transgenic mouse models of cancer have indicated that the SOI models have more features of clinical metastatic cancer (Hoffman, 1999). Cancer cell lines have been stably transfected with the jellyfish Aéquorea Victoria green fluorescent protein (GFP) to track metastases in the SOI models using bioluminescence imaging techniques. Xenograft tumors seem to yield a much higher frequency of metastases when implanted orthotopically. It is hoped that these models will increasingly be used in preclinical evaluation of potential therapeutics.

One factor in selecting the orthotopic location of the transplanted xenograft is the hormonal dependency of the tumor. Many breast cancer models (e.g., MCF-7, ZR-75-1, Br-10) and prostate cancer models (e.g., LNCaP, CWR-22) depend on estrogen or testosterone and they require supplementation with sex hormones or intact sexual organs to grow in the mouse (Brodie et al., 2003; van Weerden and Romijn, 2000). Usually hormone supplements are given by implanting time-release pellets of hormone (commercially available from, e.g., Innovative Research of America, Sarasota, FL) via trocar needle into the subcutaneous tissue of mice. The hormones in these pellets are designed to be constantly released and can last a defined period of time, from days to weeks. It is believed that it is better to have human breast cancers grow in ovariectomized female mice supplemented with extra estrogen due to the influence of fertility cycles on the primary tumor growth rate and metastasis rate (Bove et al., 2002). One should note, however, that the extra estrogen can cause toxicity or even death to the mouse, so caution and especially close monitoring of the treated mouse (e.g., by physical signs or radioimmunoassays) needs to be exerted. Combinations of castration/ovariectomy and supplementation of testosterone/estrogen in various sequences have been used to mimic hormonal blocking therapies, hormonal supplementation therapies, and different drug- and hormone-resistant conditions (Buhler et al., 2000; Clarke, 1996; Gleave et al., 2001).

10.1.4 TUMOR METASTASIS

It is widely acknowledged that tumors grown subcutaneously are less likely to metastasize than those grown in the anatomically correct or orthotopic site. Various studies have shown that transfection with human angiogenic cytokines and subcutaneous implantation of xenogeneic tumor cells with Matrigel and/or helper host cells such as fibroblasts can increase tumor growth and/or malignant potential (Eccles, 2002; Kleinman et al., 1986; Noel et al., 1993).

The simple "experimental metastasis" assays are intended to mimic the late stage of metastasis (dissemination, extravasation, and colonization). The most commonly used method is the injection of cells into the tail vein of mice. In this case, the resultant tumor colonies are most commonly confined to the lung – which harbors the first capillary bed encountered – due to mechanical trapping. Tumor cells growing in culture frequently have a good plating efficiency and are generally easy to clone; cloning efficiency is thought to be related to the ability to metastasize (Giovanella, 2002). The intravenous inoculation of cultured cell suspensions has been widely used in the hope of obtaining pulmonary metastasis. Curiously, though, this rarely happens, and the majority of cell lines do not give rise to pulmonary metastases even following intravenous injection of several million cells. Only a limited number of cell lines produce metastases, and most of these do so with low frequency. It should be noted that there is no direct correlation between lung colonization and spontaneous metastasis, and in some tumor models there are significant discrepancies between these two functions (Eccles, 2002; Welch, 1997). It is perhaps less surprising that tumor cells derived from cells that are naturally migratory – such as leukemias, lymphomas, and plasmacytomas – more readily metastasize and form colonies in multiple sites, including bone marrow, spleen, and liver.

Other methods have been used to recapitulate the process of metastasis. Cells can be introduced into the portal circulation for liver colonies or the left ventricle of the heart for bone colonies. Inoculation of cultured cells into the left ventricle has proven to be a useful route for determining the specific organotropism of the inoculated cells (Verschraegen et al., 1991). The spleen provides an alternative site for the injection of tumor cells than the portal or mesenteric vein. Injected cells pass almost immediately into the portal circulation. Inoculation of human tumor cell suspension into the mouse spleen can produce liver metastases using tumors (e.g., colorectal carcinoma) selected for liver implantation. A few metastases generated from the initial splenic implantation are isolated and cultured for further rounds of splenic innoculation, selection, and cultivation. Cell lines generated this way will metastasize exclusively to liver and cause significant amount of micrometastases (Potmesil et al., 1995). Tumor cells can be injected directly into the liver parenchyma for circumstances in which a small number of colonies are required (Chen et al., 1998). Tumor cells have also been injected directly into the pleura, peritoneal cavity, bone marrow, and brain; but there is a risk of morbidity and mortality, and quantitation of tumor burden is difficult (Eccles, 2002).

10.1.5 MONITORING TUMOR PROGRESSION AND DETERMINING EFFICACY

The growth of the primary tumor is routinely quantified by in situ caliper measurements of the three perpendicular dimensions. Various formulas are used to determine the volume of the tumor. Two of the most common are L

(length, longest diameter, in millimeters) $\times$ W (width, in millimeters) $\times$ H (height, in millimeters) $\times$ $\pi/6$ or $(L)^2 \times W \times \pi/6$. Both formulas assume that tumors are ellipsoid and both generate values that correlate well with tumor weight measured on necropsy (Euhus et al., 1986). Weight of the tumor can also be converted from measurements of two perpendicular dimensions, L and W, using the formula, $L \times (W)^2 \times 1/2$, for a prolate ellipsoid that assumes a specific gravity of 1.0 g/cm^3 (Geran et al., 1972; Plowman et al., 1997). Parameters commonly used to quantify the effects of the test agents include the following:

- Dose levels.
- Relative growth: percent treated (T)/control (C) (Grever et al., 1992).
- Growth inhibition: 1 − % T/C (Zubrod et al., 1966).
- Growth delay: % (T − C)/C (Zubrod et al., 1966).
- Net log cell kill: [(T − C)−treatment period] × 0.301/median doubling time (Goldin et al., 1979).
- Rate of partial regression (PR) or complete regression (CR) (DeVita et al., 1979).
- Percent tumor-free animals (DeVita et al., 1979).
- Increase in life span (DeVita et al., 1979).

Many researchers use computers in connection with electronic calipers, weighing balances, and telemetry devices to facilitate and automate the operations. There are various computer programs designed for in vivo tumor biology applications available commercially, e.g., *LABCAT* (www.LABCAT.com), *STUDYLOG* (www.studylog.com), and *NVIVO* (www.nvivo.com).

Histology has traditionally been used to study tumor progression and both qualitative and quantitative evaluation of the changes of various tumor markers. Immunological techniques such as ELISA, immunoblotting, immunohistochemical staining, and nucleic acid based techniques such as Southern and northern blottings, and quantitative PCR all have been widely used to study the changes of biological markers in the tumor samples. These techniques have been mostly used as end point assays. It is uncommon to have biopsies done on tumors of preclinical studies. There have been efforts to overcome such limitations by transfecting tumor cell lines that are used to generate grafted tumors with a tumor marker gene (e.g., prostate specific antigen), and then monitoring the growth of the tumor in vivo by quantifying the tumor marker in the serum of the mouse host (Bankert et al., 2002; Conway et al., 2000). However, despite such efforts, more sensitive technology to study tumor progression and metastasis in both the transgenic and xenograft models would be desired.

Recent work has established the feasibility of analyzing tumor growth and growth inhibition in live animals. In vivo imaging of cells tagged with light-emitting probes, such as green fluorescent proteins (GFP) (Hoffman, 2002) or firefly luciferase (Contag et al., 2000), has been shown to be a powerful technology that enables imaging of single tumor cells and metastases and a wide range of other biological characterizations in tumor-bearing mice. GFP- or luciferase-expressing tumors (e.g., colon, prostate, brain, liver, lung, bone, and others) were visualized externally through quantitative transcutaneous whole-body imaging. Since these technologies can be conducted in a real-time fashion on live animals, they provide particularly useful tools to study of the process of tumor progression. For example, direct observation of the metastatic process can be performed using in vivo videomicroscopy, and early steps in the metastatic process related to cell survival or extravasation have been successfully imaged with this technique (Chambers et al., 1998–1999; MacDonald et al., 2002).

Various transparent window preparations offer useful methods to allow noninvasive, continuous measurement of tumor growth, angiogenesis, blood flow, and expression of tagged genes in living tissue (Jain et al., 1997). In addition, tumor tissue oxygen content can be measured using a polarographic needle microelectrode (pO_2-Histograph, Eppendorf, Inc.) to document responses to various anticancer agents (e.g., cyclophosphamide, cisplatin) in combination with agents that aim at reversing hypoxia in tumor tissue (Teicher et al., 1997). Initial efforts are also being made to study tumor responses such as angiogenesis, extracellular volume, and microvascular permeability) to anticancer agents with MRI (Furman-Haran et al., 1998; Leach, 2001), PET (Gupta et al., 2002), ultrasound (Malich et al., 2003), CT (Kennel et al., 2000), and SPECT techniques (Waterhouse et al., 1997). The area of in vivo tumor imaging technology, which offers exciting potential for studies of cancer physiology and treatment regimens, continues to develop quite rapidly.

10.1.6 XENOGRAFT MODELS: PRACTICAL ILLUSTRATIONS

Human tumor xenograft assays were successfully exploited at the NCI to facilitate the discovery of more than a dozen clinically useful cytotoxic anticancer drugs (Plowman et al., 1997). Current anticancer drugs that were evaluated in these systems include melphalan, cytoxan, dacarbazine, BCNU, mitomycin C, cisplatin, actinomycin, doxorubicin, bleomycin, methotrexate, 5-fluorouracil, vinblastine, and paclitaxel. These drugs were evaluated in a panel of tumor lines that belong to different organ or disease types (colon, CNS, leukemia, non-small cell lung, small cell lung, melanoma, and ovary). Drugs were administered intraperitoneally, over different schedules such as:

- Once daily for consecutive 4 days.
- Once daily for consecutive 5 days.
- Once every 4 days, three times.
- Once every 7 days, three times.

At each dose or schedule level, each drug was ranked at one of five levels of efficacy:

- $0 =$ inactive, % T/C $> 40\%$
- $1 =$ tumor inhibition, % T/C range 1 to 40
- $2 =$ tumor stasis, % T/C range 0 to -49
- $3 =$ tumor regression, % T/C range -50 to -100
- $4 =$ tumor regression, % T/C range -50 to -100 with $> 30\%$ tumor-free mice.

Where T/C = change in tumor weight, each treated or control group of mice.

To screen and prioritize compounds for testing in the xenograft models, so-called hollow-fiber assays were used, in which tumor cells are cultured in sealed hollow fibers and implanted either subcutaneously or intraperitoneally in the nude mouse. After drug treatment for 6–8 days, cell survival is quantified by 3-(4,5-dimethyl thiazol-2-yl)-2,5-diphenyl tetrazolium bromide (MTT) dye conversion measurements (Hollingshead et al., 1995). The in vivo drug sensitivity profiles of these human tumor xenografts have served as worldwide benchmarks for the testing of new agents.

More recently, xenograft tumor models have been used to evaluate molecularly targeted therapies. Efforts devoted to such target-oriented drug discovery have produced some fruitful results (Hao and Rowinsky, 2002; Liu et al., 2001). One example of this effort has focused on the *ras* oncoprotein, a critical signal transduction protein that regulates cellular growth and differentiation. Inhibitors of farnesyl transferase, the enzyme that is responsible for posttranslational modification of various cellular proteins, including *ras* proteins, were developed as an indirect strategy to block the function of oncogenic *ras* in tumor cells. Preclinical tests have demonstrated that farnesyl transferase inhibitors (FTIs) can block *ras*-dependent tumor growth (Hao and Rowinsky, 2002). The results of studies of SCH 66336, a potent tricyclic FTI that is currently undergoing Phase III clinical trial evaluations, offers a leading example of the potential of this class of targeted therapeutic compounds (Liu et al., 1998). After pharmacokinetic testing in mice, compounds with superior biochemical and cellular potency were evaluated in NIH 3T3 mouse fibroblasts transformed with activated Ha-*ras* and implanted in the nude mouse. Potent dose-dependent efficacy was observed in this model. Greatly reduced inhibition was seen in cells transformed by the *mos* oncogene, which acts independently of *ras,* supporting the desired target specificity of the inhibitors (Liu et al., 1999). In addition, reduced inhibition was seen in the growth of NIH 3T3 cells transformed by a geranylgeranylated isoform of *ras,* which by virtue of its independence of farnesyl transferase was expected to render cells resistant to the tricyclic inhibitors. Lead compounds were then evaluated in a panel of xenograft models containing various *ras* mutations (H-, N-, K-, or wild type), including cancers of lung (A-549, NCI-H460), colorectal (DLD-1, HCT-116), pancreas (AsPc-1, HPAF-II, Hs700T, MIA Paca), prostate (DU-145), bladder (EJ), and melanoma (LOX). Tricyclic FTIs demonstrated dose-dependent inhibition upon oral dosing in all these models. SCH 66336 displayed efficacy in NOD-SCID mouse models bearing a panel of human astrocytoma explant xenografts (XEN01, XEN05, XEN08), (Feldkamp et al., 2001). Similar results have been reported in solid tumor xenograft models with FTIs of two other chemical classes that are being assessed currently in clinical trials, including R115777 (End et al., 2001) and BMS-214662 (Rose et al., 2001).

Another class of targeted molecular therapeutics that have been evaluated in human tumor xenografts are inhibitors of the epidermal growth factor receptor–tyrosine kinase (EGFR-TK). A variety of human tumor xenografts, including prostate (TSU-PRI, PC-3. DU-145, CWR-22), ovarian (OVCAR-3), breast (MCF-7, ZR-75-1), colon (GEO), vulal (A431), and lung carcinoma (A549, SK-LC-16, LX-1), grown in nude mice, were used to evaluate the EGFR-TK inhibitor (EGFR-TKI) – termed ZD1839, also known as Iressa (Blackledge et al., 2000; Ciardiello and Tortora, 2001; Sirotnak et al., 2002). The EGFR signaling pathway contributes to a number of processes important to tumor progression, including cell proliferation and apoptosis. EGFR is highly expressed in many tumors and is associated with poor disease prognosis. Oral administration of ZD1839 produced dose-dependent reversible growth inhibition in a wide range of tumor xenograft models. ZD1839 has also shown growth inhibitory activity against xenografts initiated from ductal carcinoma in situ tissues, indicating that EGFR inhibition may have a role in the treatment of early stage breast cancer (Morris, 2002).

Human tumor xenograft models have also been used to evaluate the combinatorial efficacy of molecular therapeutics, including FTIs and EGFR-TKIs, in combination with cytotoxic drugs (Liu et al., 1998; Nakamura et al., 2000; Shi et al., 2000) or gene therapeutic agents, such as a p53 recombinant adenovirus (Gurnani et al., 1999). Enhanced in vivo efficacy was observed when SCH 66336 was combined with paclitaxel, cytoxan, 5-Fluorouracil (5-FU), or vincristine in the human non-small cell lung cancer NCI-H460 xenograft model. Significantly greater combined efficacy for SCH 66336 and a recombinant p53 adenovirus was also observed, compared to either agent alone, in both intraperitoneal and subcutaneous DU-145 human prostate xenograft models. The EGFR-TKI ZD1839 in combination with a range of cytotoxic agents has also shown promising activities in several human tumor xenografts. ZD1839 combination therapy was associated with a significant inhibition of tumor growth and a significant increase in survival of nude mice in the GEO colon model, especially when ZD1839 was combined with paclitaxel (Ciardiello et al., 2000). ZD1839 in combination with cytotoxic agents (carboplatin, paclitaxel, or edatrexate) enhanced antitumor activity and in some cases produced tumor regression in nude mice bearing prostate tumor xenografts TSU-PrI and PC-3 (Sirotnak et al., 2000).

These examples illustrate that, in addition to offering an efficacy screening tool, xenograft tumor models can also be used to assess the action of molecular targeted therapeutics. Furthermore, xenograft models can be used not only to evaluate anticancer agents but also to facilitate target validation and proof of principle, by exploring how manipulating the presumptive therapeutic target(s) in tumor cells alters their biological response to such agents.

10.2 Transgenic Tumor Models in Drug Discovery

10.2.1 TARGET SELECTION AND VALIDATION AND PROOF OF PRINCIPLE

The generation of mice with specific genetic lesions are discussed in detail in Chapter 12. Briefly, in the early 1980s, soon after recombinant genes were first introduced into the mouse germ line, the introduction of single oncogenes (e.g., c-*myc* in mammary and SV40 T-Ag in brain) were shown to predispose transgenic mice to cancer, and subsequent work greatly expanded this technology (Adams and Cory, 1991; Van Dyke and Jacks, 2002). In the middle to late 1980s, the discovery of tumor-suppressor genes and the development of the technology to inactivate genes in the mouse germ line through homologous recombination in pluripotent embryonic stem cells brought the generation of a new class of cancer-prone strains of mice. In such knock-out mice, one or both copies of a tumor suppressor gene were mutated in the germ line. As examples of this approach, mice heterozygous for a null Rb allele developed pituitary adenomas and mice lacking p53 were shown to be predisposed to lymphomas and sarcomas (Donehower et al., 1992; Jacks et al., 1992).

Genetically engineered mice have contributed greatly to target selection and validation in recent cancer drug discovery, with hundreds of studies that have reported the alteration of cancer genes throughout the entire organism or in a specific tissue. Transgenic mouse studies have helped define cellular responses to specific genetic changes. Basic mechanisms shown to confer selective advantages to tumor cells include the disruption of major cell cycle regulatory genes resulting in aberrant proliferation (e.g., pRb), interference with cell cycle checkpoints (e.g., p53, ATM) or DNA integrity maintenance genes (e.g., mismatch repair genes, *BRCA1, BRCA2*), and inhibition of apoptotic pathways (e.g., bcl2 overexpression or p53 disruption). Many of transgenic and knock-out strains available for study are cataloged in online databases such as the Induced Mutation Registry Database maintained by the Jackson Laboratory www.jax.org/resources/documents/imr,) the Mouse Knockout and Mutation Database by Biomednet (www.research.bmn.com/mkmd), and the Mouse Repository of the Mouse Models of Human Cancer Consortium database at the National Cancer Institute (web.ncifcrf.gov/researchresources/mmhcc/default.asp). Specific genetic changes and disease phenotypes in these transgenic/knock-out mice greatly enhance the capability of tumor biologists to validate the selected molecular targets and to design proof-of-principle experiments to study the effects of inhibiting these targets.

10.2.2 PROPHYLACTIC AND THERAPEUTIC MODALITIES

Transgenic/knock-out tumor models have specific genetic changes that lead to development of natural or autochthonous tumors with well-defined penetrance and progression characteristics. These characteristics enable utility for testing prophylactic and chemoprevention regimens. For example, p53 knock-out or dominant negative mutant mice (Donehower et al., 1992; Laviguer et al., 1989), Adenomatous Polyposis Coli (APC) mutant mice (e.g., *min,* multiple intestinal neoplasia) (Moser et al., 1995), and Human Papilloma Virus, HPV-16 transgenic mice (Arbeit et al., 1996) have been widely used for carcinogenesis, toxicology, and chemoprevention studies. Other models used in chemoprevention studies include *pim-1*-transgenic mice, in which the onset of ethylnitrosourea-induced T-cell lymphomas are delayed by the synthetic retinoid fenretinide (McCormick et al., 1996), and Transgenic Adenocarcinoma of the Mouse Prostate (TRAMP) mice, in which the onset of prostate cancers driven by tissue-expression expression of SV40 large-T antigen (Gingrich and Greenberg, 1996) are delayed by the NSAID R-flurbiprofen (Wechter et al., 2000). Examples of the use of transgenic tumor models in drug testing in prophylactic and therapeutic regimens are described in the following section.

10.2.3 TRANSGENIC MODELS: PRACTICAL ILLUSTRATIONS

Transgenic/knock-out models are increasingly used in the drug discovery process. FTI testing again offers one illustration of how such models have been used in target-directed drug evaluation. WAP-*ras* transgenic mice carry an activated Ha-*ras* oncogene whose expression is driven by the breast-specific WAP promoter (Andres et al., 1987). Since the Ha-*ras* transgene is carried on the Y chromosome in this strain, only male mice develop mammary or salivary tumors. Both prophylactic and therapeutic regimens were used to evaluate the efficacy of the FTI SCH 66336 (Liu et al., 1998). For prophylactic studies, WAP-*ras* mice were enrolled in drug testing studies when they were 35 days of age, before the onset of spontaneous tumors occurred at between 6 and 9 weeks of age. For therapeutic studies, mice were enrolled on development of a palpable tumor (50 mm^3). In the prophylactic settings, SCH 66336 treatment (oral, BID) delayed tumor onset and reduced the average number of tumors per mouse as well as the average tumor weight. In therapeutic settings, significant tumor regression was seen at the highest dose level, at which animals remained tumor free throughout a significant portion of the 4-week study period after tumor regression occurred. In addition, WAP-*ras* transgenic mice were used to evaluate combination therapy (Liu et al., 1998; Shi et al., 2000). Interestingly, although 200 mg/kg Cytoxan or 10 mg/kg SCH 66336 as single agents did not result in tumor regression, the combination of both treatments resulted in significant tumor regression. Similarly, although the WAP-*ras* model was previously shown to be resistant to paclitaxel (Porter et al., 1995), combination treatment of SCH 66336-sensitized WAP-*ras* mammary tumors to paclitaxel treatment (Shi et al., 2000).

SCH 66336 was also used to treat a mouse model for acute lymphoblastic leukemia (ALL), which is driven by the *bcr-abl* p190 oncoprotein (Reichert et al., 2001). In the early leukemic phase, when circulating *bcr-abl* positive cells were first detected, mice were randomly assigned to vehicle, SCH 66336, or a nontreatment group. All animals in the control groups died of leukemia/lymphoma within a mean period of 103 days. In contrast, 80% of the drug-treated group survived without any signs of leukemia or lymphoma at the termination of treatment, after a median treatment period of 200 days. This study indicated that SCH 66336 could revert early signs of leukemia and significantly prolong survival, a conclusion that corresponded well with findings obtained in a syngeneic model of *bcr-abl*-driven disease (Peters et al., 2001).

Transgenic mice were also used to evaluate the FTI L-744,832 (Barrington et al., 1998; Kohl et al., 1995). In MMTV-Ha-*ras* transgenic mice, this compound demonstrated dose-dependent efficacy with significant tumor regression at higher dose levels. The MMTV-Ha-*ras* transgenic model was also used to determine whether FTI efficacy depended on the presence of functional p53, by interbreeding MMTV-Ha-*ras* mice with p53 knock-out mice to produce offspring that developed *ras*-expressing tumors lacking p53 function. The loss of p53 resulted in greatly accelerated tumorigenesis and more aggressive phenotypes than the tumors arising in p53 wild-type mice. Tumors expressing activated *ras* also underwent significant apoptosis and tumor regression on administration of L-744,832, despite the absence of p53. The fact that tumors from MMTV-Ha-*ras*/p53 null mice responded similarly to the FTI as did p53 wild-type tumors indicated that FTI-induced apoptosis occurred through a p53-independent mechanism. This demonstration illustrates the utility of interbreeding transgenic mice for the purpose of assessing the effects of multiple genetic defects on a drug response. In summary, transgenic and knock-out mice not only can be used as a tool for target validation but can also contribute to predicting which tumor genotypes may be responsive to a specific molecularly-targeted therapy.

10.3 Pros and Cons

No model is ideal. Arguably the most useful models would be those that reflect the natural history and histopathology of human disease, provide more accurate understanding of cellular and molecular mechanisms, and allow for discovery and development of therapeutics that are clinically effective (i.e., models that could predict human clinical response). In the cancer arena, while not all chemotherapeutic agents that test positively in mouse models are efficacious in humans, agents that are efficacious in humans are generally effective in mice (Rosenberg and Bortner, 1999). Both xenograft and transgenic models have strengths and limitations in this regard. A comparison of the main general advantages and disadvantages of each type of model is listed in Table 10.2.

10.3.1 XENOGRAFT MODELS

It is critical for a new drug to define the selectivity profile against human tumors of different organs. In this respect, human tumor xenografts are considered by many as the most relevant models (Fiegbig and Burger, 2002). In addition, xenograft models have been characterized and calibrated by many tumor biologists, and they have been widely used in large-scale screening for many years. Due to their extensive use, both the mouse hosts and the tumor cell lines are generally in the public domain and can be easily obtained. Especially for the subcutaneous models, the methods are relatively easy to set up, less labor-intensive, less costly, and reproducible. Therefore, it is relatively straight-forward to screen a large panel of xenograft models representing various cancers of different organ types to evaluate tissue sensitivity of a particular series of anticancer drugs.

Table 10.2 Comparison of Xenograft (Subcutaneous or Orthotopic) and Transgenic/Knock-Out Mouse Models

Issue	Subcutaneous Xenograft	Orthotopic Xenograft	Transgenic/Knock-Out
Availability	Cell lines easily acquired	Cell line less available	Mice less available
Set up	Relatively easy	Need surgical expertise	Need TG/KO expertise
Expense	Relatively inexpensive	More expensive	More expensive
Labor	Economic	Intensive	Intensive
Experience	Widely used and understood	Accumulating	Accumulating
Progression	Easy to monitor tumors	Not as easily monitored	Not as easily monitored
Immunity	Cannot study	Can not study	Can be studied
Gene expression	Not organ specific	Organ specific	Organ specific
Tumor host	Least relevance	Relevant	Relevant
Metastasis	Lack of natural metastasis	Can be studied	Can be studied
Target Study	Less desirable for validation	Less desirable for validation	Good for validation and proof of principle
Early events	Not well suited	Not well suited	Can be studied
Prophylaxis and prevention	Not well suited	Not well suited	Good for study

Another advantage of xenograft models is reflected in the naturally occurring mutational changes that take place during the process of carcinogenesis. Each cell line likely carries multiple mutations, and each set of mutations differ among cell lines, mirroring the clinical situation. The efficacy results may be more predictive if the anticancer candidates are tested against a panel of tumors within a given histology.

Xenograft systems are not suitable for testing agents that work through immune-based or species-specific mechanisms that involve host cell interactions. Although the implanted cancer cells are of human origin, their growth depends on the mouse host (e.g., blood and nutrient supply, infiltration of extracellular matrix, interactions with host hormones and growth factors). Lack of spontaneous metastasis is another major drawback of xenograft models, although orthotopic implants can address this issue to some extent. However, while orthotopic procedures result in an environment that is more natural for the tumor, increasing the likelihood of metastatic progression, orthotopic models are far more labor intensive and require more surgical expertise. This feature makes it impractical to perform orthotopic assays for large-scale screening.

Incorporation of green fluorescence protein or luciferase into xenografts make these models more powerful, by offering simple methods to monitor even very small and systemically distributed tumors by bioluminescence imaging techniques. It should be noted that, after the establishment of the new tumor cell line with the indicator gene, a thorough characterization needs to be conducted to make sure the new cell line still has the same phenotype as the original line. Although orthotopic xenograft models may be more representative than subcutaneous models, both share an intrinsic problem.

The tumors usually are derived from cell lines that have been selected for in vitro and in vivo growth. They grow rapidly, so they can be used in higher throughput testing in a timely fashion; however, this rapid growth certainly does not reflect the progression of clinical cancers that are typically relatively slow growing. In addition, in vitro establishment of human tumor cells may alter their genetic and biological character. Last, the interactions between xenograft tumor cells and vasculature or stromal microenvironment also are not well defined and may not reflect the interactions that occur during cancer development in human patients. These caveats must be considered when drug response studies are performed xenograft models.

10.3.2 TRANSGENIC MODELS

Transgenic and knock-out mouse models of cancer offer a more natural in vivo course of tumor development relative to xenograft models, where frank tumor cells are introduced into mice. In transgenic and knock-out models, the target molecules and pathways are better defined, thus the early events can be readily followed, making these models better suited to target selection and proof-of-principle studies. This technology also enables tumor biologists to test prophylactic regimens. Compared to xenograft models, tumors formed in transgenic and knockout mice arise spontaneously, in a stochastically fashion, in a natural environment that includes tissue-relevant tumor-stroma interactions and endocrine effects. Tumor development also occurs in an immunocompetent setting that more accurately mimics the conditions of human tumor growth than that modeled by ectopic xenograft tumors. For these reasons, some investigators have argued that the outcome of drug efficacy testing in transgenic and knock-out mice will be more predictive of clinical outcome (Rosenberg and Bortner, 1999).

However, while tumors that develop in transgenic and knock-out models share certain histopathological features of human tumors, the origin is not necessarily the same. Many knock-out mice carry the targeted gene in every cell type, although progress continues to be made on tissue-specific knockout (e.g., using the Cre/LoxP system). However, even in tissue-specific settings knockouts are driven quickly and broadly in a manner that cannot replicate the rare, stochastic gene alteration that precedes the development of human cancer. Similarly, transgenic mice also generally express the transforming oncogene in every cell of a particular lineage in a specific tissue (Rosenberg and Bortner, 1999). Nevertheless, while information gained from transgenic/knock-out studies may be transferable to human settings, one must acknowledge the empirical, species-specific differences that exist in the role of different genes in different cell types, which can lead to different mutant phenotypes in the two species (Jacks, 1996). Indeed, it remains to be proven in many cases that the strong overexpression of a single oncogene (or knockout of a suppressor gene) results in cancers whose biology accurately mimics that of the human diseases, in which multiple genetic and environmental factors contribute to tumor progression (Eccles, 2002). The increasing sophistication being brought to transgenic/knock-out technologies will help address this issue.

There remain, however, issues relating to the comparability of the subtype of tumors that arise in mice and humans, for example, a mouse mammary tumor arising in the alveolar epithelium versus a human ductal carcinoma. It is important to also note that some drugs may have pharmacodynamic or pharmacokinetic properties that will preclude their testing in mouse models. Last, other agents may be highly specific to a human target, making transgenic models unsuitable for testing in certain cases (Van Dyke and Jacks, 2002).

Theoretical issues aside, the main drawbacks of the transgenic systems used for testing targeted therapies are the variable penetrance, the relatively high variability in tumor onset relative to transplantable systems, the long latency, the development of multiple tumors per animal, and the relative inaccessibility of tumors that arise in many models (which may require imaging methods to monitor readily). All these factors necessitate the maintenance and breeding of large animal colonies that must be housed for long periods of time to produce sufficient numbers of animals at appropriate stages of tumorigenesis (Rosenberg and Bortner, 1999). The frequency and site of metastases are also unpredictable (Eccles, 2002). In some cases, tumors can be developed into cell lines and reintroduced into young mice of the transgenic strain or immunodeficient mice, providing it is an inbred genetic background (Eccles, 2002; Nielsen et al., 1994). In such syngeneic grafts, tumors are immunologically similar but may grow more reliably. Such strategies have been used for tumors developing in both the *neu* (c-*erb*-b2) protooncogene-overexpressing MMTV-*neu* and the human carcinoembryonic antigen-overexpressing transgenic strains (Eccles, 2002).

10.4 Pharmacology Issues and Efficacy Prediction

Recent developments in molecular biology and chemistry, such as genomics and bioinformatics, improvements in cloning/expression technologies, structural biology, high-throughput screening, and combinatorial chemistry, have led to a significant increase of new drugable targets. In the year 2002, there were nearly 500 molecules undergoing clinical studies for cancer treatment and this number could well reach 1000 in 2003 (Sikora, 2002). Due to the rapid increase of candidates, new methods and criteria to prioritize the most promising candidates are needed before and during clinical trials. In addition to the traditional Phase I clinical trial goals (maximal tolerated dose determination, dose recommendation for Phase II studies, and safety and pharmacokinetic evaluation), current Phase I studies also need to examine biomarkers and to stratify patients and clinical end points with these biomarkers (Wagner, 2002). Therefore, intense efforts are needed to correlate pharmacokinetic profiles and pharmacodynamic changes in the drug target with preclinical efficacy, such that adequate information can be gained in regard to the problems of dose determination and patient stratification. Current drug-discovery programs commonly incorporate such integrated approaches at early phases of the program.

Pharmacokinetic and pharmacodynamic issues in drug discovery and development are discussed in detail in Chapters 13 and 14. However, a brief consideration of these issues as they relate to the use of xenograft and transgenic models is made here. Since anticancer drugs traditionally have been given to patients by intravenous infusion, in preclinical testing most experimental agents have been given by tail vein or intraperitoneal injections. However, there is a growing trend toward developing anticancer drugs that can be administered via additional routes of administration, in particular by oral routes. Therefore, before in vivo efficacy evaluations are started, pharmacokinetic studies of compounds delivered by various dosing routes and using various formulations should be tested (e.g., saline solution, 0.4% methylcellulose, 5–20% hydroxy-propyl-betacyclodextrin). Different pharmacokinetic parameters for experimental compounds, including half-life, area under the curve (AUC), C_{max}, C_{min}, and bioavailability can all be readily established in mouse models, as discussed further in Chapters 14 and 15. For settings in which constant delivery of a compound is desired, such as through the use of the Alzet Osmotic Pump, a pharamcokinetic profile using appropriately sized pumps and compound formulations is also readily established. To determine the most suitable dose for reaching maximum in vivo efficacy, drug exposures should be compared with IC_{50} and IC_{90} concentrations required to achieve efficacy in cell-based assays. Mouse pharmacokinetic studies also provide important preliminary information for performing subsequent studies in larger animal species (Kim et al., 1999; Liu et al., 1999, 2001).

To correlate pharmacokinetic parameters with efficacy and to help determine clinical dose levels/dose responses more accurately, changes in suitable molecular markers (pharmacodynamic markers) need to be monitored carefully. Taking an example from FTI studies, several farnesylated proteins such as pre-lamin A and HDJ-2 have been used as pharmacodynamic markers of FTI activity in in tumor and normal tissues. Quantifying the inhibition in farnesylation of these markers contributed to an accurate determination of both the concentration and timing of efficacious doses, both preclinically and clinically (Adjei et al., 2000; Britten et al., 2001; Karp et al., 2001). In another example, the EGFR-TKI ZD1839 (Iressa) was used in a series of preclinical studies to surrogate markers of EGFR activity, including EGFR phosphorylation and phosphorylation of the downstream molecules MAPK, AKT, and p27KIP1 (Albanell et al., 2001). In tumor xenograft models (including head and neck carcinoma, gastric adenocarcinoma, and breast adenocarcinoma), a relationship among ZD1839 efficacy, EGFR level, and downstream markers (e.g., phosphorylated MAPK) was established. Preliminary analysis of serial skin biopsies from patients enrolled in Phase I trials confirmed that ZD1839 results in substantial changes in EGFR-dependent molecules such as phosphorylated MAPK and p27KIP1. However, the level of expression of EGFR in cells or tumors was not found to predict sensitivity to ZD1839 in terms of efficacy (Wakeling et al., 2002). Thus additional biomarkers that can specifically indicate ZD1839 sensitivity in terms of efficacy need to be further defined. The advancing developments in genomic, proteomic, molecular biology, chemistry, and imaging tools will help streamline marker identification and quantitation, promoting a more rapid rate of drug testing.

10.5 Future Perspectives

In vivo tumor models have been criticized widely for their inability to predict efficacy in humna clinical trials. However, the perceived failure of animal tumor systems to serve as accurate predictive models for human cancer does not diminish their potential utility, given the absolute necessity of precisely defining the question that the model will be used to answer (Fidler, 1982). As every investigator is aware, the art of choosing an appropriate model follows upon the art of framing the most appropriate question. One should carefully select xenograft or transgenic models that are relevant to the study objective, selecting models that represent the pathological or mechanistic setting that most closely fits the goals of the study. Well-established and/or simpler models still have their place. For example, while a bolus intravenous injection of enzymatically prepared tumor cells may not be appropriate to study the process of metastasis, this approach may be invaluable for comparing the access and activity of drugs or biological agents to the tumor colonies of relatively uniform number, size, and organ location which form (Eccles, 2002).

In the future, we believe it is likely that a combination of transgenic and xenograft models will allow investigators to reach a more complete understanding of tumor inhibition as it relates to humans, thereby facilitating the ability to discover highly disease-specific and efficacious drugs. The transgenic model can be extremely valuable for asking mechanism-related questions; however, its accessibility as a tool for large scale drug discovery remains an issue. In contrast, the xenograft model is less defined in nature but remains more accessible than the transgenic models. With increasing molecular characterization of each tumor line and the use of orthotopic methodology, the xenograft model will continue to contribute centrally to target/disease-oriented cancer research, as well as large scale drug screening. The marriage of mouse tumor models with rapidly evolving methods to profile genetic and epigenetic alterations in tumors, and to finely map genetic modifier loci, will continue to provide insight into the key pathways leading to tumorigenesis. The utility of both types of models continues to offer significant promise for identifying drug targets that are relevant to human cancer treatment (Jackson-Grusby, 2002).

Current anticancer drug discovery is focused on target-oriented and tumor cell-specific approaches, however, there is still an ongoing need for a better understanding of tumor biology. It is widely believed that the successful application of molecular cancer therapeutics will require accurate genetic profiling of tumors as well as the identification of novel, tractable, more promising, and more tumor-type-specific therapeutic targets. Enhanced efforts to identify and use pharmacodynamic markers during tumor progression and treatment would be expected to enable faster, more efficient, and more accurate studies of drug efficacy. Further advancement and availability of in vivo imaging technologies will also be helpful to the design, analysis, and interpretation of results generated from both transgenic and xenograft studies, especially for the detection of early events in tumor progression and metastasis. One benefit of these technologies will also be a reduction in the number of animals needed for experimental paradigms.

It has been proposed that it would be fruitful for the NCI to partner with the FDA and the pharmaceutical industry to play a centralized role in identifying and establishing a standardized set of transgenic and knock-out models of proven reliability and predictability to be used in preclinical cancer efficacy tests (Rosenberg and Bortner, 1999). Collaborations such as The Mouse Models of Human Cancers Consortium (MMHCC) headed by NCI and the collaboration of the NIEDHS with the FDA, International Life Science Institute (ILSI), and several major pharmaceutical companies will be crucial to validate and promote such goals. Progress in these areas are already being made, and new resources including useful Web site references (emice.nci.nih.gov), databases for cancer models (cancermodels.nci.nih.gov), and cancer images (cancerimages.nci.nih.gov) are avaibable to be freely shared by the cancer research community. A broadening of such centralized concerted efforts to include an extensive panel of well-characterized human tumor xenograft models representative of different tissue types would be a welcome addition to the collaborations being pursued. We expect that its inclusion will prove highly beneficial to accelerating successful oncology drug discovery.

References

Adams, J. M., and Cory, S. Transgenic models of tumor development. *Science* **254,** 1161–1167 (1991).

Adams, J. M., Harris, A. W., Pinkert, C. A., et al. The c-*myc* oncogene driven by immunoglobulin enhancers induces lymphoid malignancy in transgenic mice. *Nature* **318,** 533–538 (1985).

Adjei, A. A., Davis, J. N., Erlichman, C., et al. Comparison of potential markers of farnesyltransferase inhibition. *Clin. Cancer Res.* **6,** 2318–2325 (2000).

Albanell, J., Rojo, F., and Baselga, J. Pharmacodynamic studies with the epidermal growth factor receptor tyrosine kinase inhibitor ZD1839. *Semin. Oncol.* **28 (5 suppl 16),** 56–66 (2001).

Amadori, A., Belardelli, F., Cavallo, F., et al. Mind the mouse! A consensus view on the use of immunodeficient mice in immunology and oncology. *J. Immunol. Res.* **4,** 1–5 (1992).

Andres, A.-C., Schonenberger, C.-A., Groner, B., et al. Ha-ras oncogene expression directed by a milk protein gene promoter: Tissue specificity, hormonal regulation, and tumor induction in transgenic mice. *Proc. Natl. Acad. Sci. USA* **84,** 1299–1303 (1987).

Arbeit, J. M., Howley, P. M., and Hanahan, D. Chronic estrogen-induced cervical and vaginal squamous carcinogenesis in human papillomavirus type 16 transgenic mice. *Proc. Natl. Acad. Sci. USA* **93,** 2930–2935 (1996).

Bankert, R. B., Hess, S. D., and Egilmez, N. K. SCID mouse models to study human cancer pathogenesis and approaches to therapy: potential, limitations, and future directions. *Front. Biosci.* **7,** c44–62 (2002).

Barrington, R. E., Subler, M. A., Rands, E., et al. A farnesyltransferase inhibitor induces tumor regression in transgenic mice harboring multiple oncogenic mutations by mediating alterations in both cell cycle control and apoptosis. *Mol. Cell. Biol.* **18,** 85–92 (1998).

Blackledge, G., Averbuch, S., Kay, A., and Barton, J. Anti-EGF receptor therapy. *Prostate Cancer Prostatic Dis.* **3,** 296–302 (2000).

Bove, K., Lincoln, D. W., Wood, P. A., and Hrushesky, W. J. Fertility cycle influence on surgical breast cancer cure. *Breast Cancer Res. Treat.* **75,** 65–72 (2002).

Boyd, M. R. Status of the NCI preclinical antitumor drug discovery screen. In: V. T. DeVita Jr., S. Hellman, and S. A. Rosenberg, eds., *Cancer: Principles and Practice of Oncology, Updates.* Vol. 3. Philadelphia, Lippincott (1989).

Brinster, R. L., Chen, H. Y., Messing, A., et al. Transgenic mice harboring SV40 T-antigen genes develop characteristic brain tumors. *Cell* **37,** 367–379 (1984).

Britten C. D., Rowinsky, E. K., Soignet S, Patnaik A., Yao S. L., Deutsch P., Lee Y., Lobell R. B., Mazina K. E., McCreery H., Pezzuli S. and Spriggs D et al. A phase I and pharmacological study of the farnesyl protein transferase inhibitor L-778,123 in patients with solid malignancies. *Clin. Cancer Res.* **7,** 3894–3903 (2001).

Brodie, A., Jelovac, D., and Long, B. J. Predictions from a preclinical model: Studies of aromatase inhibitors and antiestrogens. *Clin. Cancer Res.* **9,** 455S–459S (2003).

Buhler, K. R., Santucci, R. A., Royai, R. A., et al. Intermittent androgen suppression in the LuCaP 23.12 prostate cancer xenograft model. *Prostate* **43,** 63–70 (2000).

Chambers, A. F., MacDonald, I. C., Schmidt, E. E., et al. Preclinical assessment of anti-cancer therapeutic strategies using in vivo videomicroscopy. *Cancer Metastasis Rev.* **17,** 263–269 (1998–1999).

Chen, L., Ter Haar, G., Hill, C. R., et al. Treatment of implanted liver tumors with focused ultrasound. *Ultrasound Med. Biol.* **24,** 1475–1488 (1998).

Ciardiello, F., Caputo, R., Bianco, R., et al. Antitumor effect and potentiation of cytotoxic drugs activity in human cancer cells by ZD-1839 (Iressa), an epidermal growth factor receptor-selective tyrosine kinase inhibitor. *Clin. Cancer Res.* **6,** 2053–2063 (2000).

Ciardiello, F., and Tortora, G. A novel approach in the treatment of cancer: Targeting the epidermal growth factor receptor. *Clin. Cancer Res.* **7,** 2958–2970 (2001).

Clark, E. A., Shultz, L. D., and Pollack, S. B. Mutations in mice that influence natural killer (NK) cell activity. *Immunogenetics* **12,** 601–613 (1981).

Clarke, R. Human breast cancer cell line xenografts as models of breast cancer: The immunobiologies of recipient mice and the characteristics of several tumorigenic cell lines. *Breast Cancer Res. Treat.* **39,** 69–86 (1996).

Clarke, R., and Dickson, R. B. Animal models of tumor onset, growth, and metastasis. In: J. R. Bertino, eds. *Encyclopedia of Cancer.* New York: Academic Press (1997).

Contag, C. H., Jenkins, K., Contag, P. R., and Negrin, R. S. Use of reporter genes for optical measurements of neoplastic disease in vivo. *Neoplasia* **2,** 41–52 (2000).

Conway, T. F. Jr., Sabel, M. S., Sugano, M., et al. Growth of human tumor xenografts in SCID mice quantified using an immunoassay for tumor marker protein in serum. *J. Immunol. Methods* **233,** 57–65 (2000).

Croy, B. A., Linder, K. E., and Yager, J. A. Primer for non-immunologists on immune-deficient mice and their applications in research. *Comp. Med.* **51,** 300–313 (2001).

Dawe, C. J., and Potter, M. Morphologic and biologica progression of a lymphoid neoplasm of the mouse in vivo and in vitro [Abstract]. *Am. J. Pathol.* **33,** 603 (1957).

DeVita, V. T. Jr., Goldin, A., Oliverio, V. T., et al. The drug development and clinical trials programs of the Division of Cancer Treatment, National Cancer Institute. *Cancer Clin. Trials* **2,** 195–216 (1979).

Donehower, L. A., Harvey, M., Slagele, B. L., et al. Mice deficient for p53 are developmentally normal but susceptible to spontaneous tumours. *Nature* **356,** 215–221 (1992).

Eccles, S. A. Models for evaluation of targeted therapies of metastatic disease. In: B. A. Teicher, ed., *Tumor Models in vancer Research.* Totowa, NJ, Humana (2002).

End, D. W., Smets, G., Todd, A. V., et al. Characterization of the antitumor effects of the selective farnesyl protein transferase inhibitor R115777 in vivo and in vitro. *Cancer Res.* **61,** 131–137 (2001).

Euhus, D. M., Hudd, C., LaRegina, M. C., and Johnson, F. E. Tumor measurement in the nude mouse. *J. Surg. Oncol.* **31,** 229–234 (1986).

Feldkamp, M. M., Lau, N., Roncari, L., and Guha, A. Isotype-specific Ras.GTP-levels predict the efficacy of farnesyl transferase inhibitors against human astrocytomas regardless of Ras mutational status. *Cancer Res.* **61,** 4425–4431 (2001).

Fidler, I. J. The role of host factors and tumor heterogeneity in the testing of therapeutic agents. In: I. J. Fidler, and R. J. White eds., *Design of Models for Testing Cancer Therapeutic Agents.* New York, Van Nostrand Reinhold (1982).

Fidler, I. J. Seed and soil revisited: Contribution of the organ microenvironment to cancer metastasis. *Surg. Oncol. Clin. North Am.* **10,** 257–269 (2001).

Fiebig, H. H., and Burger, A. M. Human tumor xenografts and explants. In: B. A. Teicher, ed., *Tumor Models in Cancer Research.* Totowa NJ, Humana Press (2002).

Fiebig, H. H., Dengler, W. A., and Roth, T. Human tumor xenografts: Predictivity, characterization, and discovery of new-anticancer agents. In: H. H. Fiebig and A. M. Burger, eds., *Relevance of Tumor Models for Anti-Cancer Drug Development. Contrib. Oncol.* Vol. 54. Basel, Karger (1999).

Furman-Haran, E., Grobgeld, D., Margalit, R., and Degani, H. Response of MCF7 human breast cancer to tamoxifen: Evaluation by the three-time-point, contrast-enhanced magnetic resonance imaging method. *Clin. Cancer Res.* **4,** 2299–2304 (1998).

Gao, J., Tombal, B., and Isaacs, J. T. Rapid in situ hybridization technique for detecting malignant mouse cell contamination in human xenograft tissue from nude mice and in vitro cultures from such xenografts. *Prostate* **39,** 67–70 (1999).

Geran, R. I., Greenberg, N. H., Macdonald N. M., and Abbott, B. J. Modified protocol for the testing of new synthetics in the L1210 lymphoid leukemia murine model in the DR&D program, DCT, NCI. *Natl Cancer Inst Monogr.* **45,** 151–153 (1977).

Gingrich, J. R., and Greenberg, N. M. A transgenic mouse prostate cancer model. *Toxicol. Pathol.* **24,** 502–504 (1996).

Giovanella, B. C. Xenotransplantation of human cell cultures in nude mice. In: B. A. Teicher, ed., *Tumor Models in Cancer Research.* Totowa, NJ, Humana (2002).

Giovanella, B. C., Vardeman, D. M., Williams, L. J., et al. Heterotransplantation of human breast carcinomas in nude mice: Correlation between nude takes, poor prognosis and overexpression of the HER-2/neu oncogene. *Int. J. Cancer* **47,** 66–71 (1991).

Giovanella, B. C., Yim, S. O., Stgehlin, J. S., and Williams, L. J. Jr. Development of invasive tumors in the "nude" mouse after injection of cultured human melanoma cells. *J. Natl. Cancer Inst.* **48,** 1531–1533 (1972).

Gleave, M. E., Miyake, H., Zellweger, T., et al. Use of antisense oligonucleotides targeting the antiapoptotic gene, clusterin/testosterone-repressed prostate message 2, to enhance androgen sensitivity and chemosensitivity in prostate cancer. *Urology* **58 (2 suppl 1),** 39–49 (2001).

Goldin, A., Schepartz, S. A., Venditti, J. M., and DeVita, V. T. Jr. Historical development and current strategy of the National Cancer Institute Drug Development Program. In: V. T. DeVita Jr. and H. Busch, eds., *Methods in Cancer Research.* Vol. 16. New York, Academic Press (1979).

Grever, M. R., Schepartz, S. A., and Chabner, B. A. The National Cancer Institute: Cancer drug discovery and development program. *Semin. Oncol.* **19,** 622–638 (1992).

Gupta, N., Price, P. M., and Aboagye, E. O. PET for in vivo pharmacokinetic and pharmacodynamic measurements. *Eur. J. Cancer* **38,** 2094–2107 (2002).

Gura, T. Systems for identifying new drugs are often faulty. *Science* **278,** 1041–1042 (1997).

Gurnani, M., Lipari, P., Dell, J., Shi, et al. Adenovirus-mediated p53 gene therapy has greater efficacy when combined with chemotherapy against human head and neck, ovarian, prostate, and breast cancer. *Cancer Chemother. Pharmacol.* **44,** 143–151 (1999).

Hanahan, D. Heritable formation of pancreatic beta-cell tumours in transgenic mice expressing recombinant insulin/simian virus 40 oncogenes. *Nature* **315,** 115–122 (1985).

Hao, D., and Rowinsky, E. K. Inhibiting signal transduction: Recent advances in the development of receptor tyrosine kinase and Ras inhibitors. *Cancer Invest.* **20,** 387–404 (2002).

Hoffman, R. Green fluorescent protein imaging of tumour growth, metastasis, and angiogenesis in mouse models. *Lancet Oncol.* **3,** 546–556 (2002).

Hoffman, R. M. Orthotopic metastatic mouse models for anticancer drug discovery and evaluation: A bridge to the clinic. *Invest. New Drugs* **17,** 343–359 (1999).

Hollingshead, M. G., Alley, M. C., Camalier, R. F., et al. In vivo cultivation of tumor cells in hollow fibers. *Life Sci.* **57,** 131–141 (1995).

Horvath, G., Ferno, M., Baldetorp, B., et al. Progression of human endometrial adenocarcinoma heterotransplanted into nude mice from hormone-sensitive to hormone resistant growth. *In Vivo* **5,** 185–190 (1991).

Jacks, T. Tumor suppressor gene mutations in mice. *Ann. Rev. Genet.* **30,** 606–636 (1996).

Jacks, T., Fazeli, A., Schmitt, E. M., et al. Effects of an Rb mutation in the mouse. *Nature* **359,** 295–300 (1992).

Jackson-Grusby, L. Modeling cancer in mice. *Oncogene* **21,** 5504–5514 (2002).

Jain, R. K., Schlenger, K., Hockel, M., and Yuan, F. Quantitative angiogenesis assays: Progress and problems. *Nature Med.* **3,** 1203–1208 (1997).

Karp, J. E., Lancet, J. E., Kaufmann, S. H., et al. Clinical and biologic activity of the farnesyltransferase inhibitor R115777 in adults with refractory and relapsed acute leukemias: A phase 1 clinical-laboratory correlative trial. *Blood* **97,** 3361–3369 (2001).

Kennel, S. J., Davis, I. A., Branning, J., et al. High resolution computed tomography and MRI for monitoring lung tumor growth in mice undergoing radioimmunotherapy: Correlation with histology. *Med. Phys.* **27,** 1101–1107 (2000).

Kerbel, R. S. What is the optimal rodent model for anti-tumor drug testing? *Cancer Metastasis Rev.* **17,** 301–304 (1999).

Killion, J. J., Radinsky, R., and Fidler, I. J. Orthotopic models are necessary to predict therapy of transplantable tumors in mice. *Cancer Metastasis Rev.* **17,** 279–284 (1999).

Kim, H., Likhari, P., Lin, C. C., and Nomeir, A. A. High-performance liquid chromatographic analysis of the anti-tumor agent SCH 66336 in cynomolgus monkey plasma and evaluation of its chiral inversion in animals. *J. Chromatogr. B. Biomed. Sci. Appl.* **728,** 133–141 (1999).

Kleinman, H. K., McGarvey, M. L., Hassell, J. R., et al. Basement membrane complexes with biological activity. *Biochemistry* **25,** 312–318 (1986).

Kohl, N. E., Omer, C. A., Conner, M. W., et al. Inhibition of farnesyltransferase induces regression of mammary and salivary carcinomas in ras transgenic mice. *Nature Med.* **1,** 792–797 (1995).

Lavigueur, A., Maltby, V., Mock, D., et al. High incidence of lung, bone, and lymphoid tumors in transgenic mice overexpressing mutant alleles of the p53 oncogene. *Mol. Cell. Biol.* **9,** 3982–3991 (1989).

Law, L. W., Dunn, T. B., Boyle, P. J., and Miller, J. H. Observations on the effect of a follic-acid antagonist on transplantable lymphoid leukemias in mice. *J. Natl. Cancer Inst.* **10,** 179–192 (1949).

Leach, M. O. Application of magnetic resonance imaging to angiogenesis in breast cancer. *Breast Cancer Res.* **3,** 22–27 (2001).

Liu, M., Bishop, W. R., Nielsen, L. L., et al. Orally bioavailable farnesyl transferase inhibitors as anticancer agents in transgenic and xenograft models. *Methods Enzymol.* **333,** 306–318 (2001).

Liu, M., Bryant, M. S., Chen, J., et al. Antitumor activity of SCH 66336, an orally bioavailable tricyclic inhibitor of farnesyl protein transferase, in human tumor xenograft models and wap-ras transgenic mice. *Cancer Res.* **58,** 4947–4956 (1998).

Liu, M., Bryant, M. S., Chen, J., et al. Effects of SCH 59228, an orally bioavailable farnesyl protein transferase inhibitor, on the growth of oncogene-transformed fibroblasts and a human colon carcinoma xenograft in nude mice. *Cancer Chemother. Pharmacol.* **43,** 50–58 (1999).

MacDonald, I. C., Groom, A. C., and Chambers, A. F. Cancer spread and micrometastasis development: Quantitative approaches for in vivo models. *Bioessays* **24,** 885–893 (2002).

Malich, A., Bohm, T., Fritsch, T., et al. Animal-based model to investigate the minimum tumor size detectable with an electrical impedance scanning technique. *Acad. Radiol.* **10,** 37–44 (2003).

McCormick, D. L., Johnson, W. D., Rao, K. V., et al. Comparative activity of N-(4-hydroxyphenyl)-all-trans-retinamide and alpha-difluoromethylornithine as inhibitors of lymphoma induction in PIM transgenic mice. *Carcinogenesis* **17,** 2513–2517 (1996).

Morris, C. The role of EGFR-directed therapy in the treatment of breast cancer. *Breast Cancer Res. Treatment* **75,** S51–55 (2002).

Moser, A. R., Luongo, C., Gould K. A., et al. ApcMin: a mouse model for intestinal and mammary tumorigenesis. *Eur J Cancer* **31A,** 1061–1064 (1995).

Mosier, D. E., Gulizia, R. J., Baird, S. M., and Wilson, D. B. Transfer of a functional human immune system to mice with severe combined immunodeficiency. *Nature* **335,** 256–259 (1988).

Nakamura, K., Yamaguchi, A., Namiki, M., et al. Antitumor activity of ER-51785, a new peptidomimetic inhibitor of farnesyl transferase: Synergistic effect in combination with paclitaxel. *Oncol. Res.* **12,** 477–484 (2000).

Nielsen, L. L. NK cells mediate the anti-tumor effects of E1-deleted, type 5 adenovirus in a human tumor xenograft model. *Oncol. Rep.* **7,** 1191–1196 (2000).

Noel, A., De Pauw-Gillet, M. C., Purnell, G., et al. Enhancement of tumorigenecity of human breast adenocarcinoma cells in nude mice by Matrigel and fibroblasts. *Br. J. Cancer* **68,** 909–915 (1993).

Peters, D. G., Hoover, R. R., Gerlach, M. J., et al. Activity of the farnesyl protein transferase inhibitor SCH66336 against BCR/ABL-induced murine leukemia and primary cells from patients with chronic myeloid leukemia. *Blood* **97,** 1404–1412 (2001).

Plowman, J., Dykes, D. J., Melinda, H., et al. Human tumor xenograft models in NCI drug development. In: B. A. Teicher, ed., *Anticancer Drug Development Guide*. Totowa NJ, Humana (1997).

Porter, G., Armstrong, L., and Nielsen, L. L. A strategy for developing transgenic assays for screening anti-neoplastic drugs which affect tubulin polymerization. *Lab. Animal Sci.* **45,** 145–150 (1995).

Potmesil, M., Vardeman, D., Kozielski, A. J., et al. Growth inhibition of human cancer metastases by camptothecins in newly developed xenograft models. *Cancer Res.* **55,** 5637–5641 (1995).

Reichert, A., Heisterkamp, N., Daley, G. Q., and Groffen, J. Treatment of Bcr/Abl-positive acute lymphoblastic leukemia in P190 transgenic mice with the farnesyl transferase inhibitor SCH66336. *Blood* **97,** 1399–1403 (2001).

Rose, W. C., Lee, F. Y., Fairchild, C. R., et al. Preclinical antitumor activity of BMS-214662, a highly apoptotic and novel farnesyltransferase inhibitor. *Cancer Res.* **61,** 7507–7517 (2001).

Rosenberg, M. P., and Bortner, D. Why transgenic and knockout animal models should be used (for drug efficacy studies in cancer). *Cancer Metastasis Rev.* **17,** 295–299 (1999).

Rydell, R., Lybak, S., Wennerberg, J., and Willen, R. Increased response to cisplatin after long-term serial passage of a squamous cell carcinoma xenograft. *In Vivo* **5,** 23–27 (1991).

Rygaard, J., and Povlsen, C. O. Heterotransplantation of a human malignant tumor to the mouse mutant "nude." *Acta Pathol. Microbiol. Scand.* **77,** 758–760 (1969).

Sausville, E. A., and Feigal, E. Evolving approaches to cancer drug discovery and development at the National Cancer Institute. *Ann. Oncol.* **10,** 1287–1292 (1999).

Schuler, W., and Bosma, M. J. Nature of the scid defect: A defective VDJ recombinase system. *Curr. Top. Microbiol. Immunol.* **152,** 55–62 (1989).

Shi, B., Yaremko, B., Hajian, G., et al. The farnesyl protein transferase inhibitor SCH66336 synergizes with taxanes in vitro and enhances their antitumor activity in vivo. *Cancer Chemother. Pharmacol.* **46,** 387–393 (2000).

Sikora, K. Surrogate endpoints in cancer drug development. *Drug Discovery Today* **7,** 951–956 (2002).

Sirotnak, F. M., She, Y., Lee, F., et al. Studies with CWR22 xenografts in nude mice suggest that ZD1839 may have a role in the treatment of both androgen-dependent and androgen-independent human prostate cancer. *Clin. Cancer Res.* **8,** 3870–3876 (2002).

Sirotnak, F. M., Zakowski, M. F., Miller, V. A., et al. Efficacy of cytotoxic agents against human tumor xenografts is markedly enhanced by coadministration of ZD1839 (Iressa), an inhibitor of EGFR tyrosine kinase. *Clin. Cancer Res.* **6,** 4885–4892 (2000).

Stewart, T. A., Pattengale, P. K., and Leder, P. Spontaneous mammary adenocarcinomas in transgenic mice that carry and express MTV/*myc* fusion genes. *Cell* **38,** 627–637 (1984).

Teicher, B. A. In vivo tumor response end points. In: B. A. Teicher, ed., *Tumor Models in Cancer Research.* Totowa, NJ, Humana Press (2002).

Teicher, B. A., Ara, G., Herbst, R., et al. PEG-hemoglobin: Effects on tumor oxygenation and response to chemotherapy. *In Vivo* **11,** 301–311 (1997).

Van Dyke, T., and Jacks, T. Cancer modeling in the modern era: Progress and challenges. *Cell* **108,** 135–144 (2002).

van Weerden, W. M., and Romijn, J. C. Use of nude mouse xenograft models in prostate cancer research. *Prostate* **43,** 263–271 (2000).

Verschraegen, C., Giovanella, B. C., Mendoza, J. T., et al. Specific organ metatases of human melanoma cells injected into the arterial circulation of nude mice. *Anticancer Res.* **11,** 529–536 (1991).

Wagner, J. A. Overview of biomarkers and surrogate endpoints in drug development. *Dis. Markers* **18,** 41–46 (2002).

Wakeling, A. E., Guy, S. P., Woodburn, J. R., et al. ZD1839 (Iressa): An orally active inhibitor of epidermal growth factor signaling with potential for cancer therapy. *Cancer Res.* **62,** 5749–5754 (2002).

Waterhouse, R. N., Chapman, J., Izard, B., et al. Examination of four 123I-labeled piperidine-based sigma receptor ligands as potential melanoma imaging agents: Initial studies in mouse tumor models. *Nucl. Med. Biol.* **24,** 587–593 (1997).

Wechter, W. J., Leipold, D. D., Murray, E. D. Jr., et al. E-7869 (R-flurbiprofen) inhibits progression of prostate cancer in the TRAMP mouse. *Cancer Res* **60,** 2203–2208 (2000).

Welch, D. R. Technical considerations for studying cancer metastasis in vivo. *Clin. Exp. Metastasis* **15,** 272–306 (1997).

Williams, S. S., Alosco, T. R., Croy, B. A., and Bankert, R. B. The study of human neoplastic disease in severe combined immunodeficient mice. *Lab. Animal Sci.* **43,** 139–146 (1993).

Zheng, L. M., Ojcius, D. M., Garaud, F., et al. Interleukin-10 inhibits tumor metastasis through an NK cell-dependent mechanism. *J. Exp. Med.* **184,** 579–584 (1996).

Zubrod, C. G., Schepartz, S., Leiter, J., et al. The chemotherapy program of the National Cancer Institute: History, analysis, and plans. *Cancer Chemother. Rep.* **50,** 349–540 (1966).

CHAPTER 11

Pharmacodynamic Assays in Cancer Drug Discovery: From Preclinical Validation to Clinical Trial Monitoring

Robert B. Lobell, Nancy E. Kohl, and Laura Sepp-Lorenzino

Molecular Cancer Therapeutics: Strategies for Drug Discovery and Development, edited by George C. Prendergast
ISBN 0-471-43202-4

Within the past decade there has been an explosion in our understanding of the molecular biology of cell growth and survival and how these highly regulated processes are altered in cancer cells. This has resulted in a shift in the development of anticancer drugs from general antiproliferative agents to those that inhibit a specific molecular target thought to play a key role in the development or maintenance of the malignant phenotype. These mechanism-based drugs may not cause toxicity at efficacious doses. Thus the accepted clinical paradigm of dosing to maximum tolerated dose (MTD) may not be optimal for these agents. Instead, the clinical goal becomes identification of a dose of the compound that produces an appropriate level of inhibition of the target. This can be accomplished through the use of pharmacodynamic (PD) assays. In addition to enabling selection of an appropriate clinical dose, PD assays enable fulfillment of a second, equally important goal, demonstration of mechanism of action.

PD assays measure the effect of the inhibitor on the target, either directly or indirectly. This can be accomplished by either comparing the results of the assay in drug-treated versus control-treated subjects or by comparing the assay readout in the same tissue from the same subject pretreatment and post-treatment. The types of PD assays can be roughly divided into two categories: those that measure a biochemical event and those that measure a biological event. A wide range of biochemical parameters have been measured, including levels and subcellular localization of cellular molecules and proteins, posttranslational modification of proteins, and modification of DNA. Techniques that have been used frequently to measure these parameters include western blotting and immunohistochemistry. Reagent availability and the assay throughput required often dictate the choice of technique. The tighter the link between the target and the biochemical event being monitored, the greater the probability that the assay is a reliable measure of the effect of the drug.

PD assays that measure a change in a biological parameter can also be used to assess the activity of an anticancer agent. Examples of biological events that have been measured include cell cycle arrest, apoptosis, and vascular permeability. The recent development of DNA microarrays capable of measuring changes in gene expression has led to the possibility of using a gene signature, a set of genes that are either upregulated or downregulated as a result of modulation of the target, as the basis for a PD assay (Sotiriou et al., 2002). It should be noted that in the past, the term *pharmacodynamics* in the field of oncology has been synonymous with measurement of myelosuppression in response to chemotherapeutic agents (examples of this abound in the literature; Fetterly et al., 2001; Mould et al., 2002). While myelosuppression can be an appropriate PD/- surrogate end point for oncology drugs, in our view, PD end points should be based on biochemical or biological effects more closely linked to drug target inhibition. In general, biological events such as hematopoietic suppression can be more distant from the target than biochemical events, requiring the function of not only the target but also additional proteins to produce the effect. It is critical, therefore, that the monitored event be as directly linked as possible to the drug target. Monitoring more than one biological parameter can increase confidence that the assay accurately reflects modulation of the target.

PD assays can and should be used early in the process of developing a mechanism-based anticancer agent. In an optimal situation, a PD assay is developed for use in in vivo tumor models, usually performed in mice, to demonstrate mechanism of action. In the course of these preclinical studies, the assay is validated. This usually consists of correlating the results from the PD assay with plasma drug levels or with an independent measure of inhibition of the target. Finally, the validated assay is used for dose selection and demonstration of mechanism of action in early clinical trials. Therefore, a PD assay should ideally be amenable to analysis of samples from both mice and humans.

The selection of a source of tissue for the assay deserves careful consideration. The tumor is clearly the most physiologically relevant tissue when evaluating the activity of an anticancer agent. While tumors are generally accessible in tumor-bearing mice, this is often not the case in cancer patients. Further complicating the use of tumor tissue in patients is the need for sequential tumor biopsies in studies that require comparison of pretreatment- and posttreatment results. This has been accomplished in several clinical studies through enrollment of patients with accessible tumors who consented to multiple biopsies (Dowlati et al., 2001a).

An alternative to performing assays on biopsied tumor samples that still allows the assay to be performed on the tumor is the use of noninvasive imaging. MRI and Doppler imaging are examples of two techniques that have been used to measure tumor vascular permeability and tumor blood flow in response to treatment (Drevs et al., 2000, 2002). In addition, PET imaging can be used to detect changes in a tumor, such as changes in glucose uptake and thymidine metabolism (Workman, 1995). The development of the appropriate equipment to be able to apply these techniques to small animals will allow the extrapolation of a validated assay in mice to humans.

A less invasive alternative to the analysis of tumor tissue is the use of nontumor tissue. The most frequently analyzed surrogate tissue used in PD assays is peripheral blood leukocytes, although assays have been developed that use plasma proteins and buccal mucosa cells (Adjei et al., 2000a, 2000b). While these tissues are easily acquired in adequate quantities from patients, they could be limiting in mice. Pooling samples from cohorts of mice for analysis can circumvent the low yield. A central issue in the use of surrogate tissues in PD assays is the ability to predict activity in tumors. Some indication of this can be gained from testing the assay in preclinical in vivo models using both tumor tissue and the surrogate tissue from the same animals. However, in one case in which such a comparison was made in patients, activity in peripheral blood mononuclear cells failed to predict activity in the tumor (Spiro et al., 1999).

The need to tailor a PD assay to the target poses several challenges. Each target or class of target may require its own assay using reagents specific to the target. In addition, knowledge not only of the biochemistry of the target but also of the cellular pathways in which it functions is essential. While this information is available for well-studied cancer targets, it may not be for newer targets, such as those identified by differential expression analyses.

In this chapter, we discuss several novel drug targets for which PD assays have played a critical role in the development of the chemotherapeutic agent, focusing on farnesyl protein transferase inhibitors (FTIs), endothelial growth factor receptor (EGFR) inhibitors (Iressa), the *bcr-abl* inhibitor, Gleevec, and vascular endothelial growth factor (VEGF) receptor inhibitors. Many other excellent examples have been published recently and are described briefly in Table 11.1.

11.1 Prenylation Inhibitors

11.1.1 FARNESYL TRANSFERASE INHIBITORS

FTIs are a class of compounds that began clinical evaluation in 1997. PD assays played an important role throughout the development of these compounds. FTI development programs were initiated with the intention of targeting the *ras* oncogene. The four isoforms of *ras,* Ki4A-*ras,* Ki4B-*ras,* Ha-*ras,* and N-*ras,* are 21-kDa GTP binding proteins that control cell proliferation by transducing signals from extracellular growth factor receptors to downstream effectors, including the *raf* and phosphatidylinositol 3-kinases (PI3Ks). While *ras* is normally activated by GTP binding and deactivated by GTP hydrolysis, approximately 30% of all human cancers have *ras* mutations that inactivate its GTPase activity, giving rise to oncogenic proteins that signal constitutively (Barbacid, 1987; Bos, 1989). Of the ras alleles, Ki-Ras is most commonly mutated in human cancers.

ras is posttranslationally modified at its C-terminus by farnesyl:protein transferase (FPTase), which catalyzes the addition of a farnesyl isoprenoid moiety to a cysteine residue at the C-terminus of the protein within an amino acid sequence known as the CA_1A_2X motif, where C is cysteine, A is typically an aliphatic amino acid, and X is typically serine or methionine. Subsequent to farnesylation, *ras* and other farnesylated proteins are subject to proteolytic cleavage of the AAX residues. Farnesylation is required for anchorage of *ras* at the plasma membrane, and without a C-terminal prenyl moiety, *ras* is incapable of carrying out its normal biological or transforming activities (Barbacid, 1987). Knowledge of the role of farnesylation in *ras* signaling activity stimulated significant interest in developing FPTase inhibitors as a potential treatment for cancer.

The first FTIs were peptidomimetic compounds designed based on the CA_1A_2X motif (James et al., 1993; Kohl et al., 1993). These compounds caused reversion of the transformed morphology of oncogenic Ha-*ras*-transformed rodent fibroblasts and inhibited their proliferation. Importantly, the compounds were shown to inhibit the activity of FPTase in intact cells using assays that were effectively cell culture-PD assays. One method used for demonstrating FPTase inhibition in cells involved labeling with $[^3H]$mevalonolactone, an intermediate in the synthesis of the farnesyl isoprenoid. Using this method, the FTI peptidomimetics were shown to

Table 11.1 Examples of Chemotherapeutic Agents/Molecular Targets Employing PD Assays in the Preclinical/Clinical Development Process[a]

Compound	Molecular Target/ Mechanism of Action	PD Assay/Results	References
FTIs (R115777, SCH66336, L-778,123)	FPTase inhibitors	Assays for inhibition of prenylation of FPTase substrates in PBMC, buccal cell, and leukemic bone marrow (see text)	Adjei et al. (2000), Britten et al. (2001), Karp et al. (2001), Lobell et al. (2002)
Iressa	EGFR kinase inhibitor	Inhibition of phosphorylated EGFR in basal keratinocytes in skin punch biopsies (see text)	Albanell et al. (2002), Baselga et al. (2002).
Gleevec	*bcr-abl* kinase inhibitor	Inhibition of phosphorylation of the *bcr-abl* adaptor protein, CRKL, in peripheral blood samples (see text)	Druker et al. (2001)
SU6668, PTK787	VEGF-R2 (KDR) kinase inhibitors	Decreased vascular permeability in response to inhibitor measured in tumors by noninvasive imaging (DCE-MRI) in preclinical models and in a Phase I trial with PTK787 (see text)	Drevs et al. (2000), Morgan et al. (2001), Pesenti et al. (2002).
Paclitaxel	Microtubule-stabilizing agent (targets mitotic spindle)	Phase I combination paclitaxel/radiotherapy trial; buccal mucosa and tumor biopsies obtained pretreatment and posttreatment; mitotic arrest induced by paclitaxel demonstrated by counting mitotic figures.	Steinberg et al. (1997)
BMS-247550	Microtubule-stabilizing agent (targets mitotic spindle)	BMS-247550 dose- and time-dependent microtubule bundle formation in PBMC demonstrated in Phase I study; drug-dependent bundle formation demonstrated in tumor biopsy from single patient.	McDaid et al. (2002)
PS341	Proteosome inhibitor	Ex-vivo 20S proteosome assay using whole blood lysates from drug-treated patients; dose-dependent inhibition of proteosome activity demonstrated in Phase I study.	Aghajanian et al. (2002)
ZD9331	TS/DNA synthesis inhibitor	ZD9331 caused time-dependent depletion of dTTP pool and elevation of dUMP pool in tumor model. dTTP and dUMP measured in tumor lysates by radioimmunoassay.	Aherne et al. (2001)

Table 11.1 (Continued)

Compound	Molecular Target/ Mechanism of Action	PD Assay/Results	References
Raltitrexed (ZD1694, Tomudex), ZD9331, 5-FU	TS/DNA synthesis inhibitor	TS inhibition causes increase in TS substrate dUMP, and the corresponding nucleoside dUrd; measured plasma levels of dUrd (by HPLC) as a surrogate marker for TS inhibition; used dUrd assay to compare TS inhibitors/dosing regimens.	Ford et al. (2002)
Gemcitabine	Gemcitabine metabolites inhibit ribonucleotide reductase/DNA synthesis	Phase-I trial in AML to optimize Gemcitabine dosing; Gemcitabine treatment caused decrease in dNTP pools and inhibited DNA synthesis in AML cells (measured ex-vivo by [^{3}H]thymidine incorporation)	Gandhi et al. (2002)
SAM486A	S-adenosyl-methionine decarboxylase inhibitor; rate-limiting enzyme in polyamine biosynthesis	Analyzed predose and postdose tumor biopsy from single patient in Phase I study; observed decrease in S-adenosyl-methionine decarboxylase activity and other markers of polyamine biosynthesis	Siu et al. (2002)
TMZ	DNA methylating agent causing O^6-methyl-guanine DNA adducts leading to DNA strand breaks	TMZ-induced DNA adducts removed by DNA repair protein O^6-alkylguanine AGT in a suicide enzymatic reaction; following depletion of AGT, DNA adducts cause mismatch-repair-dependent strand breaks and cytotoxicity; measured AGT depletion by AGT enzyme assay in PBMC ex-vivo as a PD readout for TMZ; Phase I trial comparing TMZ dosing regimens showed that AGT depletion in PBMC did not correlate with AGT depletion in tumor biopsies	Spiro et al. (2001)
BG	Binds and inactivates O^6-alkylguanine AGT; Enhances cytotoxicity of alkylating agents	In clinical trial, determined the dose of BG that causes optimal depletion of AGT in posttreatment tumor biopsies using the AGT depletion assay (see TMZ, above); as in the TMZ trial, found that AGT depletion in PBMC did not predict AGT depletion in tumor	Spiro et al. (1999)

Table 11.1 (Continued)

Compound	Molecular Target/ Mechanism of Action	PD Assay/Results	References
Dacarbazine	DNA methylating agent: produces lesions including O^6-methylguanine and 3-MeA; DNA strand breaks caused by base-excision repair pathway during 3-MeA removal.	Clinical trial using PD assays for measuring DNA strand breaks in response to Dacarbazine treatment; strand breaks in PBL measured by the alkaline comet (electrophoresis) assay; urinary 3-MeA excretion measured by ELISA as readout for repair of DNA methyl adducts; Dacarbazine treatment of 39 melanoma patients resulted in DNA strand breaks and 3-MeA excretion, with wide interpatient variation in both PD parameters; positive correlation between DNA strand breaks and 3-meA excretion	Braybrooke et al. (2000)
PZA	DNA binding agent: inhibits RNA/DNA synthesis, DNA repair; TopoI/II function	Phase I trial; measured PZA-induced DNA fragmentation by pulse-field electrophoresis in bone marrow mononuclear cells before and during infusion (before in vitro studies correlated DNA fragmentation to cell death); dose-dependent DNA fragmentation observed, correlated with extent of myelosuppression	Grem et al. (2002)
Topotecan	Topoisomerase I	Preclinical study comparing topotecan dosing schedules in an ovarian carcinoma xenograft model; measured Topo I activity in ascites tumor using enzyme assay; observed correlation between dosing schedule exhibiting optimal efficacy and level of Topo I inhibition in tumor	Guichard et al. (2001)
Topotecan	Topoisomerase I	Topo I inhibitors stabilize DNA-Topo I adducts; used immunoblotting to quantitate free Topo I and Topo I-DNA complexes. In a Phase II trial, detected DNA-Topo I adducts in tumor and normal mucosa biopsies resulting from Topotecan treatment and negatively correlated free Topo I to topotecan plasma concentrations	Liebes et al. (1998)

Table 11.1 (Continued)

Compound	Molecular Target/ Mechanism of Action	PD Assay/Results	References
Topotecan/ Etoposide	Topoisomerase I (Topotecan) and II (Etoposide)	Phase II trial in NSCLC testing combination of Topo I/II inhibitors; mutational frequency in HPRT locus in PBL, resulting from topotecan treatment measured in ex vivo assay (preclinical studies correlated HPRT mutational frequency with efficacy of Topo inhibitors); the only patient (of 19) exhibiting a tumor response was only one to show a significant change in HPRT mutational frequency	Dowlati et al. (2001)
ISIS 5132 (CGP-69846A)	c- *raf*-1 antisense oligo-deoxynucleotide	Inhibition of c- *raf*-1 gene expression in PBMCs of ISIS5132-treated patients in a Phase I trial was demonstrated using an RT-PCR assay	Stevenson et al. (1999)
Remitogen (Hu1D10)	Monoclonal antibody to cell surface antigen on B cells and B cell lymphomas.	Preclinical study showing that Hu1D10 administration caused B cell depletion in rhesus macaques	Shi et al. (2002)
IL-18	Potential tumor immunotherapy; IL-18 induces TH1 cytokines and enhances NK cell cytolytic activity	Preclinical study exploring changes in cytokine gene expression as PD end points for IL-18 administration; in mouse model, demonstrated induction of IFN-γ and GM-CSF mRNA in splenocytes (by RT-PCR assay) and circulating levels in plasma (by ELISA); human PBMC exhibited similar response to IL-18 ex vivo	Jonak et al. (2002)

[a]KDR; PBMC; CRKL; PBMC; TS; HPLC; TMZ; BG; PZA; 3-MeA; Topo; AML; AGT; PBMC; PBL; IL; TH1; NK; NSCLC; HPRT; IFN-γ; GM-CSF.

inhibit tritium incorporation into farnesylated proteins, including *ras* and the nuclear lamins (James et al., 1993). While this assay was applicable to cell culture, it was difficult to translate into PD readouts suitable for animal studies. Using an immunoblotting method, it was shown that FTIs induced a new species of Ha-*ras* with a molecular weight slightly different from the native protein (Kohl et al., 1993). Similarly, SCH44342, a nonpeptidomimetic compound identified by the Schering-Plough Corporation, induced a molecular weight shift in Ha-*ras* (Bishop et al., 1995). These "band shifted" forms of *ras* produced upon FTI treatment were confirmed to be nonfarnesylated forms of *ras* by Triton X-114 partitioning, a method that distinguishes prenylated and unprenylated proteins based on their partitioning into the detergent phase (Overmeyer and Maltese, 1992). The FTI-induced band shift in Ha-*ras* represented a PD readout, showing that, in cultured cells, the target enzyme was inhibited. These results strengthened the correlation between FPTase inhibition and the antiproliferative effects of FTIs in cell culture. Furthermore, the band-shift observed in Ha-*ras* and other farnesylated proteins represented a PD readout that could monitor FPTase inhibition in animal models and in clinical trials.

Once inhibitor activity has been demonstrated in cultured cells, the typical drug development process involves demonstration of activity in animal models. While FTIs followed this typical route of drug discovery, PD assays did not play a prominent role in the early animal studies with FTIs. For example, in 1994, the Merck peptidomimetic L-739,749 was shown to be active in a nude-mouse xenograft tumor model (Kohl et al., 1994). The compound inhibited tumor growth by Rat1 cells transformed with the Ha-*ras* oncogene, but not by Rat1 cells transformed by v-*raf,* a result consistent with cell culture studies. In a follow-up study, a related Merck peptidomimetic, L-744,832, caused dramatic tumor regression in a Ha-*ras* transgenic mouse-model in which overexpression of oncogenic *ras* causes mammary tumors (Kohl et al., 1995).

Similarly, SCH66336, a compound developed by the Schering-Plough Corporation and tested in clinical trials, was characterized extensively against human tumor xenografts in nude mice and was also shown to cause tumor regression in a Ha-*ras* transgenic mouse-model (Liu et al., 1998). However, none of these studies used PD assays to correlate FPTase inhibition to the inhibition of tumor growth.

Several preclinical in vivo studies with FTIs have reported PD data. One of the first direct demonstrations of FPTase inhibition in a mouse tumor model involved the peptidomimetic inhibitor B956 developed by the Eisai Corporation (Nagasu et al., 1995). In this study, B956 inhibited the growth of xenograft tumors formed by Ha-*ras*-transformed NIH-3T3 cells. Furthermore, B956 inhibited the localization of Ha-*ras* to the plasma membrane, as demonstrated by cellular fractionation studies on tumor lysates. Thus B956 treatment caused the expected biochemical effect in the cell: prevention of Ha-*ras* farnesylation, resulting in the inability of the protein to localize to the plasma membrane. In another study involving xenograft tumors formed by oncogenic Ha-*ras*-transformed NIH-3T3 cells, concentrations of a different FTI, FTI-276, that inhibited tumor growth were shown to inhibit prenylation of oncogenic Ha-*ras* using the band-shift assay (Sun et al., 1995). A third example of preclinical PD monitoring of an FTI involved R115777, an FTI evaluated in clinical trials by Janssen Pharmaceutica. R115777 was evaluated in an efficacy study using the human breast tumor line, MCF7, in the xenograft model (Kelland et al., 2001), and in this case, the farnesylated protein, lamin A, served as a marker of FPTase inhibition. Immunoblotting with an antibody specific for prelamin A, an unprocessed form of the protein that lacks farnesylation and retains the C-terminal CAAX motif, showed that R115777 treatment caused a dose-dependent increase in the accumulation of unfarnesylated prelamin A in tumors. The doses of R11577 that caused the greatest accumulation of prelamin A also caused the greatest antitumor efficacy.

As FTIs proceeded through the preclinical development process, it became apparent that their mechanism of action did not solely depend on inhibition of *ras* processing. In general, cancer cells transformed with oncogenic Ha-*ras* were found to be more sensitive to FTIs than cells harboring oncogenic forms of Ki-*ras* or N-*ras* (Nagasu et al., 1995; Prendergast and Gibbs, 1994; Sepp-Lorenzino et al., 1995).Ultimately, these results were explained by studies showing that FTI treatment inhibits the prenylation of oncogenic Ha-*ras* while Ki-*ras* and N-*ras* remain prenylated in FTI-treated cells (Rowell et al., 1997; Whyte et al., 1997). While all three *ras* isoforms are substrates for FPTase in vitro, Ki-*ras* and N-*ras* are also substrates for geranylgeranyl:protein transferase type I (GGPTase-I), a prenyltransferase related to FPTase (Zhang et al., 1997). Thus while Ki-*ras* and N-*ras* are normally farnesylated in cells, they are subject to cross-prenylation by GGPTase-I in FTI-treated cells (Rowell et al., 1997; Whyte et al., 1997). An oncogenic form of Ki-*ras* with a genetically altered CAAX motif that results exclusively in its geranylgeranylation can transform rodent cells (Cox et al., 1992; Kato et al., 1992), suggesting that geranylgeranylated Ki-*ras* produced upon FTI treatment is still capable of cellular transformation. These findings led to the suggestion that the antiproliferative effect of FTIs against tumor cells harboring oncogenic N-*ras* or Ki-*ras* may be due to inhibition of other farnesylated proteins. RhoB, a *ras*-related protein, has received considerable attention as an FPTase substrate that might be critical to the antiproliferative activity of FTIs (Du and Prendergast, 1999; Du et al., 1999; Prendergast, 2000). CENP-E, a centromere associated kinesin (Ashar et al., 2000; Crespo et al., 2001), and PRL-1/PTP-CAAX, a protein tyrosine phosphatase (Diamond et al., 1994; Cates et al., 1996), are other farnesylated proteins that might also be relevant to the mechanism of action of FTIs.

Since the FPTase substrates critical to the antiproliferative mechanism of action of FTIs were not well defined, it was unclear which farnesylated protein to choose as a PD marker for clinical monitoring. Therefore, we and others followed an approach in which the primary goal of the PD studies was not to demonstrate a correlation between efficacy and inhibition of the chosen FPTase substrate marker but rather to demonstrate conclusively that the target, FPTase, was inhibited. To this end, two farnesylated proteins, human DJ2 protein (HDJ2) and prelamin A were identified as FPTase substrates suitable as PD markers because they are not cross-prenylated by GGPTase-I and their prenylation status can be readily determined by immunoblotting (Adjei et al., 2000a; Lobell et al., 2001).

Clinical trials with R115777 from Janssen Pharmaceutica, SCH66336 from Schering Plough, L-778,123 from Merck, and BMS-214662 from Bristol Meyers Squibb have been reported. Some clinical responses have been observed in early phase clinical trials, though results from later stage trials are still pending (Haluska et al., 2002). In several of these trials, PD assays were employed to monitor FPTase inhibition. While these assays provided definitive evidence of target inhibition in these trials, no clear relationship between efficacy and FPTase inhibition can be drawn at this time.

PD monitoring played an integral role in the design of the clinical trials with the FTI, L-778,123. This compound has dual inhibitory activity against

FPTase and GGPTase-I and was selected, in part, for its ability to inhibit the prenylation of Ki-*ras* (see below). Before initiation of the clinical trials with L-778,123, a PD assay for monitoring FPTase inhibition was validated in animal models (Lobell et al., 2002). This assay monitored FPTase inhibition in peripheral blood mononuclear cells (PBMC) by HDJ2 immunoblotting. Using an FTI related in structure to L-778,123 (Lobell et al., 2002), it was shown that HDJ2 prenylation is inhibited in a dose-dependent manner in white blood cells obtained from FTI-treated mice. In addition, infusion of FTI-1 in mice for up to 2 weeks caused a continuous, stable level of unprenylated HDJ2. Comparable results were seen with PBMCs from dogs treated by continuous infusion with both FTI-1 and L-778,123. By 2 days after discontinuation of the infusion, HDJ2 prenylation returned to predose levels. While these studies did not attempt to correlate HDJ2 prenylation inhibition to antitumor efficacy, the studies did establish that plasma concentrations of both FTI-1 and L-778,123 that caused maximal inhibition of HDJ2 in PBMC were in the range of concentrations required for inhibition of proliferation of a variety of human tumor cell lines in cell culture (unpublished data). Thus it was concluded that HDJ2 prenylation in PBMC provides a reasonable PD marker for FPTase inhibition and that the assay could provide quantitative information suitable for establishment of a dose-response relationship in humans.

The first clinical trial with L-778,123 involved continuous infusion of the compound for 7 days every 3 weeks in patients with advanced solid malignancies, with doses ranging from 35 to 1120 mg/m^2/day (Britten et al., 1999, 2001). The study was designed to determine the maximum tolerated dose of the compound and to determine the extent of inhibition of FPTase as a function of drug dose. Blood samples were drawn before, during, and after infusion for PD analysis. No objective evidence for tumor regression was observed in any of the 25 patients who received L-778,123. Dose-limiting toxicities, consisting of prolongation of the electrocardiographic QTc interval in some patients (toxicity unrelated to the FPTase activity of the compound) as well as grade 4 thrombocytopenia, were observed at the highest dose, 1120 mg/m^2/day. The next lowest dose, 560 mg/m^2/day, was established as the MTD, with mild to moderate myelosuppression observed in 2 of 12 treatment courses. Dose-dependent inhibition of HDJ2 prenylation in PBMC by L-778,123 was observed, with an apparent plateau level of prenylation inhibition observed above the 560 mg/m^2/day dose. The findings of this study guided a second Phase I study where L-778,123 was infused for 2- or 4-weeks (Rubin et al., 2000) at doses ranging from 140 to 840 mg/m^2/day. Here as well, a dose-response relationship between L-778,123 dose and HDJ2 prenylation inhibition in PBMC was evident. Again, no objective tumor responses were noted in this study. Although clinical responses with L-778,123 were not observed in these monotherapy trials with L-778,123, more promising results were observed in a Phase I trial of L-778,123 in combination with radiotherapy (Hahn et al., 2002). In this case, three of six patients with non-small cell lung cancer (NSCLC) had a complete response to treatment. This result suggests that further studies with FTIs in combination with other agents may be warranted.

SCH66336 is an orally administered FTI that has been evaluated in several clinical trials as monotherapy and in combination with other chemotherapeutic agents. Several Phase I studies with SCH66336 involving dose-escalation to determine the MTD and to assess inhibition of FPTase have been reported (Adjei et al., 2000b; Awada et al., 2002). In the first study with SCH66336, gastrointestinal toxicity was dose limiting at 400 mg SCH66336 BID and 1 of 20 patients exhibited a partial response. Inhibition of lamin A processing was followed as a PD marker for FPTase using an immunohistochemical assay for detection of prelamin A in fixed buccal smears obtained predose and 12 h after the final dose. Buccal mucosa was chosen as a source tissue because of its ease of isolation and the rapid turnover of the tissue and also because lamin A could not be detected in PBMC. A dose-dependent trend was observed, with 60, 67, 75, and 100% of the patient samples exhibiting prelamin A in the buccal tissue at the 200, 300, 350, and 400 mg SCH66336 doses, respectively (Adjei et al., 2000b). This result demonstrated that the MTD of SCH66336 occurred in the context of significant inhibition of FPTase. These PD studies served as an important foundation for ongoing clinical studies with this agent.

R115777 has been tested in many clinical trials, both alone and in combination with other agents. One of the most promising FTI clinical studies to date involved the evaluation of R115777 in a Phase I single agent trial in patients with refractory and relapsed acute leukemias (Karp et al., 2001). In this study, patients received oral R115777 twice daily for 21 days at doses ranging from 100 to 1200 mg. Of the 34 patients, 10 responded, including 2 with complete remissions. Complete and partial tumor responses were seen across all dose levels, with no clear dose-response relationship. Dose-limiting toxicity (central neurotoxicity evidenced by ataxia) was observed at the 1200 mg dose, and 7 of 19 patients displayed myelosuppression that was not dose limiting. This study also included several different types of PD analyses. Before dosing, and at day 8 in the dosing regimen, leukemic bone marrow samples were obtained for PD analysis from patients in the 100–900 mg dose groups. In this case, FPTase enzyme activity was assessed using bone marrow cell lysates. Consistent levels of enzyme inhibition ranging from 50 to 95% were observed in patient samples from the 300–900 mg dose groups, but no dose-response relationship was observed. Although this type of direct enzyme assay may be complicated by a variety of technical issues, including dissociation of the drug from the enzyme during the cell isolation procedure, these results provided evidence of FPTase inhibition in the patient samples. Additional PD analyses in this study employing immunoblotting for prelamin A and band shift assays for unprenylated HDJ2 provided further evidence for FPTase inhibition. Working with leukemic bone marrow samples obtained before treatment and from treatment days 8–21, consistent inhibition of HDJ2 was observed in most patients receiving 600 or 900 mg doses but not at lower doses. Results with prelamin A were less interpretable due to its inconsistent detection in pretreatment samples. A third type of PD measurement was made in these clinical samples by monitoring the phosphorylation state of the mitogen-activated protein kinase (MAPK) and extracellular signal-regulated kinase (ERK), via immunoblotting with phospho-ERK-specific antibodies. Since ERKs are phosphorylated in response to a *ras*-mediated signaling pathway, the phosphorylation of these kinases can be sensitive to FPTase inhibition, as indicated by cell culture studies (James et al., 1994; Lerner et al., 1995). Of the 22 patients evaluated, 8 displayed constitutive phosphorylation of ERK, and 4 of these patients displayed decreased ERK phosphorylation in response to R115777 treatment. Decreases in ERK phosphorylation were observed in patient samples at the 300 and 600 mg doses but not at the 100 mg dose.

Taken together, the three types of PD analysis performed in this study provide qualitative evidence for inhibition of FPTase in response to R115777, with PD effects observed in all three types of assays at the 600 mg dose or greater. This study is also significant because malignant cells, and not a surrogate tissue, were used for the PD analysis. However, since clinical responses were observed at all dose levels, the data set from this Phase I trial was insufficient to draw any relationship between efficacy and FPTase inhibition as measured by PD monitoring.

Recently reported abstracts indicate that HDJ2 prenylation in PBMC has been employed as a PD assay in several other clinical trials involving R115777 (Alsina et al., 2002; Kurzrock et al., 2002; Morrow et al., 2002). Of 18 patients with myelodysplastic syndrome 6 responded to treatment with R115777, yet all patients displayed inhibition of HDJ2 prenylation in PBMC lysates. Similarly, 6 of 12 patients with multiple myeloma responded to R115777, while all patients displayed inhibition of HDJ2 prenylation in PBMC lysates.

In summary, the PD analyses with L-778,123, SCH66336, and R115777 indicate that these agents all inhibit their molecular target in their respective clinical studies. These PD assays have been invaluable to the interpretation of the clinical responses, or lack thereof, to these agents. As stated, the complex biological mechanism of action of FTIs made it difficult to identify PD correlates of efficacy before commencement of clinical trials, and it is not surprising that the PD assays that were employed to follow FPTase inhibition in the clinic have not shown correlation to clinical response. Having demonstrated significant FPTase inhibition at the MTD, SCH66336 and R115777 proceeded into further clinical trials and the final outcome of these studies is pending.

11.1.2 FTI-GGTI COMBINATION THERAPY

As mentioned above, the peptidomimetic FTI, L-744,832, and SCH66336 caused dramatic tumor regression in Ha-*ras* transgenic mouse-models (Kohl et al., 1995; Liu et al., 1998). This impressive efficacy was likely due to inhibition of Ha-*ras* prenylation and, therefore, to a direct inhibitory effect on the transforming function of the Ha-*ras* oncoprotein in this model system.

However, in transgenic mouse models where either wild-type N-*ras* or oncogenic Ki-*ras* was overexpressed, L-744,832 inhibited mammary tumor growth but did not cause regression. By band shift assays, it was shown that L-744,832 inhibited the prenylation of Ha-*ras* but not N-*ras* in tumor tissue from N-*ras* transgenic mice (Mangues et al., 1998), and similarly, L-744,832 inhibited the prenylation of HDJ2, but not Ki-*ras* in tumors from Ki-*ras* transgenic mice (Omer et al., 2000). The lack of inhibition of Ki-*ras* and N-*ras* prenylation by L-744,832, and the less dramatic antitumor response in these models is likely due to the cross-prenylation of these oncoproteins by GGPTase-I, as was seen in cell culture models (Rowell et al., 1997; Whyte et al., 1997).

Our laboratory and other investigators at Merck have explored whether cancers involving oncogenic Ki-*ras* can be effectively targeted by inhibition of both FPTase and GGPTase-I. Specific inhibitors of GGPTase-I that could be used in combination with an FTI (Huber et al., 2001) and compounds that have dual inhibitory activity against FPTase and GGPTase-I were identified (Bergman et al., 2001; Lobell et al., 2002; Nguyen et al., 2002). GGPTase-I inhibitors (GGTIs) have also been identified by the laboratory of Sebti and Hamilton (2000) and have been shown to inhibit the growth of tumor lines in culture and in nude mouse xenograft models (Sun et al., 1998, 1999). Band shift studies demonstrated that FTI-GGTI combinations can inhibit the prenylation of Ki-*ras* in cultured cells (Lerner et al., 1997; Sun et al., 1998). Furthermore, combination FTI-GGTI treatment inhibits MAPK signaling in cells that overexpress Ki-*ras* (Mazet et al., 1999) and causes the apoptosis of tumor cells in culture to a greater extent than is seen with either agent alone (Lobell et al., 2001).

The band shift PD assay was critical to the assessment of whether FTI-GGTI combinations can be used to target Ki-*ras* prenylation in animal models (Lobell et al., 2001). Selective FTIs, GGTIs, and dual inhibitors of both enzymes were administered to nude mice by continuous infusion in a xenograft model involving a human pancreatic tumor line expressing an oncogenic form of Ki-*ras*. Prenylation of marker proteins was examined in lysates of both tumor tissue and normal tissue. In addition to studying the inhibition of prenylation of oncogenic and normal Ki-*ras* in this model, HDJ2 prenylation was studied as a marker for FPTase inhibition, and Rap1A prenylation was followed as a marker for GGPTase-I inhibition. As expected, FTI infusion caused inhibition of HDJ2, GGTI infusion caused inhibition of Rap1A prenylation, and co-administration of FTI and GGTI caused the inhibition of prenylation of Ki-*ras* in both normal and tumor tissue. However, GGTIs were found to be poorly tolerated, and extended infusion (for up to 3 days) caused lethality. The lethality of GGTIs was observed in a dose-dependent manner and correlated with the extent of GGPTase-I inhibition as shown by the inhibition of Rap1A prenylation. In addition, the doses of GGTI that were required for inhibition of Ki-*ras* when co-infused with an FTI, caused lethality if infused for 3 days. Dual FTI-GGTI inhibitors also caused lethality when infused for 3 days at doses that caused inhibition of Ki-*ras* prenylation. While we found that shorter infusion (24 h) of a dual FTI-GGTI inhibitor blocked Ki-*ras* prenylation, this treatment was no more efficacious than an FTI at inhibiting growth of the xenograft. By performing a thorough PD analysis of FTI-GGTI combinations, we were able to conclude definitively that while these combinations can inhibit the prenylation, and presumably the function of the intended molecular target (Ki-*ras*), the utility of the combination treatment was limited by toxicities inherent to the inhibition of GGPTase-I.

As mentioned, L-778,123 was evaluated in clinical trials. L-778,123 was selected for clinical evaluation in part because it is a potent FTI and contains GGTI activity sufficient to inhibit the prenylation of Rap1A as well as Ki-*ras* in cell culture models (Bell, 2000; Bergman et al., 2001; Buser et al., 2001; Huber et al., 2001). In a dog model involving continuous infusion, L-778,123 inhibited the prenylation of the FPTase substrate HDJ2, and at high doses, inhibited the prenylation of the GGPTase-I substrate Rap1A in PBMC. However, while Ki-*ras* protein was detectable in the dog PBMC, inhibition of Ki-*ras* prenylation was not observed even at the highest dose (Lobell et al., 2002). In the clinical trials with L-778,123, GGPTase-I inhibition was observed in some patients at the MTD but not at lower doses, as measured by inhibition of Rap1A prenylation in PBMC (Lobell et al., 2002). However, similar to the dog studies, Ki-*ras* prenylation inhibition was not observed in any patient samples. These PD analyses imply that the intended target of L-778,123, Ki-*ras,* was not inhibited in these clinical studies. However, in preclinical mouse models, while we were able to demonstrate inhibition of Ki-*ras* in normal tissues exposed to FTI-GGTI combinations, PBMC were not examined, so it remains possible that PBMC are not a suitable tissue for detecting inhibition of prenylation of Ki-*ras*. Nonetheless, the toxicity observed in preclinical studies with FTI-GGTI combinations suggests that further pursuit of this strategy is not warranted. In summary, it can be seen that PD analyses played a central role in evaluating whether FTI-GGTI combinations could be used for targeting the Ki-*ras* oncoprotein and in reaching the conclusion that the approach is not viable.

11.2 Tyrosine Kinase Inhibitors

The development of several tyrosine kinase inhibitors has also been guided by the use of validated PD assays. Tyrosine kinases are often deregulated in a wide variety of malignancies, either via co-expression of activating ligands, gene amplification, or overexpression; via the acquisition of activating mutations; or via chromosomal translocations (Blume-Jensen and Hunter, 2001). Malignant transformation due to kinase activation depends on the presence of a catalytically active kinase, which led to the development of a number of therapeutic approaches aimed at blocking kinase activity (Fabbro et al., 2002). In this section, we discuss how PD assays have been used in the development of tyrosine kinase inhibitors to determine biological responses and, when appropriate, evaluate mechanisms of resistance.

11.2.1 IRESSA: AN EPIDERMAL GROWTH FACTOR RECEPTOR INHIBITOR

Epidermal growth factor receptor (EGFR) represents an attractive target for drug development. EGFR overexpression and autocrine pathway activation by co-expression of its activator, transforming-growth factor α (TGFα), is frequently observed in many human cancers and represents a negative prognostic indicator for colorectal and gastric cancers (Nicholson et al., 2001; Arteaga, 2002). The agents targeting EGFR that are furthest along in clinical development are the small molecule tyrosine kinase inhibitors Iressa (AstraZeneca) and Tarceva (OSI/Genentech/Roche), and the monoclonal antibody IMC-C225/Erbitux (Imclone/BMS) (Ciardiello and Tortora, 2002).

Iressa, also known as ZD1839 or Gefitinib, is a selective and potent inhibitor of EGFR. (Barker et al., 2001). It blocks receptor autophosphorylation and activation of a proliferative and survival signaling cascade in cell culture, resulting in inhibition of tumor growth and causing regressions in a wide variety of xenograft models (Bianco et al., 2002; Ciardiello et al., 2000; Sirotnak et al., 2000; Wakeling et al., 2002). Iressa was also shown to potentiate the cytotoxic effects of a number of chemotherapeutic agents and of radiation therapy (Bianco et al., 2002; Ciardiello et al., 2000; Sirotnak et al., 2000) and to have antiangiogenic effects (Ciardiello et al., 2001). *her*-2 overexpressing tumors were also shown to be particularly sensitive to this agent and additive effects were observed when given in combination with anti-*her*-2 therapies (Baselga, 2002; Moasser et al., 2001; Moulder et al., 2001; Normanno et al., 2002). Iressa is currently being studied as monotherapy and in combination with various cytotoxic regimes in NSCLC, gastric, colorectal, breast, and hormone-resistant prostate cancer (Baselga and Averbuch, 2000; Ciardiello and Tortora, 2002).

Phase I clinical studies incorporated PD assays to identify an optimal biological dose indicative of EGFR suppression. Although the desired pharmacologic target for Iressa is activated EGFR in the tumor, clinical studies employing repeated tumor biopsies are not practical in most patients in the target cancer populations. Therefore, investigators turned to other tissues, in which EGFR activation could be evaluated more readily. As indicated by its name, EGFR is physiologically expressed in the epidermis, being highly active in the basal layer and in the outer root sheath of hair follicles (Jost et al., 2000). EGFR activation mediates keratinocyte proliferation and survival, important for skin replenishment. As keratinocytes differentiate and migrate toward the more superficial strata, EGFR activity diminishes. Pharmacological suppression of EGFR activity impaired keratinocyte proliferation in culture systems leading to apoptosis and premature terminal differentiation (Peus et al., 1997), recapitulating the skin phenotype observed in mice genetically deficient in EGFR (Miettinen et al., 1995; Threadgill et al., 1995).

Drug effects on keratinocytes were evaluated in skin punches (or biopsies) in cancer patients participating in two Phase I clinical trials (Albanell et al., 2002; Baselga et al., 2002). Skin biopsies were taken from clinically normal skin 2 weeks before initiation of therapy and near the completion a 28-day drug cycle. Specimens were analyzed by immunohistochemistry for activated (phosphorylated, pEGFR) and total EGFR, for EGFR signaling partners, and for proliferative, survival and differentiation markers. Pretreatment pEGFR was highest in basal keratinocytes, which also expressed high levels of pERK, a downstream mediator of EGFR signaling. The basal layer had a high proliferative index, as shown by Ki67 staining, and low expression of the cell cycle inhibitor p27, which was increased in more differentiated suprabasal layers. Treatment with Iressa nearly completely suppressed pEGFR immunoreactivity in basal keratinocytes, without affecting the levels of total EGFR expression. pERK activity and Ki67 were also reduced, and cells showed an increase in expression of p27. This cyclin-dependent kinase inhibitor had been shown to mediate the antiproliferative effects of anti-EGFR agents in both keratinocytes and in tumor cells, and together with pSTAT3, to be critical for keratinocyte differentiation (Hauser et al., 1998; Sano et al., 1999). Furthermore, as was observed in preclinical models, elevated p27 expression in basal keratinocytes from treated patients was also accompanied by high levels of pSTAT3 and by expression of keratin 1, a specific epithelial differentiation marker (Albanell et al., 2002; Baselga et al., 2002). Reduced proliferation rates and premature differentiation of basal keratinocytes resulted in thinning of the uppermost layer of the epidermis, the stratum corneum, a clear histopathological defect observed in treated patients. These cellular changes provide an explanation for the acne-like rash that occurred in 55–65% of the patients (Albanell et al., 2002; Baselga et al., 2002).

These PD studies unambiguously demonstrated that Iressa effectively suppressed EGFR activity in a physiologic cellular target, reproducing the skin phenotype observed in mice deficient in EGFR. However, most of the histological changes were an all-or-none phenomenon, with no detectable dynamic range. The only statistically significant inverse correlates were between Ki67 and dose, and activated pERK and steady-state plasma concentrations. Even at its lowest dose (150 mg/day), Iressa severely blocked EGFR activation in keratinocytes. However, this presumed lack of linearity may be explained by the fact that even at this low dose, the targeted plasma levels were over the IC_{90} for inhibition of EGFR-dependent proliferation in vitro (Herbst et al., 2002; Wakeling et al., 2002).

Although pharmacologic suppression of tumoral pEGFR can only be determined directly via serial biopsy, the fact that clinical benefit was observed at all dose levels and in several tumor types, suggests that skin pEGFR levels may be predictive of tumor target modulation and responsiveness (Baselga et al., 2002; Herbst et al., 2002; Ranson et al., 2002). Ongoing clinical trials include tumoral pEGFR analysis in patients consenting to undergo serial needle biopsies. In preparation for these studies, Rojo et al. (2002) analyzed drug effects on various histological readouts in a mouse xenograft efficacy study and demonstrated that antitumor efficacy with Iressa correlates with inhibition of tumoral pEGFR and pERK and decreases in Ki67 in a dose-dependent manner.

An important lesson learned from Iressa phase I studies was the demonstration that molecular and clinical efficacy can be achieved at doses well below the MTD, which guided phase II dose selection to ensure biological suppression in the setting of improved tolerability. Although trials are ongoing, the appearance of skin reactions is being taken as an indirect measure of mechanism-based biological activity which is also observed with the EGFR kinase inhibitor Tarceva (Hidalgo et al., 2001) and the blocking monoclonal antibody IMC-C225/Erbitux (Busam et al., 2001).

11.2.2 GLEEVEC: A *bcr-abl* AND *kit* INHIBITOR

Gleevec, also known as Imanitib, STI-571 or Glivec (Novartis Pharma), has revolutionized the treatment of chronic myelogenous leukemia (CML) and gastrointestinal stromal tumors (GISTs) and represents the first successful example of targeted therapy (Sawyers, 2002). Gleevec is a selective kinase inhibitor with equipotent inhibitory activity against the platelet-derived growth factor receptor (PDGFR) and the hematopoietic tyrosine kinases *abl* and *kit* (Buchdunger et al., 2002; Druker, 2002; Manley et al., 2002). *abl* and *kit* contribute to the transforming events leading to the pathogenesis of several solid cancers and leukemias (Rubin et al., 2001).

abl is activated in nearly all CML patients as a consequence of a balanced chromosomal translocation between the long arms of chromosomes 9 and 22 in hematopoietic stem cells. As a result of this translocation, also known as the Philadelphia chromosome, a chimeric protein, *bcr-abl,* is expressed in which the tyrosine kinase activity of *abl* is constitutively activated (Sawyers, 1999; Druker, 2002a and 2002b). Expression of *bcr-abl* confers a proliferative and survival advantage to early myeloid precursors, which no longer depend on cytokines and are insensitive to apoptotic stimulation. This population is subject to massive clonal expansion and terminally differentiated neutrophils are released into circulation. Eventually, accumulation of secondary hits leading to progressive loss in differentiation changes the course of the disease. In the accelerated phase there are increasing numbers of circulating immature myeloid and lymphoid cells, which eventually progresses into an acute blastic leukemia (Sawyers, 1999).

Preclinical murine studies evaluated the role of *bcr-abl* in tumorigenesis and demonstrated that catalytically-active *bcr-abl* alone is sufficient to cause CML (Daley et al., 1990; Heisterkamp et al., 1990; Kelliher et al., 1990). Inhibition of *bcr-abl* tyrosine kinase activity by genetic or pharmacological means prevented disease, validating the development of kinase inhibitors for CML (Lugo et al., 1990).

In cell culture, Gleevec exhibited cytotoxic activity against Philadelphia positive CML cells and *bcr-abl*-transfectants under conditions that made growth dependent on the activity of the oncogene (Druker et al., 1996). Exposure to drug resulted in G1 arrest and apoptosis and was accompanied at the molecular level by dephosphorylation of pERK and of the protein adaptor Crk-like (CRKL), a specific substrate of the *bcr-abl* oncogene that couples it to the phosphatidylinositol 3-kinase (PI3K) pathway (Dan et al., 1998; Feller, 2001; Sattler et al., 1996; Senechal et al., 1996). CRKL is the most prominent tyrosine phosphorylated protein detected in circulating tumor cells from CML patients (Nichols et al., 1994; Oda et al., 1994; ten Hoeve et al., 1994). Changes in CRKL tyrosine phosphorylation affect the protein's electrophoretic mobility. These observations provided the basis for the development of a PD assay for clinical use, which unlike solid tumors is easily applicable to peripheral blood (Druker et al., 2001; Gorre and Sawyers, 2002).

Phase I trials of Gleevec started in June 1998 in CML patients in the chronic phase of the disease. CKRL phosphorylation was evaluated in buffy coat lysates from peripheral blood samples taken before and 2 h after the second dose of Gleevec (Druker et al., 2001). PD effects were not observed at the lowest dose levels, but with dose-dependent changes observed from 85 mg to 250–750 mg/day doses. Daily doses > 300 mg showed clinical efficacy; complete hematological responses were observed in 53 of 54 patients, and cytogenetic responses were observed in 29. The drug was well tolerated, and a MTD was not reached. Unlike conventional chemotherapeutic development, safety parameters did not drive Phase II dose selection, but rather, the selected dose was based on PD proof of molecular target modulation that correlated with clinical efficacy.

Responses have been durable in patients with chronic phase CML but not in the advanced stages. PD analyses helped understand the mechanism of clinical relapse. Gorre and Sawyers (2002) conducted a longitudinal study in nine patients who exhibited disease progression and resistance, and they carried out correlative ex vivo studies with the clinical samples. In all nine patients, resistance to Gleevec was associated with loss of inhibition of CRKL phosphorylation, indicative of reactivation of *bcr-abl* signaling. In three patients this was due to overexpression of *bcr-abl*; in the other six a single point mutation (Thr-315-Ile) was selected for that impaired drug binding to *bcr-abl* without affecting catalytic activity (Gorre and Sawyers, 2002; Manley et al., 2002). These results have been confirmed and expanded by others (Hochhaus et al., 2002; Nimmanapalli and Bhalla, 2002), and based on this knowledge, a new generation of *bcr-abl* inhibitors are being developed to target Gleevec-refractory CML (Huang et al., 2002; La Rosee et al., 2002; Warmuth et al., 2003; Wisniewski et al., 2002).

Gleevec is also efficacious in GIST and represents the selected treatment for chemoresistant, inoperable, or metastatic GIST (Dagher et al., 2002; Demetri et al., 2002; van Oosterom et al., 2002). Activating mutations in kit have been detected in the majority of GIST patients, and cell lines carrying these mutations are susceptible to inhibition by Gleevec in vitro (Rubin et al., 2001; Tuveson et al., 2001). Clinical efficacy led to the approval of Gleveec for the treatment of metastatic and/or unresectable GISTs (Dagher et al., 2002; Demetri et al., 2002).

Unlike hematopoietic disorders, solid tumors are difficult to evaluate directly for PD purposes. Nevertheless, serial needle biopsies from a liver metastasis from a GIST patient were obtained pretreatment and 1 and 2 months after daily doses of Gleevec (Joensuu et al., 2001). Treatment reduced the density of tumor cells with clear histopathological features of tumor death, but without signs of inflammation or necrosis. *kit*-expressing tumor cells exhibited a clear decrease in Ki67 staining, whereas endothelial cells – not targeted by Gleevec – showed no signs of damage. These histological changes were correlated with measurements of tumor metabolism as determined by [^{18}fluoro]-2-deoxy-D-glucose (^{18}FDG) (PET). This technique evaluates the rate of anaerobic glycolysis, a hallmark of cancer cell physiology, and constitutes a reliable test to differentiate normal tissue from malignant lesions (Van den Abbeele and Badawi, 2002). Gleevec treatment for 1 month normalized the metabolic rate of metastatic lesions that exhibited ^{18}FDG-uptake indistinguishable from normal tissues. It is interesting that, these metabolic changes preceded anatomic responses evaluated by conventional MRI scanning and CT, which were clearly observed after 8 weeks of treatment (Joensuu et al., 2001; Van den Abbeele and Badawi, 2002). This highlights the value of metabolic PET for early assessment of tumor response.

11.2.3 KDR INHIBITORS: IMAGING TECHNIQUES TO EVALUATE ANGIOGENESIS

Noninvasive imaging techniques, such as MRI, PET, and CT, have become standard tests for diagnosis and disease management by providing morphometric measurements of the tumor. In addition, protocols are being developed to determine pharmacological effects on tumor cell physiology before macroscopic effects are evident, as discussed above for GIST. As discussed here, clinical development of antiangiogenic agents are exploiting the use of imaging techniques to evaluate drug effects on vascular parameters (Libutti et al., 1999; Neeman et al., 2001; Padhani and Husband, 2001).

Small molecule inhibitors and biological agents are under development to impair blood supply to the tumor. Tumor angiogenesis is critical for tumor growth and metastasis. Given that oxygen diffusion in tissues is limited to several hundred micrometers, tumors cannot grow beyond 1 mm in diameter in the absence of new vessel growth. Hypoxia in the tumor leads to expression of angiogenic factors, of which vascular endothelial growth factor (VEGF) mediates the first and committed step in neo-angiogenesis (Ferrara, 2002). Several therapeutic approaches have been developed to target VEGF and its mitogenic receptor VEGF-receptor-2 or kinase-insert domain containing receptor (KDR) (Sepp-Lorenzino and Thomas, 2002).

The net result of targeting the KDR pathway is inhibition of endothelial cell proliferation and survival, which translates into a net decrease in microvascular density after chronic treatment. In addition, inhibition of KDR activity results in decreased vascular permeability, a parameter that can be modulated acutely. Both of these processes can be studied using novel imaging techniques that, in addition to being noninvasive, can evaluate the tumor in its entirety and provide morphometric as well as functional information. One such technique is dynamic contrast-enhanced (DCE) MRI, which involves the rapid administration of a gadolinium-based contrast agent followed by rapid analysis of signal intensity as a function of time. Low molecular weight contrast agents, such as gadopentate dimeglumide (Gd-DTPA), are widely used in the clinic, but due to their small size, they readily extravasate to the interstitial space. For these agents, measurements of perfusion, blood volume, and microvascularity are made immediately after injection while the contrast agent is mainly intravascular. Larger contrast agents are under development to overcome this limitation and were shown to efficiently map microvessel density in preclinical models (van Dijke et al., 1996, 2002).

On the other hand, small contrast agent leakage into the extravascular space can provide a functional readout for vascular permeability. Increased vascular permeability and extravasation of plasma proteins is one of the earliest functional consequences of activation of KDR by VEGF (Dvorak, 2002; Gille et al., 2001; Zeng et al., 2001). VEGF-secreting tumors are very permeable (leaky) and can be easily visualized by DCE-MRI due to tumor accumulation of contrast agent in the tumor interstitial space. Changes in tumor enhancement have been employed as early readouts for KDR suppression by specific kinase inhibitors. In a study reported by Packard et al. (2002) a KDR kinase inhibitor from Merck decreased permeability in rat brain GS-9L tumors detectable as early as 2 h posttreatment. In this study, tumor-bearing animals were imaged before and after drug treatment. Following the algorithms of Tofts and Kermode (1991), contrast agent enhancement was translated into a measure of permeability. The KDR inhibitor caused a statistically significant decrease in permeability that was correlated with inhibition of tyrosine-phosphorylated KDR in animal tissues. Similar results were observed by Pesenti et al. (2002) with the KDR kinase inhibitor SU6668 (Sugen Inc./Pharmacia Inc.) in HT29 human colon carcinoma xenografts by DCE-MRI using albumin-Gd-DTPA to quantify fractional plasma volume and endothelial permeability. DCE-MRI performed before and 1–14 days after treatment with SU6668 at doses that inhibit tumor growth, showed a decrease in endothelial permeability at 24 h of treatment.

Novartis Pharma AG validated DCE-MRI in a mouse model of renal cell carcinoma (Drevs et al., 2000, 2002) and used changes in vascular permeability as a measure of molecular efficacy in a phase I study with PTK787 (Morgan et al., 2001). Of 46 patients enrolled in two Phase I trials 39 were monitored by DCE-MRI, and 29% exhibited dose-dependent decreases in permeability ranging from 29 to 58%, after 48 h of treatment. Changes in permeability were correlated with dose and pharmacokinetics, and in a subset of colon cancer patients with liver metastases, with clinical response. Although these are early results from a small number of patients, they lend support to the continued use of noninvasive imaging techniques to quantify angiogenic activity in tumors.

11.3 Summary

PD assays measuring drug target inhibition are critical to the development of novel chemotherapeutics directed against molecular targets and pathways. These assays facilitate the drug-discovery process and guide decisions in both the preclinical and clinical settings. Preclinically, PD assays can identify compounds with chemical properties suitable for target inhibition in animal models, and can provide direct evidence that the desired biological response, typically antitumor activity, is due to target inhibition and not to unknown off-target activities in the test compound. Similarly, PD assays can help determine whether toxicities observed in preclinical models are linked to target inhibition. This was seen in the evaluation of FTI-GGTI combination therapy targeting Ki-*ras,* in which PD assay directly linked toxic effects in mice with GGPTase-I inhibition in vivo. Clinically, deployment of PD assays in Phase I dose-escalation trials can be critical in the selection of appropriate doses for later stage trials. This is particularly apparent in the case of Gleevec and Iressa, with which effective target modulation occurred at drug levels that were below the MTD. Furthermore, PD assays can be invaluable in the early phases of clinical development when efficacy data may be limited. This is exemplified by the FTIs, for which definitive PD data showing inhibition of the drug target gave clinical investigators the confidence to proceed with further trials knowing that the administered dose is indeed modulating the target. PD assays may rely on standard methods, including immunoblotting, and employ tissues, such as PBMC, that can be readily obtained. In the future, newer technologies (such as gene expression profiling and imaging methods that permit measurement of target inhibition in the tumor without the need for invasive biopsies) may form the basis of PD assays that guide the selection of the next generation of chemotherapeutic agents.

References

Adjei, A. A., Davis, J. N., Erlichman, C., et al. Comparison of potential markers of farnesyltransferase inhibition. *Clin. Cancer Res.* **6,** 2318–2325 (2000a).

Adjei, A.A., Erlichman, C., Davis, J. N., et al. A phase I trial of the farnesyl transferase inhibitor SCH66336: Evidence for biological and clinical activity. *Cancer Res.* **60,** 1871–1877 (2000b).

Aghajanian, C., Soignet, S., Dizon, D. S., et al. A phase I trial of the novel proteasome inhibitor PS341 in advanced solid tumor malignancies. *Clin. Cancer Res.* **80,** 2505–2511 (2002).

Aherne, G. W., Hardcastle, A., Ward, E., et al. Pharmacokinetic/pharmacodynamic study of ZD9331, a nonpolyglutamatable inhibitor of thymidylate synthase, in a murine model following two curative administration schedules. *Clin. Cancer Res.* **7,** 2923–2930 (2001).

Albanell, J., Rojo, F., Averbuch, S., et al. J. Pharmacodynamic studies of the epidermal growth factor receptor inhibitor ZD1839 in skin from cancer patients: Histopathologic and molecular consequences of receptor inhibition. *J. Clin. Oncol.* **20,** 110–124 (2002).

Alsina, M., Overton, R., Belle, N., et al. Farnesyl transferase inhibitor FTI-R115777 is well tolerated, induces stabilization of disease and inhibits farnesylation and oncogenic/tumor survival pathways in patients with advanced multiple myeloma (Abstract). *Proc. Am. Assoc. Cancer Res.* **43,** 1000 (2002).

Arteaga, C. L. Overview of epidermal growth factor receptor biology and its role as a therapeutic target in human neoplasia. *Semin. Oncol.* **29,** 3–9 (2002).

Ashar, H. R., James, L., Gray, K., et al. Farnesyl transferase inhibitors block the farnesylation of CENP-E and CENP-F and alter the association of CENP-E with the microtubules. *J. Biol. Chem.* **275,** 30451–30457 (2000).

Awada, A., Eskens, F. A. L. M., Piccart, M., et al. Phase I and pharmacological study of the oral farnesyltransferase inhibitor SCH66336 given once daily to patients with advanced solid tumours. *Eur. J. Cancer* **38,** 2272–2278 (2002).

Barbacid, M. *Ras* genes. *Ann. Rev. Biochem.* **56,** 779–827 (1987).

Barker, A. J., Gibson, K. H., Grundy, W., et al. Studies leading to the identification of ZD1839 (Iressa): An orally active, selective epidermal growth factor receptor tyrosine kinase inhibitor targeted to the treatment of cancer. *Bioorg. Med. Chem. Lett.* **11,** 1911–1914 (2001).

Baselga, J. Combined anti-egf receptor and anti-Her2 receptor therapy in breast cancer: A promising strategy ready for clinical testing. *Ann. Oncol.* **13,** 8–9 (2002).

Baselga, J., and Averbuch, S. D. ZD1839 (Iressa) as an anticancer agent. *Drugs* **60,** 33–40 (2002).

Baselga, J., Rischin, D., Ranson, M., et al. Phase I safety, pharmacokinetic, and pharmacodynamic trial of ZD1839, a selective oral epidermal growth factor receptor tyrosine kinase inhibitor, in patients with five selected solid tumor types. *J. Clin. Oncol.* **20,** 4292–4302 (2002).

Bell, I. M. Inhibitors of protein prenylation 2000. *Exp. Opin. Ther. Patents* **10,** 1813–1831 (2000).

Bergman, J. M., Abrams, M. T., Davide, J. P., et al. Aryloxy substituted *n*-arylpiperazinones as dual inhibitors of farnesyltransferase and geranylgeranyltransferase-I. *Biorg. Med. Chem. Lett.* **11,** 1411–1415 (2001).

Bianco, C., Tortora, G., Bianco, R., et al. Enhancement of antitumor activity of ionizing radiation by combined treatment with the selective epidermal growth factor receptor-tyrosine kinase inhibitor ZD1839 (Iressa). *Clin. Cancer Res.* **8,** 3250–3258 (2002).

Bishop, W. R., Bond, R., Petrin, J., et al. Novel tricyclic inhibitors of farnesyl protein transferase. *J. Biol. Chem.* **270,** 30611–30618 (1995).

Blume-Jensen, P., and Hunter, T. Oncogenic kinase signaling. Nature **411,** 355–365 (2001).

Bos, J. L. *Ras* oncogenes in human cancer: A review. *Cancer Res.* **49,** 4682–4689 (1989).

Braybrooke, J. P., Houlbrook, S., Crawley, J. E., et al. Evaluation of the alkaline comet assay and urinary 3-methyladenine excretion for monitoring DNA damage in melanoma patients treated with dacarbazine and tamoxifen. *Cancer Chemother. Pharmacol.* **45,** 111–119 (2000).

Britten, C. D., Rowinsky, E. K., Soignet, S., et al. A phase I and pharmacologic study of the farnesyl protein transferase inhibitor L-778,123 in patients with solid malignancies. *Clin. Cancer Res.* **7,** 3894–3903 (2001).

Britten, C. D., Rowinsky, E. K., Yao, S.-L., et al. The farnesyl protein transferase (FPTase) inhibitor L-778,123 in patients with solid cancers. *Proc. Am. Soc. Clin. Oncol.* **18,** 155A (1999).

Buchdunger, E., O'Reilly, T., and Wood, J. Pharmacology of Imatinib (STI571). *Eur. J. Cancer* **38 (suppl 5),** S28–S36 (2002).

Busam, K. J., Capodieci, P., Motzer, R., et al. Cutaneous side-effects in cancer patients treated with the antiepidermal growth factor receptor antibody C225. *Br. J. Dermatol.* **144,** 1169–1176 (2001).

Buser, C. A., Dinsmore, C. J., Fernandes, C., et al. High-performance liquid chromatography/mass spectrometry characterization of Ki4b-Ras in PSN-1 cells treated with the prenyltransferase inhibitor L-778,123. *Anal. Biochem.* **290,** 126–137 (2001).

Cates, C. A., Michael, R. L., Stayrook, K. R., et al. Prenylation of oncogenic human PTP-CAAX protein tyrosine phosphatases. *Cancer Lett.* **110,** 49–55 (1996).

Ciardiello, F., Caputo, R., Bianco, R., et al. Antitumor effect and potentiation of cytotoxic drugs activity in human cancer cells by ZD-1839 (Iressa), an epidermal growth factor receptor-selective tyrosine kinase inhibitor. *Clin. Cancer Res.* **6,** 2053–2063 (2000).

Ciardiello, F., Caputo, R., Bianco, R., et al. Inhibition of growth factor production and angiogenesis in human cancer cells by ZD1839 (Iressa), a selective epidermal growth factor receptor tyrosine kinase inhibitor. *Clin Cancer Res.* **7,** 1459–1465 (2001).

Ciardiello, F., and Tortora, G. Anti-epidermal growth factor receptor drugs in cancer therapy. *Expert Opin. Investig. Drugs* **11,** 755–768 (2002).

Cox, A. D., Hisaka, M. M., Buss, J. E., and Der, C. J. Specific isoprenoid modification is required for function of normal, but not oncogenic, ras protein. *Mol. Cell. Biol.* **12,** 2606–2615 (1992).

Crespo, N. C., Ohkanda, J., Yen, T., et al. The farnesyltransferase inhibitor, FTI-2153, blocks bipolar spindle formation and chromosome alignment and causes prometaphase accumulation during mitosis of human lung cancer cells. *J. Biol. Chem.* **276,** 16161–16167 (2001).

Dagher, R., Cohen, M., Williams, G., et al. Approval summary: Imatinib mesylate in the treatment of metastatic and/or unresectable malignant gastrointestinal stromal tumors. *Clin. Cancer Res.* **8,** 3034–3038 (2002).

Daley, G. Q., Van Etten, R. A., and Baltimore, D. Induction of chronic myelogenous leukemia in mice by the p210*bcr-abl* gene of the Philadelphia chromosome. *Science* **247,** 824–830 (1990).

Dan, S., Naito, M., and Tsuruo, T. Selective induction of apoptosis in Philadelphia chromosome-positive chronic myelogenous leukemia cells by an inhibitor of Bcr-Abl tyrosine kinase, CGP57148. *Cell Death Differ.* **5,** 710–715 (1998).

Demetri, G. D., von Mehren, M., Blanke, C. D., et al. Efficacy and safety of Imatinib Mesylate in advanced gastrointestinal stromal tumors. *N. Engl. J. Med.* **347,** 472–480 (2002).

Diamond, R. H., Cressman, D. E., Laz, T. M., et al. PRL-1, a unique nuclear protein tyrosine phosphatase, affects cell growth. *Mol. Cell. Biol.* **14,** 3752–3762 (1994).

Dowlati, A., Haaga, J., Remick, S. C., et al. Sequential tumor biopsies in early phase clinical trials of anticancer agents for pharmacodynamic evaluation. *Clin. Cancer Res.* **7,** 2971–2976 (2001a).

Dowlati, A., Levitan, N., Gordon, N. H., et al. Phase II and pharmacokinetic/pharmacodynamic trial of sequential topoisomerase I and II inhibition with topotecan and etoposide in advanced non-small-cell lung cancer. *Cancer Chemother. Pharmacol.* **47,** 141–148 (2001).

Drevs, J., Hofmann, I., Hugenschmidt, H., et al. Effects of PTK787/ZK 222584, a specific inhibitor of vascular endothelial growth factor receptor tyrosine kinases, on primary tumor, metastasis, vessel density, and blood flow in a murine renal cell carcinoma model. *Cancer Res.* **60,** 4819–4824 (2000).

Drevs, J., Muller-Driver, R., Wittig, C., et al. PTK787/ZK222584, a specific vascular endothelial growth factor-receptor tyrosine kinase inhibitor, affects the anatomy of the tumor vascular bed and the functional vascular properties as detected by dynamic enhanced magnetic resonance imaging. *Cancer Res.* **62,** 4015–4022 (2002).

Druker, B. J. Imatinib and chronic myeloid leukemia: Validating the promise of molecularly targeted therapy. *Eur. J. Cancer* **38 (suppl 5),** S70–S76 (2002a).

Druker, B. J. Inhibition of the Bcr-Abl tyrosine kinase as a therapeutic strategy for CML. *Oncogene* **21,** 8541–8546 (2002b).

Druker, B. J., Talpaz, M., Resta, D. J., et al. Efficacy and safety of a specific inhibitor of the Bcr-Abl tyrosine kinase in chronic myeloid leukemia. *N. Engl. J. Med.* **344,** 1031–1017 (2001).
Druker, B. J., Tamura, S., Buchdunger, E., et al. Effects of a selective inhibitor of the Abl tyrosine kinase on the growth of Bcr-Abl positive cells. *Nature Med.* **2,** 561–566 (1996).
Du, W., Lebowitz, P. F., and Prendergast, G. C. Cell growth inhibition by farnesyltransferase inhibitors is mediated by gain of geranylgeranylated RhoB. *Mol. Cell. Biol.* **19,** 1831–1840 (1999).
Du, W., and Prendergast, G. C. Geranylgeranylated RhoB mediates suppression of human tumor cell growth by farnesyltransferase inhibitors. *Cancer Res.* **59,** 5492–5496 (1999).
Dvorak, H. F. Vascular permeability factor/vascular endothelial growth factor: A critical cytokine in tumor angiogenesis and a potential target for diagnosis and therapy. *J. Clin. Oncol.* **20,** 4368–4380 (2002).
Fabbro, D., Ruetz, S., Buchdunger, E., et al. Protein kinases as targets for anticancer agents: From inhibitors to useful drugs. *Pharmacol. Ther.* **93,** 79–98 (2002).
Feller, S. M. CRK family adaptors-signalling complex formation and biological roles. *Oncogene* **20,** 6348–6371 (2001).
Ferrara, N. Role of vascular endothelial growth factor in physiologic and pathologic angiogenesis: Therapeutic implications. *Semin. Oncol.* **29,** 4–10 (2002).
Fetterly, G. J., Tamburlin, J. M., and Straubinger, R. M. Paclitaxel pharmacodynamics: Application of a mechanism-based neutropenia model. *Biopharm. Drug Dispos.* **22,** 251–261 (2001).
Ford, H. E. R., Mitchell, F., Cunningham, D., et al. Patterns of elevation of plasma 2′-deoxyuridine, a surrogate marker of thymidylate synthase (TS) inhibition, after administration of two different schedules of 5-fluorouracil and the specific TS inhibitors Raltitrexed (Tomudex) and ZD9331. *Clin. Cancer Res.* **8,** 103–109 (2002).
Gandhi, V., Plunkett, W., Du, M., et al. Prolonged infusion of Gemcitabine: Clinical and pharmacodynamic studies during a phase I trial in relapsed acute myelogenous leukemia. *J. Clin. Oncol.* **20,** 665–673 (2002).
Gille, H., Kowalski, J., Li, B., et al. Analysis of biological effects and signaling properties of Flt-1 (VEGFR-1) and KDR (VEGFR-2). A reassessment using novel receptor-specific vascular endothelial growth factor mutants. *J. Biol. Chem.* **276,** 3222–3230 (2001).
Gorre, M. E., and Sawyers, C. L. Molecular mechanisms of resistance to STI571 in chronic myeloid leukemia. *Curr. Opin. Hematol.* **9,** 303–307 (2002).
Grem, J. L., Harold, N., Keith, B., et al. A phase I pharmacologic and pharmacodynamic study of pyrazoloacridine given as a weekly 24-hour continuous intravenous infusion in adult cancer patients. *Clin Cancer Res.* **8,** 2149–2156 (2002).
Guichard, S., Montazeri, A., Chatelut, E., Hennebelle, I., Bugat, R., and Canal, P. Schedule-dependent activity of Topotecan in OVCAR-3 ovarian carcinoma xenograft: Pharmacokinetic and pharmacodynamic evaluation. *Clin. Cancer Res.* **7,** 3222–3228 (2001).
Haluska, P., Dy, G. K., and Adjei, A. A. Farnesyl transferase inhibitors as anticancer agents. *Eur. J. Cancer* **38,** 1685–1700 (2002).
Hahn, S. M., Bernhard, E. J., Regine, W., et al. A phase I trial of the farnesyltransferase inhibitor L-778,123 and radiotherapy for locally advanced lung and head and neck cancer. *Clin. Cancer Res.* **8,** 1065–1072 (2002).
Hauser, P. J., Agrawal, D., Hackney, J., and Pledger, W. J. Stat3 activation accompanies keratinocyte differentiation. *Cell Growth Differ.* **9,** 847–855 (1998).
Heisterkamp, N., Jenster, G., ten Hoeve, J., et al. Acute leukaemia in *bcr-abl* transgenic mice. *Nature* **344,** 251–253 (1990).
Herbst, R. S., Maddox, A. M., Rothenberg, M. L., et al. Selective oral epidermal growth factor receptor tyrosine kinase inhibitor ZD1839 is generally well-tolerated and has activity in non-small-cell lung cancer and other solid tumors: Results of a phase I trial. *J. Clin. Oncol.* **20,** 3815–3825 (2002).
Hidalgo, M., Siu, L. L., Nemunaitis, J., et al. Phase I and pharmacologic study of OSI-774, an epidermal growth factor receptor tyrosine kinase inhibitor, in patients with advanced solid malignancies. *J. Clin. Oncol.* **19,** 3267–3279 (2001).
Hochhaus, A., Kreil, S., Corbin, A. S., et al. Molecular and chromosomal mechanisms of resistance to Imatinib (STI571) therapy. *Leukemia* **16,** 2190–2196 (2002).
Huang, M., Dorsey, J. F., Epling-Burnette, P. K., et al. Inhibition of Bcr-Abl kinase activity by PD180970 blocks constitutive activation of Stat5 and growth of CML cells. *Oncogene* **21,** 8804–8816 (2002).
Huber, H. E., Robinson, R. G., Watkins, A., et al. Anions modulate the potency of geranylgeranyl-protein transferase I inhibitors. *J. Biol. Chem.* **276,** 24457–24465 (2001).
James, G. L., Brown, M. S., Cobb, M. H., and Goldstein, J. L. Benzodiazepine peptidomimetic BZA-5B interrupts the map kinase activation pathway in H-*ras*-transformed Rat-1 cells, but not in untransformed cells. *J. Biol. Chem.* **269,** 27705–27714 (1994).

James, G. L., Goldstein, J. L., Brown, M. S., et al. Benzodiazepine peptidomimetics: Potent inhibitors of Ras farnesylation in animal cells. *Science* **260,** 1937–1942 (1993).

Joensuu, H., Roberts, P. J., Sarlomo-Rikala, M., et al. Effect of the tyrosine kinase inhibitor STI571 in a patient with a metastatic gastrointestinal stromal tumor. *N. Engl. J. Med.* **344,** 1052–1056 (2001).

Jonak, Z. L., Trulli, S., Maier, C., et al. High-dose recombinant interleukin-18 induces an effective TH1 immune response to murine MOPC-315 plasmacytoma. *J. Immunother.* **25,** S20–S27 (2002).

Jost, M., Kari, C., and Rodeck, U. The EGF receptor – An essential regulator of multiple epidermal functions. *Eur. J. Dermatol.* **10,** 505–510 (2000).

Karp, J. E., Lancet, J. E., Kaufmann, S. H., et al. Clinical and biologic activity of the farnesyltransferase inhibitor R115777 in adults with refractory and relapsed acute leukemias: A phase I clinical-laboratory correlative trial. *Blood* **97,** 3361–3369 (2001).

Kato, K., Cox, A. D., Hisaka, M. M., et al. Isoprenoid addition to *ras* protein is the critical modification for its membrane association and transforming activity. *Proc. Natl. Acad. Sci. USA* **89,** 6403–6407 (1992).

Kelland, L. R., Smith, V., Valenti, M., et al. Preclinical antitumor activity and pharmacodynamic studies with the farnesylprotein transferase inhibitor R115777 in human breast cancer. *Clin. Cancer Res.* **7,** 3544–3550 (2001).

Kelliher, M. A., McLaughlin, J., Witte, O.N., and Rosenberg, N. Induction of a chronic myelogenous leukemia-like syndrome in mice with *v-abl* and *bcr-abl*. *Proc. Natl. Acad. Sci. USA* **87,** 6649–6653 (1990).

Kohl, N. E., Mosser, S. D., deSolms, S. J., et al. Selective inhibition of *ras*-dependent transformation by a farnesyltransferase inhibitor. *Science* **260,** 1934–1937 (1993).

Kohl, N. E., Omer, C. A., Conner, M. W., et al. Inhibition of farnesyltransferase induces regression of mammary and salivary carcinomas in *ras* transgenic mice. *Nature Med.* **1,** 792–797 (1995).

Kohl, N. E., Wilson, F. R., Mosser, S. D., et al. Protein farnesyltransferase inhibitors block the growth of *ras*-dependent tumors in nude mice. *Proc. Natl. Acad. Sci. USA* **91,** 9141–9145 (1994).

Kurzrock, R., Wilson, E. F., and Sebti, S. M. Clinical activity and inhibition of protein farnesylation in FTI-R115777 treated patients with myelodysplastic syndrome (MDS [Abstract]). *Proc. Amr. Assoc. Cancer Res.* **43,** 1000 (2002).

La Rosee, P., Corbin, A. S., Stoffregen, E. P., et al. Activity of the Bcr-Abl kinase inhibitor PD180970 against clinically relevant bcr-abl isoforms that cause resistance to Imatinib Mesylate (Gleevec, STI571). *Cancer Res.* **62,** 7149–7153 (2002).

Lerner, E. C., Qian, Y., Blaskovich, M. A., et al. Ras CAAX peptidomimetic FTI-277 selectively blocks oncogenic *ras* signaling by inducing cytoplasmic accumulation of inactive ras-raf complexes. *J. Biol. Chem.* **270,** 26802–26806 (1995).

Lerner, E. C., Zhang, T. T., Knowles, D. B., et al. Inhibition of the prenylation of K-*ras,* but not H- or N-*ras,* is highly resistant to CAAX peptidomimetics and requires both a farnesyltransferase and a geranylgeranyltransferase I inhibitor in human tumor cell lines. *Oncogene* **15,** 1283–1288 (1997).

Libutti, S., Choyke, P., Carrasquillo, J., et al. Monitoring responses to antiangiogenic agents using noninvasive tests. *Cancer J. Sci. Am.* **5,** 252–256 (1999).

Liebes, L., Potmesil, M., Kim, T., et al. Pharmacodynamics of topoisomerase I inhibition: Western blot determination of topoisomerase I and cleavable complex in patients with upper gastrointestinal malignancies treated with topotecan. *Clin. Cancer Res.* **4,** 545–557 (1998).

Liu, M., Bryant, M., Chen, J., et al. Antitumor activity of SCH66336, an orally bioavailable tricyclic inhibitor of farnesyl protein transferase, in human tumor xenograft models and wap-*ras* transgenic mice. *Cancer Res.* **58,** 4947–4956 (1998).

Lobell, R. B., Liu, D., Buser, C. A., et al. Preclinical and clinical pharmacodynamic assessment of L-778,123, a dual inhibitor of farnesyl:protein transferase and geranylgeranyl:protein transferase type-I. *Mol. Cancer Ther.* **1,** 747–758 (2002).

Lobell, R. B., Omer, C. A., Abrams, M. T., et al. Evaluation of farnesyl:protein transferase and geranylgeranyl:protein transferase inhibitor combinations in preclinical models. *Cancer Res.* **61,** 8758–8768 (2001).

Lugo, T. G., Pendergast, A. M., Muller, A. J., and Witte, O. N. Tyrosine kinase activity and transformation potency of *bcr-abl* oncogene products. *Science* **247,** 1079–1082 (1990).

Mangues, R., Corral, T., Kohl, N. E., et al. Antitumor effect of a farnesyl transferase inhibitor in mammary and lymphoid tumors over-expressing n-*ras* in transgenic mice. *Cancer Res.* **58,** 1253–1259 (1998).

Manley, P. W., Cowan-Jacob, S. W., Buchdunger, E., et al. A selective tyrosine kinase inhibitor. *Eur. J. Cancer* **38 (suppl 5),** S19–27 (2002).

Mazet, J. L., Padieu, M., Osman, H., et al. Combination of the novel farnesyltransferase inhibitor RPR130401 and the geranylgeranyltransferase-I inhibitor GGTI-298 disrupts MAP kinase activation and G1-S transition in Ki-*ras*-over-expressing transformed adrenocortical cells. *FEBS Lett.* **460,** 235–240 (1999).

McDaid, H. M., Mani, S., Shen, H.-J., et al. Validation of the pharmacodynamics of BMS-247550, an analogue of epothilone B, during a phase I clinical study. *Clin Cancer Res.* **8,** 2035–2043 (2002).

Miettinen, P. J., Berger, J. E., Meneses, J., et al. Epithelial immaturity and multiorgan failure in mice lacking epidermal growth factor receptor. *Nature* **376,** 337–341 (1995).

Moasser, M. M., Basso, A., Averbuch, S. D., and Rosen, N. The tyrosine kinase inhibitor ZD1839 (Iressa) inhibits Her2-driven signaling and suppresses the growth of Her2-overexpressing tumor cells. *Cancer Res.* **61,** 7184–7188 (2001).

Morgan, B., Ball, H., Buchert, M., et al. Resonance imaging as a surrogate marker of efficacy in a phase I trial of a VEGF receptor tyrosine kinase inhibitor. *Proc. Am. Assoc. Cancer Res.* **41,** 587a (2001).

Morrow, M., Holden, S. N., Cohen, R., et al. Correlative biological assays in a phase I study of the farnesyl transferase inhibitor (FTI) R115777 administered with capecitabine. *Proc. Am. Assoc. Cancer Res.* **43,** 338 (2002).

Mould, D., Holford, N., Schellens, J., et al. Population pharmacokinetic and adverse event analysis of topotecan in patients with solid tumors. *Clin. Pharm. Ther.* **71,** 334–348 (2002).

Moulder, S. L., Yakes, F. M., Muthuswamy, S. K., et al. Epidermal growth factor receptor (Her1) tyrosine kinase inhibitor ZD1839 (Iressa) inhibits Her2/Neu (ErbB2)-overexpressing breast cancer cells *in vitro* and *in vivo*. *Cancer Res.* **61,** 8887–8895 (2001).

Nagasu, T., Yoshimatsu, K., Rowell, C., et al. Inhibition of human tumor xenograft growth by treatment with the farnesyl transferase inhibitor B956. *Cancer Res.* **55,** 5310–5314 (1995).

Neeman, M., Provenzale, J., and Dewhirst, M. Magnetic resonance imaging applications in the evaluation of tumor angiogenesis. *Sem. Rad. Oncol.* **11,** 70–82 (2001).

Nguyen, D. N., Stump, C. A., Walsh, E. S., et al. Potent inhibitors of farnesyltransferase and geranylgeranyltransferase-I. *Bioorg. Med. Chem. Lett.* **12,** 1269–1273 (2002).

Nichols, G. L., Raines, M. A., Vera, J. C., et al. Identification of CRKL as the constitutively phosphorylated 39-kD tyrosine phosphoprotein in chronic myelogenous leukemia cells. *Blood* **84,** 2912–2918 (1994).

Nicholson, R. I., Gee, J. M., and Harper, M. E. EGFR and cancer prognosis. *Eur. J. Cancer* **37,** S9–15 (2001).

Nimmanapalli, R., and Bhalla, K. Mechanisms of resistance to Imatinib Mesylate in Bcr-Abl-positive leukemias. *Curr. Opin. Oncol.* **14,** 616–620 (2002).

Normanno, N., Campiglio, M., De, L. A., et al. Cooperative inhibitory effect of ZD1839 (Iressa) in combination with Trastuzumab (Herceptin) on human breast cancer cell growth. *Ann. Oncol.* **13,** 65–72 (2002).

Oda, T., Heaney, C., Hagopian, J. R., et al. CRKL is the major tyrosine-phosphorylated protein in neutrophils from patients with chronic myelogenous leukemia. *J. Biol. Chem.* **269,** 22925–22928 (1994).

Omer, C. A., Chen, Z., Diehl, R. E., et al. Mouse mammary tumor virus-Ki-*ras*B transgenic mice develop mammary carcinomas that can be growth-inhibited by a farnesyl:protein transferase inhibitor. *Cancer Res.* **60,** 2680–2688 (2000).

Overmeyer, J. H., and Maltese, W. A. Isoprenoid requirement for intracellular transport and processing of murine leukemia virus envelope protein. *J. Biol. Chem.* **267,** 22686–22692 (1992).

Packard, S., Moore, J., Hill, S., et al. Dynamic magnetic resonance imaging of vascular permeability changes in 9l/lacZ gliosarcoma bearing rats due to KDR kinase inhibition. *Proc. Am. Assoc. Cancer Res.***42,** 5350a (2002).

Padhani, A., and Husband, J. Dynamic contrast-enhanced MRI studies in oncology with an emphasis on quantification, validation and human studies. *Clin. Radiol.* **56,** 607–620 (2001).

Pesenti, E., Degrassi, A., Marzola, P., et al. Evaluation of the efficacy of an anti-angiogenesis compound SU6668 in a xenograft tumor model by dynamic contrast enhanced-MRI (DCE-MRI). *Proc. Am. Assoc. Cancer. Res.* **42,** 4446a (2002).

Peus, D., Hamacher, L., and Pittelkow, M. R. EGF-receptor tyrosine kinase inhibition induces keratinocyte growth arrest and terminal differentiation. *J. Invest. Dermatol.* **109,** 751–756 (1997).

Prendergast, G. C. Farnesyltransferase inhibitors: Antineoplastic mechanism and clinical prospects. *Curr. Opin. Cell Biol.* **12,** 166–173 (2000).

Prendergast, G. C., and Gibbs, J. B. Regulatory interactions in Ras signal transduction: New targets for anti-cancer intervention? *Bioessays* **16,** 187–191 (1994).

Ranson, M., Hammond, L. A., Ferry, D., et al. ZD1839, a selective oral epidermal growth factor receptor-tyrosine kinase inhibitor, is well tolerated and active in patients with solid, malignant tumors: Results of a phase I trial. *J. Clin. Oncol.* **20,** 2240–2250 (2002).

Rojo, F., Albanell, J., Anido, J., et al. Dose dependent pharmacodynamic effects of ZD1839 (Iressa) correlate with tumor growth inhibition in BT-474 breast cancer xenografts. *Proc. Am. Assoc. Cancer Res.* **42,** 3893A (2002).

Rowell, C. A., Kowalczyk, J. J., Lewis, M. D., and Garcia, A. M. Direct demonstration of geranylgeranylation and farnesylation of Ki-Ras *in vivo*. *J. Biol. Chem.* **272,** 14093–14097 (1997).

Rubin, B. P., Singer, S., Tsao, C., et al. Kit activation is a ubiquitous feature of gastrointestinal stromal tumors. *Cancer Res.* **61,** 8118–8121 (2001).

Rubin, E., Abbruzzese, J. L., Morrison, B. W., et al. Phase I trial of the farnesyl protein transferase inhibitor L-778,123 on a 14- or 28-day dosing schedule. *Proc. Ann. Meet. Am. Soc. Clin. Oncol.* **19,** 178a (2000).

Sano, S., Itami, S., Takeda, K., et al. Keratinocyte-specific ablation of Stat3 exhibits impaired skin remodeling, but does not affect skin morphogenesis. *EMBO J.* **18,** 4657–4668 (1999).

Sattler, M., Salgia, R., Okuda, K., et al. The proto-oncogene product p120Cbl and the adaptor proteins CRKL and c-Crk link c-Abl, p190Bcr-Abl and p210Bcr-Abl to the phosphatidylinositol-3' kinase pathway. *Oncogene* **12,** 839–846 (1996).

Sawyers, C. L. Chronic myeloid leukemia. *N. Engl. J. Med.* **340,** 1330–1340 (1999).

Sawyers, C. L. Disabling Abl-perspectives on abl kinase regulation and cancer therapeutics. *Cancer Cell* **1,** 13–15 (2002).

Sebti, S. M., and Hamilton, A. D. Farnesyltransferase and geranylgeranyltransferase I inhibitors in cancer therapy: Important mechanistic and bench to bedside issues. *Exp. Opin. Invest. Drugs* **9,** 2767–2782 (2000).

Senechal, K., Halpern, J., and Sawyers, C. L. The CRKL adaptor protein transforms fibroblasts and functions in transformation by the *bcr-abl* oncogene. *J. Biol. Chem.* **271,** 23255–23261 (1996).

Sepp-Lorenzino, L., Ma, Z., Rands, E., et al. A peptidomimetic inhibitor of farnesyl: Protein transferase blocks the anchorage-dependent and -independent growth of human tumor cell lines. *Cancer Res.* **55,** 5302–5309 (1995).

Sepp-Lorenzino, L., and Thomas, K. Antiangiogenic agents targeting vascular endothelial growth factor and its receptors in clinical development. *Expert Opin. Investig. Drugs* **11,** 1447–1465 (2002).

Shi, J. D., Bullock, C., Hall, W. C., Wescott, V., Wang, H., Levitt, D. J., and Klingbeil, C. K. *In vivo* pharmacodynamic effects of HU1D10 (Remitogen), a humanized antibody reactive against a polymorphic determinant of HLA-DR expressed on B cells. *Leukemia and Lymphoma* **43,** 1303–1312 (2002).

Sirotnak, F. M., Zakowski, M. F., Miller, V. A., et al. Efficacy of cytotoxic agents against human tumor xenografts is markedly enhanced by coadministration of ZD1839 (Iressa), an inhibitor of EGFR tyrosine kinase. *Clin. Cancer Res.* **6,** 4885–4892 (2000).

Siu, L. L., Rowinsky, E. K., Hammond, L. A., et al. A phase I and pharmacokinetic study of SAM486A, a novel polyamine biosynthesis inhibitor, administered on a daily-times-five every-three-week schedule in patients with advanced solid malignancies. *Clin. Cancer Res.* **8,** 2157–2166 (2002).

Sotiriou, C., Powles, T. J., Dowsett, M., et al. Gene expression profiles derived from fine needle aspiration correlate with response to systemic chemotherapy in breast cancer. *Breast Cancer Res.* **4,** R3 (2002).

Spiro, T. P., Gerson, S. L., Liu, L., et al. O6-benzylguanine: A clinical trial establishing the biochemical modulatory dose in tumor tissue for alkyltransferase-directed DNA repair. *Cancer Res.* **59,** 2402–2410 (1999).

Spiro, T. P., Liu, L., Majka, S., et al. Temozolomide: The effect of once- and twice-a-day dosing on tumor tissue levels of the DNA repair protein O6-alkylguanine-DNA-alkyltransferase. *Clin. Cancer Res.* **7,** 2309–2317 (2001).

Steinberg, L., Hassan, M., Olmsted, L., et al. A phase I trial of radiotherapy and simultaneous 24-h Paclitaxel in patients with locally advanced head and neck squamous cell carcinoma. *Semin. Oncol.* **24,** S19–51 (1997).

Stevenson, J. P., Yao, K.-S., Gallagher, M., et al. Phase I clinical/pharmacokinetic and pharmacodynamic trial of the c-*raf*-1 antisense oligonucleotide Isis 5132 (CGP 69846a). *J Clin. Oncol.* **17,** 2227 (1999).

Sun, J., Blaskovich, M. A., Knowles, D., et al. Antitumor efficacy of a novel class of non-thiol-containing peptidomimetic inhibitors of farnesyltransferase and geranylgeranyltransferase I: Combination therapy with the cytotoxic agents Cisplatin, Taxol, and Gemcitabine. *Cancer Res.* **59,** 4919–4926 (1999).

Sun, J., Qian, Y., Hamilton, A. D., and Sebti, S. M. Both farnesyltransferase and geranylgeranyltransferase I inhibitors are required for inhibition of oncogenic k-ras prenylation but each alone is sufficient to suppress human tumor growth in nude mouse xenografts. *Oncogene* **16,** 1467–1473 (1998).

Sun, J., Qian, Y., Hamilton, A. D., and Sebti, S. M. Ras CAAX peptidomimetic FTI-276 selectively blocks tumor growth in nude mice of a human lung carcinoma with K-*ras* mutation and p53 deletion. *Cancer Res.* **55,** 4243–4247 (1995).

ten Hoeve, J., Arlinghaus, R. B., Guo, J. Q., et al. Tyrosine phosphorylation of CRKL in Philadelphia$^+$ leukemia. *Blood* **84,** 1731–1736 (1994).

Threadgill, D. W., Dlugosz, A. A., Hansen, L. A., et al. Targeted disruption of mouse EGF receptor: Effect of genetic background on mutant phenotype. *Science* **269,** 230–234 (1995).

Tofts, P. S., and Kermode, A. G. Measurement of the blood-brain barrier permeability and leakage space using dynamic MR imaging. Fundamental concepts. *Magn. Reson. Med.* **17,** 357–367 (1991).

Tuveson, D. A., Willis, N. A., Jacks, T., et al. STI571 inactivation of the gastrointestinal stromal tumor c-kit oncoprotein: Biological and clinical implications. *Oncogene* **20,** 5054–5058 (2001).

Van den Abbeele, A. D., and Badawi, R. D. Use of positron emission tomography in oncology and its potential role to assess response to Imatinib Mesylate therapy in gastrointestinal stromal tumors (GISTs). *Eur. J. Cancer* **38 (suppl 5),** S60–S65 (2002).

van Dijke, C. F., Brasch, R. C., Roberts, T. P., et al. Mammary carcinoma model: Correlation of macromolecular contrast-enhanced MR imaging characterizations of tumor microvasculature and histologic capillary density. *Radiology* **198,** 813–818 (1996).

van Dijke, C. F., Mann, J. S., Rosenau, W., et al. Comparison of MR contrast-enhancing properties of Albumin-(Biotin)10-(Gadopentetate)25, a macromolecular MR blood pool contrast agent, and its microscopic distribution. *Acad. Radiol.* **9,** S257–S260 (2002).

van Oosterom, A. T., Judson, I. R., Verweij, J., et al. Update of phase I study of Imatinib (STI571) in advanced soft tissue sarcomas and gastrointestinal stromal tumors: A report of the EORTC soft tissue and bone sarcoma group. *Eur. J. Cancer* **38,** S83–S87 (2002).

Wakeling, A. E., Guy, S. P., Woodburn, J. R., et al. ZD1839 (Iressa): An orally active inhibitor of epidermal growth factor signaling with potential for cancer therapy. *Cancer Res.* **62,** 5749–5754 (2002).

Warmuth, M., Simon, N., Mitina, O., et al. Dual-specific Src and Abl kinase inhibitors, PP1 and CGP76030, inhibit growth and survival of cells expressing imatinib mesylate-resistant Bcr-Abl kinases. *Blood* **101,** 664–672 (2003).

Whyte, D. B., Kirschmeier, P., Hockenberry, T. N., et al. K- and N-ras are geranylgeranylated in cells treated with farnesyl protein transferase inhibitors. *J. Biol. Chem.* **272,** 14459–14464 (1997).

Wisniewski, D., Lambek, C. L., Liu, C., et al. Characterization of potent inhibitors of the Bcr-Abl and the c-Kit receptor tyrosine kinases. *Cancer Res.* **62,** 4244–4255 (2002).

Workman, P. Bottlenecks in anticancer drug discovery and development: *In vivo* pharmacokinetic and pharmacodynamic issues and the potential role of PET. In: D. Komar, ed., *PET for Drug Development and Evaluation*. Dordrecht, Kluwer (1995).

Zeng, H., Sanyal, S., and Mukhopadhyay, D. Tyrosine residues 951 and 1059 of vascular endothelial growth factor receptor-2 (KDR) are essential for vascular permeability factor/vascular endothelial growth factor-induced endothelium migration and proliferation, respectively. *J. Biol. Chem.* **276,** 32714–32719 (2001).

Zhang, F. L., Kirschmeier, P., Carr, D., et al. Characterization of Ha-Ras, N-Ras, Ki-Ras4A and Ki-Ras4B as in vitro substrates for farnesyl protein transferase and geranylgeranyl protein transferase type I. *J. Biol. Chem.* **272,** 10232–10239 (1997).

CHAPTER 12

Pharmacokinetic and Toxicology Issues in Cancer Drug Discovery and Development

Pamela A. Benfield and Bruce D. Car

Molecular Cancer Therapeutics: Strategies for Drug Discovery and Development, edited by George C. Prendergast
ISBN 0-471-43202-4

12.1 Importance of Pharmacokinetics and Toxicity Studies in Drug Development

The testing and marketing of drugs is subject to regulatory control to ensure that the consumer enjoys reasonable expectations of safety and therapeutic efficacy. In the United States the relevant regulatory agency is the Food and Drug Administration (FDA). This body imposes strict controls on drug testing, manufacturing, and marketing processes. For a new biological or chemical entity to be tested in humans the FDA requires that an investigational new drug application or IND be submitted for approval. The primary purpose of the IND is to ensure that new therapeutics have been appropriately tested preclinically and that a protocol has been established to allow safe testing of the agent in humans. For this purpose it is not required that the agent be completely safe since all drugs have accompanying side effects. It is, however, required that the positive benefits of the agent are likely to outweigh these negative side effects.

In the United States clinical testing of agents is divided into three principal phases (see Chapter 14). The purpose of Phase I trials is to determine the fate of the agent once it is administered to the patient (the manner in which it is absorbed, metabolized, and excreted), to establish tolerability and a safe starting dose for further efficacy trials to be carried out in Phases II and III. Phase I trials usually involve a small number of healthy volunteers. However, as will be discussed in more detail later, Phase I trials for new therapies in oncology have traditionally been carried out in cancer patients themselves. Phase II and Phase III trials examine efficacy in more detail and routinely involve testing in patient populations. Safety issues are monitored at all phases of clinical testing and an inappropriate side-effect profile can prevent the successful progress of an agent through clinical testing or impact the need for more extensive safety testing.

If clinical testing of a new agent demonstrates efficacy following rigorous evaluation of Phase III trial data, then a new drug application (NDA) needs to be submitted and approved by the FDA before a drug can be registered and marketed in the United States. New agents are evaluated by the Center for Drug Evaluation and Research (CDER). Historically, novel biologic agents were evaluated by the Center for Biologics Evaluation and Research (CBER), the function of which has more recently been rolled into CDER to ensure more uniform treatment of small molecules and biologics. For this application to be successful it is necessary, among other things, to show that the new agent is efficacious in humans with a sufficient therapeutic margin and enhanced activity over currently available therapies to represent an advance for the patient consumer.

Official IND and NDA approval is necessary for a new drug to be marketed in the United States. Similar regulatory approvals are necessary for marketing approval in other countries throughout the world. To industry, this drug development process represents the most time-consuming and expensive part of the new drug-discovery process. Depending on the agent and the therapeutic indication it can take up to 10 years from the first identification of a novel agent to its final approval as a marketed product. In addition, clinical trials can involve long-term testing in large patient populations to amass efficacy data of appropriate statistical significance. These trials can represent a considerable financial investment. Additional anticancer drug indications (e.g., different tumor targets) are generally evaluated in postregistration clinical studies collectively referred to as Phase IV. Each application for a new drug indication is submitted with a supplementary NDA (sNDA).

Satisfaction of safety and efficacy considerations is the major hurdle to overcome in clinical development. Toxicity issues account for the major reason for the failure of new agents early in development. Lack of sufficient efficacy is often the reason for drug failure at late stages. Each of these criteria has to be assessed in the context of drug exposure levels. As a result, pharmacokinetics (PK) and toxicity issues are of major concern to industry in its pursuit of new cancer therapeutics as they are for any new pharmaceutical agent.

12.2 Differences in Drug Discovery for Cancer and Other Therapeutic Areas

Typically, in the development of a novel therapeutic it is important to determine the level of drug exposure necessary to achieve efficacy. Efficacy can be evaluated based on modulation of the drug's intended target or, if the precise mechanism is unknown or difficult to monitor, efficacy can be determined based on achievement of a desired therapeutic effect, for example, lowering of blood pressure or modulation of a disease marker. During preclinical development, drug dosing above these effective levels is performed to determine at what drug exposure undesirable or unacceptable toxic effects are encountered. The ratio of the effective blood plasma concentrations to the those drug concentrations associated with the highest nontoxic dose is considered the drug's therapeutic margin. Ideally this therapeutic margin (also known as therapeutic index) should be as large as possible and consistent across species. This increases the likelihood that the drug can be dosed in the clinic at a level that will afford the patient maximum therapeutic benefit with minimal risk. To ensure accuracy of evaluation of therapeutic indices in preclinical species, the ability to measure the relevant efficacy or pharmacodynamic end point in at least one of the toxicology species is critical.

It is commonly the case that anticancer therapies cannot be dosed to full potential therapeutic efficacy because toxic side effects become limiting. These therapies frequently have little or no therapeutic margin, and treatment is a constant battle between dosing sufficient drug to give patient benefit without causing unacceptable and sometimes life-threatening toxicity. These factors significantly affect the relative role PK and toxicity issues play in anticancer drug development and have meant that traditionally toxicity considerations have played a dominant role.

Cancer is a devastating and acutely life-threatening disease. Cancer cells have lost normal growth control and frequently gain the ability to escape the primary disease site and populate other locations in the body, a process known as metastasis. The dispersed nature of end-stage disease drives the need for systemic therapy and makes surgical approaches ineffective. Particularly in end-stage metastatic disease, life expectancy is short, and the tolerance for effective agents with toxic side effects is higher than in almost any other disease.

Cancer is also a complex multifactorial disease. Tumors are frequently heterogeneous and tumor cells are notoriously genetically unstable and capable of sustaining mutations leading to drug resistance. As a consequence, it has proved virtually impossible to define single specific molecular targets likely to have selective impact on a broad range of cancer cells. Many classic anticancer therapies probably have multiple mechanisms of action, and combination therapy is normal practice in oncology. Because cancer cells are variants of normal cells, molecular targets are likely not specific to the cancer cell, increasing the potential for unwanted side effects. Consequently broadly effective therapies have had a very high potential for generating significant undesirable toxicity.

Currently, many commonly used anticancer therapeutics represent broadly cytotoxic agents. These agents were frequently discovered using cell-based cytotoxicity assays, and the exact molecular mechanism, or mechanisms of action are frequently unknown. Furthermore, the cytotoxic effect of these compounds is rarely selective for cancer cells and certain normal cells, frequently those in proliferating tissues, such as the bone marrow and gastrointestinal tract, are also targeted. Thus for traditional anticancer agents it has commonly been accepted that these molecules have no clearly defined mechanism of action and that they are expected to have significant toxicity and very narrow therapeutic margins.

During the development of most therapeutics, the agent is dosed to achieve a desired exposure of the drug to the molecular target. This dosing is driven by the PK properties of the compound as determined both in preclinical studies and in Phase I trials. Because safety considerations in compound testing are stringent, these Phase I trials are carried out in healthy volunteers.

For anticancer agents dosing has not traditionally been driven by PK considerations. Rather agents have been dosed based on the maximum dose that can be administered without incurring unacceptable toxicity. This dose is termed the maximum tolerated dose (MTD). Preclinical toxicity studies are used to determine the starting dose to be used in humans, and this dose is usually set as 1/10 the LD_{10} in rodents (i.e., 1/10 the dose in which 10% of dosed rodents die). An MTD is then determined in Phase I studies in humans using a variety of dose escalation strategies starting with this predetermined starting dose. Given the significant toxic liability associated with these agents, Phase I testing is not carried out in normal healthy volunteers but rather in a cancer patient population often represented by end-stage patients who have failed other currently available therapies.

Some of the perceived advantages of the above considerations for developing drugs for cancer, as compared to other therapeutics, are as follos:

- Frequently, new agents can be advanced into the clinic with much less preclinical PK and safety testing and with a greater tolerance for toxic side effects.
- Despite the fact that all agents tested in humans start in Phase I, cancer drugs are typically tested in a terminal group of pateints, and it may be possible to get some indications of efficacy at an early stage in the clinical trial process (before Phase II). Of course, a potential disadvantage is that as the number of patients tested in a Phase I trial is usually very small, the results are likely not statistically significant and indeed may be misleading. In addition, the advanced stage of disease in many Phase I patients may decrease the likelihood of a clinical response to selective mechanism-based therapies, perhaps discouraging further development of an agent that might be effective in a selected patient population.
- Since risk–benefit considerations permit the evaluation of relatively toxic molecules in cancer patients, progression to Phase II is simpler and clinical development for cancer agents is cheaper.

12.3 Introduction to Pharmacokinetic Issues

PK is the study of the fate of a drug when it is delivered to an organism (Rowland and Tozer, 1995). It encompasses evaluation of absorption (A) and distribution (D) of the drug into the various organs and tissues, the metabolism (M) of that drug and the ultimate elimination (E) of the drug from the body (ADME). From the pharmaceutical industry drug-discovery point of view, the organism of relevance is humans. However, initial PK studies are ordinarily conducted in a variety of surrogate species. Preclinical PK studies usually have two objectives. First, these studies aim to provide insights into how the drug will be absorbed distributed, metabolized, and excreted in humans as required by the FDA. Second, these studies are important for providing drug exposure data relevant to the interpretation of preclinical efficacy and toxicity studies.

12.3.1 ABSORPTION

Therapeutic agents may be delivered by a variety of routes, including intravenous (IV), intraperitoneal (IP), oral (PO), intranasal, transdermal, etc. When a compound is delivered by a non-IV route, it usually needs to be absorbed into the bloodstream for distribution to the site of action (e.g., from the gastrointestinal tract for an oral agent). The extent of this absorption is described as the drug's bioavailability (F). During early preclinical studies with a new agent a dosing route may be chosen based on experimental convenience, but ultimately an agent is desired that is suitable for dosing by a commercially acceptable route. Frequently this leads to a search for an orally bioavailable drug with a PK to support a convenient dosing frequency (e.g., once-a-day dosing).

For anticancer therapies, the need to deliver efficacious levels of drugs with a high degree of toxicity has led to increased flexibility in terms of drug delivery options. Historically, a large fraction of cancer drugs have been delivered by the IV route. Many classic agents have poor PK and short half-lives, and the IV delivery route provides the most effective way of controlling compound plasma levels. This route of delivery allows direct access of the compound to the bloodstream and short-circuits any problems with intestinal absorption. It can also help protect the upper gastrointestinal tract from the effect of less specific cytotoxic agents. Furthermore, with this delivery route it is easier to control plasma drug levels and avoid high peak to trough ratios, which may be problematic for a low therapeutic index drug. Obviously, this route of delivery is not the most desirable or convenient for the patient, although continuous IV delivery systems are available for outpatient use. IV administration of drugs is associated with toxicology issues peculiar to this route, including the potential for rapid onset of type I hypersensitivity reactions – rapid onset of hemodynamic alterations such as lowering of blood pressure, vascular and cutaneous irritation at the injection site, and hemolysis of red cells.

In cancer therapy, the need for efficacious agents has meant that IV delivery is acceptable and commonly used. However, the economic pressures of the current health-care environment favor IV agents that can be delivered by this route on a short infusion schedule that does not require hospitalization. This sometimes means that dosing schedules are used that do not necessarily reflect the optimum for the agent. For example, despite the fact that Taxol has been shown to be optimally efficacious when given as a long-term infusion (Eisenhauer and Vermorken, 1998; Huizing et al. 1997), it is mostly delivered in the clinic using a 1-h infusion. As will be discussed later, specific safety issues need to be addressed for the evaluation of IV agents.

Obviously, oral delivery represents a preferable drug delivery route for the patient, and newer versions of cancer drugs are emerging with improved oral bioavailability. As newer agents emerge with improved therapeutic margins, competitive pressures to discover and develop oral agents will likely increase.

12.3.2 DISTRIBUTION

Once the drug is absorbed into the bloodstream it is typically distributed to the tissues and metabolized to break-down products, which are eliminated from the system. The measure of how well a drug is distributed outside the vascular space is represented by the volume of distribution, V_d. Drugs that are well distributed to tissues or otherwise sequestered outside the bloodstream have a high V_d. Typically, for a drug to be pharmacologically effective it is necessary for sufficient active agent to reach the target mechanism for enough time to have therapeutic benefit. For an anticancer drug to be effective, usually it needs to reach the tumor site. Exceptions to this might include antiangiogenic drugs, for which the site of action is the tumor vasculature, and immunomodulatory agents For many agents, it is also important that the therapy be able to access intracellular targets and permeate the plasma and internal membrane systems. Given the variable nature of tumor vasculature, its irregular growth, and its propensity for thrombosis, areas of relatively poor perfusion exist in tumors for which penetration by antitumor agents may be limited.

Ideally, tumor levels of a test agent should be monitored. This is possible in preclinical models, but monitoring of tumor drug levels in the clinic usually requires biopsy of tumor tissue, which for solid tumors is not attractive to patients unless it is done as part of a planned surgery. Even in this case, evaluation of multiple drug doses and multiple sampling over time is not possible. An exception to this would be the testing of agents designed to treat hematological malignancies. For some targets, surrogate plasma markers can provide a readout of target inhibition in the tumor and noninvasive techniques are emerging that may be applicable in certain cases. Nevertheless plasma drug levels are commonly used as a means of estimating likely tumor drug exposure.

12.3.2.1 Plasma Protein Binding

It is important, that drug efficacy is frequently related to plasma levels of *free* drug – that is, drug that is not bound to serum proteins or other nonspecific sites. Within the bloodstream are several components that can sequester drug substance. Small molecule drugs bind frequently with moderate affinity to serum albumin and occasionally to α-acidic glycoprotein and other globulins. In addition, drug may become bound to or taken up by the cellular components in blood and effectively removed from circulation. In assessing drug exposure and its relationship to efficacy and toxicity, it is important to correct plasma drug levels so that they reflect these levels of free drug. In this regard, it is important to note that plasma protein binding can vary among species, making it important to evaluate these factors using blood components from the species of relevance – the preclinical test species and humans.

Plasma protein binding considerations are especially important for agents with narrow therapeutic margins. If an agent binds to a plasma protein component with low serum concentrations, such as α-acidic glycoprotein, then a disproportionate increase in free drug with dose could be observed once the binding component becomes saturated. This behavior is referred to as *hockey stick kinetics* and can be problematic for a highly toxic agent. One novel anticancer agent in development for which this was an issue was the protein kinase C (PKC)/cell cycle inhibitor UCN01, which is highly bound to α-acidic glycoprotein (Akinaga et al. 2000). Significant toxicity problems were encountered as the binding to this component became saturated. This issue is further complicated by significant interpatient variation in plasma protein levels, particularly in an aging and diseased patient population.

12.3.3 METABOLISM

Living organisms have evolved mechanisms to eliminate foreign substances, frequently after prior enzymatic processing or metabolism. It is important, particularly for agents with significant toxic liability, that the drug be eliminated effectively from the system. Clearance (Cl) values provide a measure of how rapidly an agent is eliminated from the body. Lack of effective clearance could be dangerous in the event of accidental overdose or could lead to drug accumulation upon repeated dosing. This could be particularly problematic for an anticancer agent with a narrow therapeutic index. It is desirable for the rate of clearance to be compatible with a convenient dosing regimen to augment patient compliance and acceptance. Of course, it is also important that the metabolic side products of a drug are not toxic and are also efficiently cleared.

Typically, agents are metabolized before clearance. The major sites of metabolism in mammals are the liver, the gut, and the kidney, although metabolism can also occur in a variety of secondary tissues (e.g., lung). Major players in drug metabolism in the liver are the components of the cytochrome P450 reductase system, but other important drug-metabolizing enzymes are represented by the glucuronidases, sulfotransferases, and esterases.

The quantitatively most important site of drug metabolism is the liver. Significant interpatient variation can be observed in metabolic capacity. For example, there is genetic variation in the population with respect to cytochrome P450 expression. Also cancer patients tend to represent an older patient population already compromised by debilitating disease where metabolism by liver and other organs may be impaired, particularly if the tumor has metastasized to these sites. This is, of course, an important consideration with low therapeutic index agents for which it may become easy to overdose the patient. An example of a commonly used anticancer drug for which interpatient variation in metabolism can be an important factor in patient exposure is provided by 5-fluorouracil (5FU) (see below). For this drug, there is is significant interpatient variation in the major enzyme involved in the drug's catabolism – namely, dihydropyrimidine dehydrogenase (DPD) (Etienne et al. 1994).

12.3.3.1 Exploiting Metabolism to Gain Tumor Selectivity

Metabolic conversion of drug substances can sometimes be used to patient advantage. If certain metabolizing enzymes are unique or enriched in tumor tissue, these can be exploited to convert an inactive drug precursor, or pro-drug, to the active substance specifically in tumor tissue. This approach has been used to generate pro-drugs of 5FU, for example, capecitabine (Xeloda) (Reigner et al. 2001). Other examples of antitumor agents that exploit tumor-selective metabolic pathways are etoposide phosphate (Budman et al. 1994) and peptide-linked cytotoxic agents. In the latter case, the linked peptide is designed to have a protease cleavage site, which is preferentially cleaved in the target tumor.

Inhibition of metabolic pathways can also be used to increase the exposure of patients to agents that have short intrinsic half-lives. For example, it has been suggested that inhibition of DPD might be a strategy to increase patient exposure to 5FU (Milano and McLeod, 2000). Of course, this has to be done with care to balance the increased therapeutic benefit with the increased toxic liability of higher drug exposure.

12.3.3.2 Metabolic Interference in Combination Therapy

Another important issue that needs to be evaluated with respect to anticancer drug metabolism is the possibility that a molecule may interfere with the metabolism of co-administered agents. For example, if a drug is an inhibitor of a cytochrome P450 enzyme, the drug may cause an increased exposure to a second co-administered agent if that agent is metabolized by the same cytochrome P450 pathway. This issue is particularly relevant to anticancer therapies that are often administered with a variety of additional agents. Since some of the additional agents may have significant toxic potential, interference with the P450 pathways can increase the dose of the additional agent. Examples of anticancer drugs that are also inhibitors of cytochrome P450 enzymes include certain aromatase inhibitors used in the clinic (Buzdar et al. 2002).

12.3.4 ELIMINATION

Free drug and metabolic side products are ultimately eliminated from the body. The major routes of elimination are bile/feces, urine, sweat, tears, and milk. The efficiency of elimination is described by the compound half life or $t_{1/2}$. The half-life of drug is the time taken for the concentration of the drug in the blood or plasma to decrease to half its value. For highly toxic anticancer agents, elimination fluids may have toxic potential that may need to be addressed for patients and health-care providers.

12.4 Determination of Compound PK

12.4.1 PRECLINICAL PK STUDIES

Compound PK is usually determined by taking blood samples at various times after compound administration and determining plasma concentrations as a function of time. Typically high-sensitivity analytical assays are developed to quantify both the levels of parent drug and major metabolites. Various parameters may be used to quantify exposure. Commonly reported parameters are the maximum plasma concentration ($C_{\max}$), the area under the plasma concentration time curve (AUC) – both as protein bound and free fractions –

the volume of distribution (V_d), and the elimination half-life $t_{1/2}$. Efficacy and toxicity have to be related to compound exposure and may be related to different PK parameters. For example, toxicity may be determined by exceeding a particular plasma concentration; it may be related to total drug exposure, or AUC, or to the time the drug is present above a certain threshold concentration. Similar considerations are important in determining efficacy exposure relationships. Threshold-based toxicities such as interactions with cardiac ion channels are typically C_{max} related, while efficacy and toxicities such as hepatotoxicity are more commonly related to AUC.

Preclinical PK studies usually include directed in vivo PK studies and in vitro metabolism studies with isolated enzymes and tissue extracts. For the prediction of human PK, there is no agreed on "best" surrogate species. Experimental animals may have biologies that either mimic or differ from humans, depending on the organ system in question. The preclinical species best matching human drug metabolism pathways are generally also best in accurate prediction of human exposure.

Different enzymes of the cytochrome P450 system, which includes many members, may be used in the hepatic metabolism of different small molecule agents. The cytochrome P450 reductase 3A4 (CYP 3A4) is the most commonly employed pathway for metabolism of small molecule drugs. P450 profiles can vary among species and even among individuals within a species. The genetic basis for polymorphisms in cytochrome P450 expression are important and relatively well understood for humans. Typically, species used for preclinical PK testing include rodents (rats and mice), dogs, and in some cases primates (rhesus monkey, cynomologus monkey, and chimpanzee). The use of higher nonhuman primates, such as the chimpanzee, for drug testing is strictly regulated to avoid undue pain or stress to the animal. Therefore, the highly toxic nature of many anticancer drugs frequently precludes their testing in these animals. PK studies may be carried out in additional species, if needed, to support evaluation of preclinical efficacy and toxicity studies.

During early phases of drug discovery, it may be important to screen a cohort of related compounds to search for ones with PK properties suitable for delivery in preclinical and clinical studies. To increase the throughput of compounds, it is often possible to dose several compounds simultaneously (so-called n in 1 studies) and determine which compound results in the best exposure. The feasibility of this approach requires analytical methods that can distinguish between the individual components tested and that the compounds tested show little interference in metabolism, uptake, or transport. Ideally, the compound showing acceptable PK properties for preclinical PK and safety testing will also be the compound with optimal predicted PK in humans, although unfortunately this may not always be the case.

PK studies should also aim at determining whether compound exposure levels increase linearly with increasing dose and over what dose range this occurs. Nonlinear PK can have a variety of causes. If hepatocellular compound concentrations significantly exceed the capacity for metabolism, pathway saturation may occur. Thereafter, drug levels will increase disproportionately to dose. Similarly, depending on the plasma protein binding component, it may be possible to saturate plasma protein binding with a similar outcome. Nonlinear PK can be especially problematic for classical cytotoxic anticancer drugs, for which the therapeutic window is narrow, leading to harmful and potentially lethal drug levels being achieved.

12.4.2 SUGGESTED SCHEME FOR PRECLINICAL EVALUATION OF A NOVEL ANTICANCER AGENT

A general scheme for PK evaluation of a novel anticancer agent is suggested below. Obviously, specific considerations may require modification of this scheme to suit individual agents. However, this scheme provides minimal guidelines for the factors that should be considered.

1. First, develop an analytic assay for the agent and, if possible, its likely metabolites. This assay needs to be sensitive enough to detect the agent at therapeutic levels and preferably at levels at least 10- to 100-fold below the therapeutic level. Commonly used assays for this purpose use a combination of high performance liquid chromatography (HPLC) and mass spectrometry. For a biologic agent, an immunoassay may represent a suitable alternative. In some cases, a bioassay can be used to detect the presence of active agent with high sensitivity, but these assays are rarely convenient for high throughput and cannot distinguish between parent drug and active metabolites.
2. Determine what species are going to be used for preclinical efficacy and toxicity studies, and if necessary, an additional species most likely to be predictive of responses in humans, if known. For anticancer drugs, the choice for efficacy species is typically the mouse, and for toxicity species typically the mouse, rat, and dog. An important consideration in the selection of preclinical species is that at least one should possess a measurable surrogate for the desired pharmacodynamic end point. Defining a surrogate for efficacy and toxicity in the same species yields the most accurate and predictive estimates of therapeutic index.
3. Evaluate plasma protein binding in the efficacy species, the toxicity species, and humans. This is most accurately done by equilibrium dialysis. An estimate of plasma protein binding can be achieved by spiking a bioassay with physiological concentrations of major plasma protein components, serum albumin and α-acidic glycoprotein, and looking for shifts in IC_{50} or K_i values in the assay for the drug target protein.
4. Conduct single-dose IV PK studies in the efficacy species and the toxicity species. Identify metabolic products, if possible, and determine the likely route of clearance of the agent. Determine the levels of agent in plasma and (if possible) the tumor that are associated with efficacy (either antitumor efficacy or antitarget efficacy). In addition, determine the levels of agent associated with toxicity, so that efficacy and toxicity can be related to compound exposure. Efficacy studies should encompass experimentation with different dosing schedules, to maximize the therapeutic window (i.e., the dose that maximizes efficacy and minimizes toxicity).
5. If a route of delivery other than IV is preferred, either for preclinical or clinical testing (e.g., oral), determine the PK of the test agent by that route in

the efficacy and toxicity species. For example, IP delivery is frequently more convenient for multiple dosing to rodents, whereas PO delivery is clearly preferable in the clinic. Oral bioavailability can be predicted to some degree in the absence of formal oral PK studies by the use of $CaCO_2$ intestinal cell permeability assays. Such assays may be useful for screening libraries of compounds for those most likely to have oral bioavailabilty, but they are not fully predictive and results need to be confirmed with formal oral PK studies.

6. Perform experiments aimed at predicting the likely metabolism in humans. For agents that are metabolized by the liver, this can be achieved by performing in vitro metabolism studies with human liver slices or human liver microsome fractions. Known inhibitors of metabolic enzymes may be used to determine which specific metabolic route is preferred (e.g., inhibitors of individual P450 enzymes for compounds metabolized in the liver).

7. Test multiple dose levels to determine if the PK properties vary in a linear or nonlinear manner with dose.

8. If efficacy of the agent is achieved without significant toxicity, perform PK studies at elevated levels of the agent where toxicity is seen (TK studies). These studies allow one to evaluate at what drug exposures toxicity is seen and what the toxicities are likely to be (see below).

12.4.3 CLINICAL DETERMINATION OF PK

Drug PK issues may be identified first in Phase I clinical trials. As indicated above, the FDA requires that studies be performed preclinically to address the likely absorption, distribution, metabolism, and excretion of the agent. Once a compound enters the clinic, PK parameters can be determined directly in humans by sampling from dosed human test subjects. In these early trials, the effects of drug dosing with food and effects of gender and age are typically evaluated.

As stated earlier, PK evaluation has not been historically widely used in guiding the clinical administration of anticancer drugs. Agents have been introduced on a milligram per square meter basis to a dose deemed to represent the MTD based on Phase I clinical studies. The problem with this approach is that given the inherent interpatient variability in metabolism and PK, an agent may either be dosed insufficiently, meaning the patient is not effectively treated, or overdosed, meaning that the patient may be put at risk for life-threatening toxicity. More recently, several studies have been performed to collect PK data from dosed patients and relate outcomes to drug exposure. This approach allows assessment of exposure levels that are associated with limiting toxicities. Once this is known, agents can be dosed to a predetermined exposure as determined by compound plasma levels or an appropriate surrogate thereof. This approach is termed *pharmacologically guided dosing,* and it is increasingly being employed to dose cancer patients (Van den Bongard et al. 2000).

The adequate resolution of PK issues is typically important in the development and delivery of anticancer agents. Frequently, drugs with a narrow therapeutic index are delivered to an older patient population already compromised by a debilitating disease, which may in many cases affect patient metabolic function. The liver, a highly active metabolic site, is also a common site for metastasis. Wide variations in interpatient drug metabolism is to be expected, and monitoring PK parameters may have an important role in ensuring that individual patients receives maximum therapeutic benefit with minimum toxic compromise. In practice, toxic exposure is frequently measured indirectly, by monitoring surrogates such as hematologic assessment of hematopoiesis and the serum transaminase activities for hepatotoxicity.

12.5 Pharmacogenomics

Genetic variation in the patient population can profoundly influence the efficacy and toxicity of a therapeutic. Variations in the target itself or in metabolic pathways important for the activation of prodrugs can affect drug efficacy. Similarly, variations in metabolic capacity or in components of targets responsible for negative side effects can modify the expression of toxicity. For anticancer drugs that are toxic by nature, this should be a particularly important issue, because the therapeutic indices of the drugs and the margin for error is so small.

Metabolic variation can result from genetic variation in components of cytochrome P450 enzyme complexes. In drugs metabolized solely through cytochrome P450 2D6, a polymorphism with low enzyme activity in a significant minority of the population may result in high and potentially toxic plasma drug concentrations (Ingelman-Sundberg, 2002). Variations in DPD levels between patients dictates toxic responses to 5-FU (Etienne et al., 1994). Genetic polymorphisms in the cardiac IKr (internal potassium rectifying) current, hERG (human ether-a-go-go-related gene), an ion channel critical for myocardial depolarization, potentiate the effects of may drugs that interact with this channel, leading to QT interval prolongation and an enhanced propensity for developing potentially fatal arrhythmias (Escande, 2000).

To date, the study of the role these factors play in dosing cancer patients with cytotoxic drugs has been relatively limited. Genotyping of common drug metabolizing and cardiac ion channel polymorphisms may allow more effective and safer dosing of individual patients. The field of pharmacogenomics, introduced at only a superficial level here, is emerging as an exceptionally significant area of drug development, as the emerging era of individualized medicine develops in this century.

12.6 Toxicity Issues

Toxicology issues play a critical role in the testing of potential new anticancer agents. Despite ongoing debate about their clinical predictive value, classically novel anticancer agents are tested preclinically in rodent-derived efficacy models. These are most typically xenograft models, in which human tumor-derived material or cell lines are propagated in immune-compromized mice (nude or SCID mice), or murine tumor models (e.g., L1210 or B16 melanoma), potentially including transgenic mouse tumor models (see Chapters 10 and 11). Rarely are agents broadly curative in these models, and toxicity to the host frequently limits the ability to dose the agent to full potential efficacy. Dose scheduling studies are used to determine empirically a dosing regimen that minimizes toxicity and maximizes efficacy – that is, to optimize the therapeutic index. These scheduling studies are sometimes used as a guide to possible scheduling in the clinic.

12.6.1 PRECLINICAL TOXICOLOGY STUDIES

The objective of preclinical safety studies is to guide dosing of patients in the clinic and to help define a likely safe starting dose to use in Phase I studies. The toxicities observed in preclinical testing might also give some indication of those that might be observed in dosed human patients. Broadly speaking, anticancer agents tend to be grouped into two categories: cytotoxic agents and noncytotoxic agents. Examples of the former category include DNA-damaging agents, antimicrotubule agents, and antimetabolites. These agents constitute the mainstay of classical anticancer therapy. Examples of the latter category include steroid hormone receptor antagonists such as antiestrogens and antiandrogens.

One of the principal concerns of the regulatory agencies in allowing testing of novel therapeutics in humans is that this testing be done with the minimal safety liability for the patient. Before an agent enters Phase I clinical trials, it is required that an appropriate series of toxicology and safety pharmacology studies be performed preclinically. Likely toxicity associated with administration of the agent needs to be defined, and it is important that the reversibility of these toxicities be assessed. Typically, these studies are carried out in at least two mammalian species, including one rodent and one nonrodent species. First-line studies include single-dose acute toxicity studies. However, longer term multiple-dose studies are usually required to support the longer-term efficacy studies that are carried out in Phases II and III. Toxicology studies usually involve dose escalation experiments to determine the side effects of the compound, sometimes at supratherapeutic doses; and these may provide guidance for specific testing to be performed during Phase I trials.

In addition to standard safety studies, particular attention should be paid to assessing potential toxicities that may be suggested by the agent's mechanism of action, chemical structure, and profile of cross-reactivity with nonspecific targets and by experience with related agents. The choice of preclinical test species may be strongly dictated by the PK profile of the compound. For example, it may be desirable, if possible, to test a compound preclinically in a species that generates a similar profile of potentially toxic metabolites to those expected in humans.

Single-dose toxicology studies must also be completed before an agent can enter Phase I trials. This work typically involves dose escalation studies aimed at determining the *no effect level* of the drug, the dose at which toxicity is first observed, and the nature of that toxicity. These studies help define a safe starting dose for Phase I clinical trials, and they provide some guidance to the toxicities that might be observed in the clinic. In all cases, the toxicity data must be interpreted with reference to drug exposure, as determined in accompanying PK and TK studies.

More extensive toxicity and repeat dose studies are frequently required to support longer-term efficacy testing in the clinic in Phases II and III. The exact format of these studies depend on the drug being tested and its likely dosing in clinical practice.

12.6.2 SAFETY PHARMACOLOGY STUDIES

In general, safety pharmacology studies may be performed as separate studies or as part of toxicology studies. These studies need to be completed before a compound can enter Phase I clinical trials. In vitro target selectivity panels are typically contracted to one of several vendors (e.g., MDS Pharma, CEREP, Nova), which assess off-target activities against a variety of specific molecules. Additional selectivity criteria are usually added for a specific target. Furthermore, compounds are evaluated for their ability to inhibit the cardiac myocyte potassium, K^+ rectifying current through interactions with the hERG channel in conveniently transfected cell lines. Typically, an in vitro evaluation of combined ion channel effects (the action potential duration) is conducted on isolated Purkinje fibers of dogs or rabbits. In vivo evaluations of the function of major vital organ systems – the cardiovascular system, the cental nervous system (CNS), and the respiratory system – are also undertaken. In addition, these studies may include assessment of the effect of the agent on the gastrointestinal system and the renal system. Typically, safety pharmacology studies aimed at assessing impact on the cardiovascular and central nervous systems are performed in dogs. The test agent is administered at progressively increasing bolus doses, encompassing the anticipated effective exposure and exposure levels above this level. Cardiovascular function is monitored by electrocardiography (ECG), cardiac contractility by telemetry, and blood pressure, and CNS function by monitoring overt behavioral effects. Gastrointestinal and renal studies are typically performed in rats. Gastrointestinal function is monitored by evaluating the effect of the test agent on the rate of passage of a carbon marker through the gastrointestinal system.

The safety pharmacology studies listed above are important for anticancer agents. Acute adverse effects on the CNS or the cardiovascular system may be less well tolerated or even acutely fatal in the clinic, compared to the generally fully reversible hematopoietic or gastrointestinal toxicities that develop several days after treatment. Safety pharmacology findings, such as relatively potent inhibition of hERG channels, are, as for all toxicities, considered in the context of risk assessment for the particular indication. The early identification of such liabilities frequently leads to a strategy to reduce this liability in back-up molecules.

12.6.3 GENOTOXICITY, REPRODUCTIVE TOXICITY AND ADDITIONAL STUDIES

In general, cytotoxic anticancer agents are considered unsafe for dosing to pregnant women. Such agents also induce sterility in males and are frequently mutagenic and carcinogenic. The evaluation of reproductive system toxicity, genotoxicity, and carcinogenicity – routinely performed for other therapeutic agents – is usually not undertaken for cytotoxic anticancer agents.

Although these types of toxicity may not negatively affect the potential for use of an agent in the oncology clinic, they may still need to be assessed for other reasons. Anticancer drugs that act by damaging DNA will score positive in genotoxicity and probably in carcinogenicity studies as required by the FDA for new agents. Although this does not affect the use of these agents as anticancer drugs in the clinic, it does have important implications for the protocols for manufacturing the agent and for handling the agent, including by members of the health-care profession and to protect nonpatients from the agent's harmful effects.

The regulation of noncytotoxic agents and novel anticancer therapies suitable for long-term administration will predictably move closer to that for other pharmaceutic agents that pose less risk.

12.6.4 CLINICAL TOXICOLOGY STUDIES

It is widely accepted that cytotoxic anticancer agents will likely have a high level of toxicity to the organism and, as a result, a low therapeutic index. The clinical oncology community has many years of experience with drugs of this type, and in many cases it has determined empirically how to dose these agents effectively. For this type of agent, dose escalation studies are used in Phase I to define the clinical MTD for subsequent use in more detailed Phase II efficacy studies. Choosing the clinical starting dose for Phase I involves a consideration of the toxicity that the agent has shown preclinically, in both rodent and nonrodent test species. The manner in which this information can be used to determine the clinical starting dose is outlined in Figure 12.1, which offers a general guide for starting dose selection for cytotoxic agents in cancer patients.

First, the dose of the agent that is severely toxic or lethal to 10% of animals (SD_{10}) of the chosen rodent species is determined. In preclinical testing, agents are usually dosed on a milligram per kilogram basis. However, in the clinic anticancer agents are dosed on a milligram per square meter basis. Therefore, the rodent SD_{10} is converted to the milligram per square meter equivalent (see www.fda.gov/cder/cancer/animalframe.htm) and one tenth of this dose is administered to the nonrodent test species. If this dose is not severely toxic to the nonrodent test species of choice and if there is no reason to believe that the rodent is an inappropriate species for prediction of human toxicity, then one tenth the rodent SD_{10} is used as the clinical starting dose for Phase I studies in humans. However, if one tenth the SD_{10} is severely toxic to the nonrodent species of choice, then a lower dose may need to be used

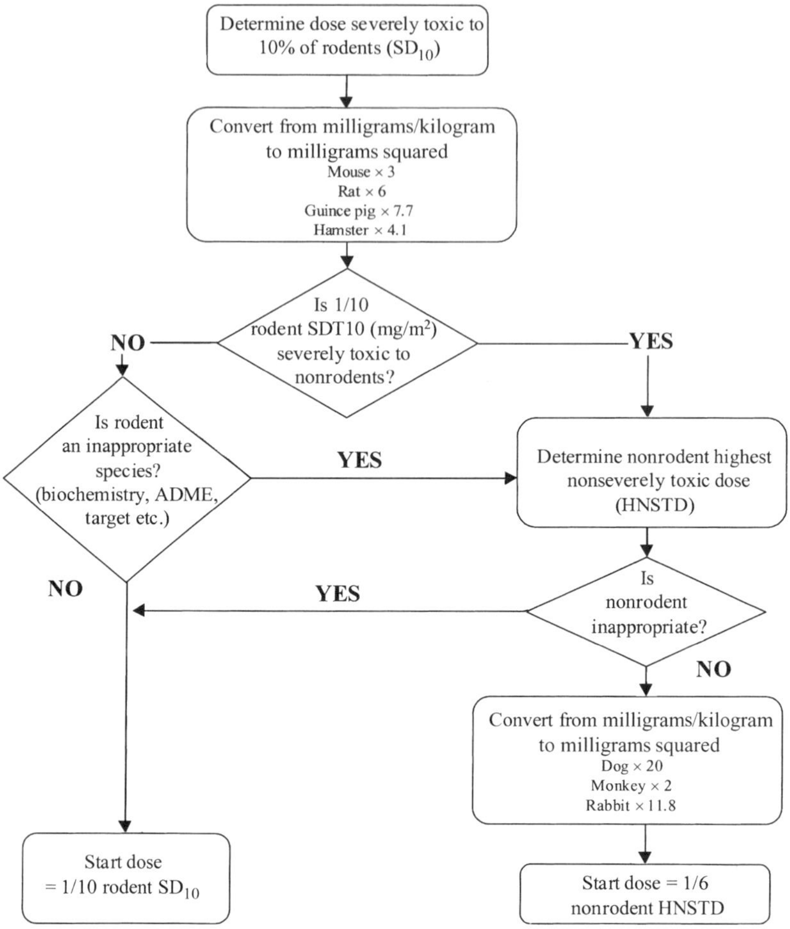

Figure 12.1 Flow chart for suggested preclinical safety assessment of a novel anticancer agent.

as the starting dose in Phase I. Provided there is again no reason to believe that the nonrodent species is an inappropriate predictor of likely toxicity in humans, the highest dose that can be given to this species without incurring serious toxicity – the MTD or highest nonseverely toxic dose (HNSTD) – is determined. The starting dose for Phase I studies in the clinic is then set at one sixth the nonrodent HNSTD. Using this kind of empiric formula in establishing Phase I doses is unique to the development of anticancer agents.

In these studies it is important to chose preclinical safety species most appropriate for prediction of likely toxicity in humans. Sometimes this is a difficult judgment call. However, it may be important to exclude, for example, a species in which the metabolism of the tested agent is substantially different from that predicted in humans. To assess target-mediated toxicity, it is also important that the agent be capable of interacting similarly with the target in both humans and the test species of choice. The specific considerations might be different for each individual test agent, but the goal is to maximize the possibility that the toxicity likely to be encountered when the drug is administered to patients will be uncovered in preclinical testing.

12.6.5 COMMON TOXICITIES ASSOCIATED WITH CYTOTOXIC ANTICANCER DRUGS

Certain types of toxicity are common in classical cytotoxic anticancer drugs. Typical targets for adverse effects are the proliferating cells of the hematopoietic system and the gastrointestinal tract. These toxicities are common and, although undesirable, can be readily managed in the clinic. As a result, these types of toxicity do not generally mean that an agent cannot advance through clinical trials and gain FDA approval. Indeed, researchers at the U.S. National Cancer Institute (NCI) evaluate the effect of potential new therapies on primary human hematopoietic cell cultures as a means of estimating toxic drug levels in humans and predicting likely therapeutic margins in patients.

However, other types of toxicity can be problematic if they might lead to a poor risk–benefit assessment. Examples of more problematic toxicities, including for cytotoxic anticancer therapies, include effect on the cardiovascular system and the CNS. Potential for negative effect on critical organ systems is evaluated in preclinical safety pharmacology studies, frequently carried out in dogs.

12.6.6 TOXICOLOGY AND NONCYTOTOXIC ANTICANCER DRUGS

For noncytotoxic agents, the conduct of safety studies and the mechanism for starting dose selection for Phase I studies could arguably be different from that described above, approaching protocols applied to agents being tested for other therapeutic areas. For noncytotoxic agents, which includes many of the more molecular-targeted therapeutics, the level of whole organism toxicity expected may be much lower than for standard cytotoxic drugs. Therefore, it makes little sense to dose noncytotoxic agents in the clinic at the MTD, since this might represent a substantial overdose with respect to target inhibition, increasing side effects without enhancing efficacy. Nevertheless, these agents have been – to date at least – typically dosed using approaches developed for standard cytotoxic therapies. As discussed further below and in Chapter 14, new approaches must be evaluated to test novel molecular-targeted therapies appropriately in the clinic.

12.6.7 PRECLINICAL ASSESSMENT OF COMMON TOXICITIES OF ANTICANCER DRUGS

In the typical preclinical evaluation of toxicity for an anticancer agent, groups of male and female animals are dosed, generally at half-log intervals (e.g., 10, 30, 100 mg/kg/day) by the desired route for clinical exposure with a form of drug that is identical to that planned for use in the clinic. Effects in drug-dosed animals are compared to control groups dosed by the same route with vehicle.

Table 12.1 Common Toxicities Displayed by Cytotoxic Anticancer Agents

Target Tissue	Histogical Alteration
Bone marrow	Depletion of hematopoietic lineages Replacement with adipocytes, myelofibrosis
Peripheral blood	Reticulocytopenia (observed on stained blood films) Neutropenia (observed on stained blood films)
Spleen, lymph nodes	Lymphoid depletion
Small intestine	Necrosis of crypt enterocytes
Skin	Loss of hair from follicles – telogen appearance
Testis	Degeneration of seminiferous tubules Loss of spermatogonia, spermatids, and sperm

Blood plasma samples are generally taken on day 1 and at the end of study to document drug exposure. Body weight changes, behavioral changes (e.g., nervousness, aggression, emesis, seizures), alterations in appearance, ECG parameters, and food consumption are monitored. The eyes are examined for retinal and other ophthalmologic effects. Animals are observed especially closely to the C_{max} of drug for potentially adverse effects. At the end of the study, blood and urine are collected for serum chemistry, hematology, and urinalysis. Organs are weighed. In addition, a selection of tissues representing all major organ systems are fixed in 10% buffered formalin and processed for microscopic evaluation by trained veterinary histopathologists. Particularly close attention is paid to common target systems affected by anticancer agents, including bone marrow, gastrointestinal tract, metaphyseal growth plates, and skin and reproductive systems. If the drug is dosed by the IV or SC routes, the site of administration is collected to assess local irritation.

Table 12.1 shows the changes commonly associated with administration of a cytotoxic anticancer agent. These effects derive both from the targeting of rapidly dividing cells and the stress of high-level systemic toxicity.

12.7 Examples of PK and Toxicity Issues of Common Anticancer Therapies

12.7.1 DNA DAMAGING AGENTS

Among the most commonly used anticancer agents are those that directly cause damage to DNA. Among these agents are the anthracycline antibiotics (doxorubicin, daunorubicin, idarubicin), alkylating agents (cyclophosphamide, bleomycin), and the platinum-based anticancer drugs (*cis*-platin, carboplatin, oxaloplatin). Each of these agents is believed to work primarily through the formation of DNA adducts. These adducts, once formed interfere with enzymatic processes on the DNA template and cause interruption of transcription, replication, and repair. Commonly the generation of stalled replication complexes leads to DNA strand breakage, which in turn causes cell death. It has been estimated that as little as a single unrepaired strand breakage may be sufficient for cell kill. For each of these agents additional mechanisms have been proposed which may contribute to cellular toxicity.

The common target of these agents is DNA. Therefore, to be effective they must be able not only to enter cells but also permeate the nuclear compartment. DNA adducts, once formed, may linger for some time, particularly in nonreplicating cells, before the are converted into a potentially toxic strand-break upon cell cycle entry. As a result, the toxic effect of the agent may be manifest long after free drug has been cleared from the peripheral circulation. This may make it difficult to relate PK parameters determined from peripheral drug levels to efficacy and toxicity outcomes. In general, with this class of molecule it appears that total drug exposure or AUC correlates with increased DNA damage; therefore, increasing both efficacy and certain toxicities.

All cells, in particular proliferating cells, are subject to the toxic effect of this class of agent. In particular, hematopoietic and gastrointestinal toxicities are reflected in side effects like nausea and vomiting. Alopecia (baldness) is also common with cytotoxic compounds. Despite these general effects on proliferating cells, some degree of tumor cell selectivity is believed to result from the defects in checkpoint control pathways in tumor cells.

Cells can be protected from the damaging effects of these agents by endogenous DNA repair pathways. Repair defects common in tumor cells can be exploited to provide tumor cell and even tumor type selectivity.

12.7.1.1 Anthracyclines

Anthracycline antibiotics that are used commonly as cancer chemotherapeutic agents include doxorubicin, epirubicin, and idarubicin. These agents constitute the mainstay of a wide range of treatment protocols used in clinical practice. Epirubicin and idarubicin are second-generation compounds developed to provide an improved toxic side-effect profile relative to doxorubicin. These agents intercalate between DNA base pairs and inhibit DNA replication and transcription. In addition, it has been reported that they may exert further cytotoxic effects by inhibition of cytochrome c oxidase activity, free radical formation, lipid peroxidation, chelation of iron, and generation of reactive oxygen species that lead to oxidative stress.

The anthracycline antibiotics are largely metabolized by the liver by aldo-keto reductase and by cytochrome P450 reductase. Biliary elimination and fecal excretion of parent compound and metabolites provide the major routes of clearance. The major toxic side effects of these agents are hematological and cardiotoxicity, which may be both acute and chronic. A significant time lag occurs between the time of administration of the drug and the nadir in neutrophil counts. Neutrophil counts < 1000/uL render a patient susceptible to bacterial infection. For this reason, patients receiving such cytotoxic agents now routinely receive recombinant granulocyte colony stimulating factor (G-CSF) to stimulate neutrophil proliferation and improve recovery.

Chronic cardiotoxicity with this class of cytotoxic agents represents a serious side effect that can in some cases be lethal, due to the development of a cumulative dose-dependent congestive cardiomyopathy. Cytochrome P450-generated aglycone metabolites are believed to play an important role in the development of cardiotoxicity. For doxorubicin there appears to be a correlation between the AUC for the drug and drops in leukocyte and platelet counts. In addition, a correlation has been shown between the AUC for the active metabolite doxorubicinol and decreases in neutrophil and platelet counts. In contrast, the occurrence of cardiotoxicity appears to be more correlated with C_{max} and can be ameliorated to some extent by dosing the drug with a prolonged intravenous infusion. Unfortunately, the use of a prolonged infusion increases the severity of mucositis and bone marrow suppression, which then become the dose-limiting toxicities (Danesi et al., 2002).

12.7.1.2 Alkylating Agents

Chemotherapeutic alkylating agents include cyclophosphamide and bleomycin. Of these agents, cyclophosphamide is the most commonly used, playing an important role particularly in the treatment of breast cancer. This agent requires bioactivation before it can exert its cytotoxic effect (Ayash et al., 1992). Cyclophosphamide is metabolized in the liver and renally excreted. Myelosuppression and cardiotoxicity represent toxic side effects of cyclophosphamide. No relationship has been shown between the AUC for this agent and myelosuppression (Lichtman et al., 1993). However, it does appear that higher concentrations of active metabolites contribute to cytotoxicity and cardiotoxicity (Ayash et al., 1992).

12.7.1.3 Platinum-Based Drugs

Platinum-based drugs include cisplatin, carboplatin, and oxaloplaton. These agents form platinum-based adducts on DNA that interfere with DNA replication and transcription. Cisplatin and carboplatin differ in the nature of the leaving group that is released on adduct formation; however, they result in the formation of the same adduct on DNA. Oxaliplatin not only has a different leaving group from the two agents above but also forms a structurally different DNA adduct. These adducts are repaired by different repair pathways, and this can affect tumor selectivity of the different platinum-based drugs. Carboplatin and oxaloplatin display less toxicity than cisplatin and a different spectrum of antitumor activity (Levi et al., 2000; O'Dwyer et al., 2002).

12.7.2 AGENTS TARGETING ENZYMES INVOLVED IN DNA METABOLISM

DNA damaging agents other than those listed above also have clinical utility in cancer therapy. DNA damage can be inflicted either directly or indirectly by inhibition of enzymes involved in DNA metabolism. Examples of currently used therapies that have enzyme targets in this category include 5FU, which targets metabolic pathways important for thymidine synthesis, thereby causing disruption of DNA replication; gemcitabine, which is a nucleoside analog that also disrupts DNA synthesis; inhibitors of DNA topoisomerase I (e.g., irinotecan) (Mathijssen et al., 2000) and DNA topoisomerase II (e.g., etoposide) (Desoize et al., 1990); and other enzymes that control DNA topology during transcription and replication. By interfering with DNA replication, inhibition of each of these targets primarily leads to generation of DNA strand breaks in replicating cells, ultimately resulting in cell death.

Agents that damage DNA tend to have preferential effect on replicating cells and have relatively little intrinsic selectivity for tumor cells over normal cells. Toxicities experienced with each of these agents are similar to those seen with other DNA damaging agents and commonly include hematopoietic toxicities, gastrointestingal toxicities, and alopecia. Despite the fact that these agents have enzyme targets, they are usually dosed to an empirically determined MTD in a manner similar to that followed for direct DNA damaging agents. Relatively few studies have addressed the relationship between efficacy, toxicity, and target enzyme inhibition.

12.7.2.1 5-Fluorouracil

The antitumor activity of 5FU is usually attributed to inhibition of thymidylate synthase. 5FU itself has to be delivered by the IV route, since the drug has highly variable gastrointestinal absorption. The drug has a narrow therapeutic index, because it has no tumor-specific distribution. Precursors of 5FU have been developed, for example, the compound capecitabine (Xeloda), which has the advantage of being suitable for oral administration. In addition, these derivatives exploit tumor-specific pathways to direct 5FU generation preferentially in tumor tissue. Capecitabine takes advantage of elevated levels of cytidine deaminase and thymidine phosphorylase in colorectal tumors, as these are key enzymes in the conversion of this precursor to active 5FU (Reigner et al., 2001).

The rate-limiting enzyme in 5FU catabolism is DPD. This enzymc is present in both tumor and healthy tissue but is especially high in liver. DPD levels can vary in human populations, and there are rare individuals who totally lack this enzyme. Diminished DPD activity can have serious negative patient consequences, given the intrinsic narrow therapeutic index of the drug, potentially leading to serious gastrointestinal toxicity, myelosuppression, and neurological toxicities, which in some cases can be life-threatening (Etienne et al., 1994). Conversely, it has been suggested that carefully monitored inhibition of DPD may represent a strategy to prolong exposure to 5FU and increase drug efficacy.

12.7.2.2 Gemcitabine

Gemcitabine is a nucleoside analog; its toxicity strongly depends on dose scheduling. Prolonged infusion is preferred for optimal efficacy; weekly bolus dosing leads to increased toxicity. Myelosuppression is the dose-limiting toxicity. Like other nucleoside analogs (cytaribine, fludaribine), gemcitabine is a prodrug that is phosphorylated by nucleoside salvage pathways, producing a triphosphate metabolite (nucleotide) that is incorporated into DNA. Gemcitabine is highly metabolized in solid tumor cells (Johnson, 2000).

12.7.2.3 Topoisomerase Inhibitors

Topisomerase inhibitors target the DNA topoisomerases that relieve torsional stress arising during DNA replication, repair, and transcription. Drugs have been developed to target both topoisomerase I (camptothecin/irinotecan) and topoisomerase II (etoposide, VP16). Typically, these drugs work by binding to DNA-bound topoisomerase and trap the enzyme in a DNA-bound complex. When cells replicate, this trapped complex results in DNA strand breakage and ultimately cell death. As a result, these agents have a toxicity profile that is similar to more direct DNA-damaging agents, with some additional unique features. For example, for Irinotecan the toxic side effects include diarrhea, fever, and shortness of breath (Mathijssen et al., 2001).

12.7.3 ANTIMICROTUBULE AGENTS

A third class of agents, generally considered to fall into the category of cytotoxic agents, is represented by microtubule poisons. These agents include the taxanes, such as Taxol (paclitaxel), and Taxotere (docetaxel), and other microtubule poisons such as vinblastine. By interfering with microtubule function, these agents interrupt the mitotic process and cause cell death, again with little selectivity for tumor cells. As expected, these agents have hematopoietic and gastrointestinal toxicity, but additional toxicities are also seen, including peripheral neuropathy, cardiotoxicity (rarely; only for paclitaxel), and fluid retention (only for docetaxel). Unlike docetaxel, paclitaxel shows nonlinear pharmacokinetics, meaning that a disproportionate change in plasma concentration and AUC is seen with increasing dose.

The first-generation taxanes are insoluble in aqueous solution, and significant negative reactions to the vehicles used for drug administration have been recorded. It has been shown that paclitaxel has more efficacy with prolonged exposure. However dosage of the compound in the clinic has drifted to shorter infusion times driven by convenience and economic considerations. Different toxicities are observed with different schedules. Myalgia and neuropathy were more prominent with short-term infusions; neutropenia and mucositis were more prominent with longer-term infusions.

Taxanes are highly protein bound ($> 90\%$), metabolized by cytochrome P450 enzymes, and excreted in the bile (Clarke and Rivory, 1999; Vaishampayan et al., 1999).

12.7.4 NONCYTOTOXIC CHEMOTHERAPEUTIC AGENTS

Chemotherapeutic agents also include therapies that are not generally broadly cytotoxic. These therapies commonly target specific functions that are important for the survival of tumor cells. As such, they tend to have more intrinsic selectivity for the tumor cell over normal cells and less accompanying toxicity. Conversely, they tend not to be as broadly active against multiple tumor types. Among these agents are those that target steroid hormones that are important for survival of reproductive tract tumors (e.g., breast, ovary, prostate). In addition, this category includes agents that target signal transduction pathways that are critical for malignant transformation and survival of maligantly transformed cells.

Given that these agents have a defined molecular target and a much more desirable safety profile, the potential exists to dose these agents based on target occupancy or inhibition. In other words, dosing can proceed to "target efficacy" rather than to MTD used traditionally. Nevertheless, in practice even antiestrogens are probably dosed to levels that exceed those needed to completely occupy the estrogen receptor in tumor cells, where the therapeutic effect is thought to be required.

12.7.5 STEROID HORMONE RECEPTOR MODULATORS

12.7.5.1 Antiestrogens

Antiestrogenic agents are believed to work by binding to the estrogen receptor (ER). The ER a ligand-activated transcription factor that is important for controlling the expression of genes necessary for appropriate functioning and survival of cells in the female reproductive tract. Estrogens also play a beneficial role in bone development and in cardioprotection. Upon binding ligand, receptor conformation is believed to change, allowing it to bind to protein co-activators that co-operate with the receptor to drive specific activation or inhibition of various genes. Co-activator populations may differ among tissues, so an estrogen receptor modulator may be estrogenic or antiestrogenic in different tissues, depending on the co-activator population distribution.

Mechanism-driven side effects of estrogen modulators can include inhibition of beneficial effects of estrogen, such as bone protection and cardioprotection or stimulation of unwanted estrogenic effects such as uterine epithelial cell proliferation or hot flashes. This has led to the search for selective estrogen receptor modulators (SERMs), molecules that block the proliferative action of estrogen on breast tumor tissue but maintain the positive effects for cardioprotection and prevention of osteoporosis (Buzdar and Horrobagyi, 1998; Taras et al., 2000).

Steroid hormone receptor modulators are free of the more adverse hematological and gastrointestinal toxicities associated with standard cytotoxic chemotherapeutic agents. Their more improved safety profile makes these agents suitable for long-term use in a chemopreventive mode. For example, tamoxifen is administered as a chemopreventive agent to women at high risk for developing breast cancer.

12.7.5.2 Aromatase Inhibitors

Aromatase inhibitors block estrogen biosynthesis by targeting the enzyme aromatase, an enzyme in the sex hormone pathway that is involved in conversion of androstenedione to the female hormone estradiol. Aromatase inhibitors are effective against hormone-dependent breast cancers and other estrogen-dependent tumors with little systemic effect on proliferating tissues. Exemestane and formestane are steroidal type I aromatase inhibitors that bind irreversibly to the aromatase target. Anastrazole and letrazole are nonsteroidal, reversible competitive inhibitors of the enzyme. Potential toxic side effects of these agents result from interference in steroid metabolism, including effects on plasma lipid levels, bone metabolism, and adrenosteroidogenesis. However, individual agents within this class differ in their PK properties and in their specific side-effect profile. For example, letrazole and exemestane appear to have different effects on plasma lipids. These agents also differ in the time it takes for them to reach steady-state dosing conditions. Anastrazole and exemestane reach steady state within 7 days, whereas letrazole takes 60 days to reach steady-state plasma levels. Adverse events occurring during exposure to long half-life drugs, such as letrazole, are particularly dangerous and difficult to manage, due to very slow compound clearance. In addition, results for letrazole suggest a nonlinear relationship between dosing and efficacy for this agent. The potential for drug–drug interactions exist with these agents, since cytochrome P450 inhibition has been reported: Anastrazole inhibits CYP1A2, CYP2C8/9, and CYP3A4; and letrazole strongly inhibits CYP2A6 and less strongly inhibits CYP3A4 (Buzdar et al., 2002).

12.7.5.3 Immune Therapies and Antibody Therapies

A variety of immunotherapies have been explored as potential anticancer therapeutics. These include agents designed to modulate host immune reactions to tumors, and antibody therapies directed to specific targets. Historically, antibody approaches have suffered from significant toxicity that results from host reactions. More recently, the generation of "humanized" monoclonal antibodies, which are made up mainly of human amino acid sequences, has allowed the successful development of a new generation of antibody therapeutics. These biomolecules have side-effect profiles that are in general different from those observed with standard cytotoxic agents, making them especially suitable for use in combination regimens. Monoclonal antibodies that are now used clinically include Herceptin, which targets the *her-2/neu* receptor of the epidermal growth factor receptor family (used to treat *her-2*-positive breast cancer), and Rituximab (Mabthera), which targets the CD20 ligand (used to treat non-Hodgkin lymphoma).

Antibody therapies are dosed by the IV route. Given that antibodies have long circulating half-lives, often well in excess of 1 week and perhaps up to months, these agents can be dosed at intervals that make them attractive for outpatient use. Herceptin for example is dosed on a 7- to 10-day cycle. Side effects that have been reported for Herceptin include fever, chills, nausea, and vomiting. More worrisome, Herceptin has been reported to have cardiotoxic liabilities. This has unfortunately limited the use of this agent in breast cancer patients previously treated with the commonly used anthracyclines, which themselves carry a cardiotoxic liability (Slamon et al., 2001). Side effects associated with rituximab include allergic reactions, flu-like symptoms, nausea, and vomiting. Some lowering of blood pressure has also been reported.

12.8 Tumor Selectivity Engineered by Tumor Site Drug Delivery

The toxicities of many traditional cytotoxic anticancer therapies stem from the fact that they lack little intrinsic selectivity for the tumor cell. For example, DNA-damaging anticancer drugs are very effective at killing proliferating cells, but they also target normal cells in the bone marrow and intestinal tract. This offers significant medical opportunities to design variants of effective existing therapies, for which the goal is to reduce the accompanying toxicity of the agent by physically targeting it specifically to the tumor cells. Several such examples of tumor-targeted cytotoxic therapies are being explored currently in clinical trials.

The oldest example of a directed cytotoxic agent that causes DNA damage is radiation, by which the cytotoxic principal (γ-irradiation) can be physically directed to the tumor. Biochemical approaches have also been used to target and thereby improve the therapeutic window of existing agents. A variety of such strategies exist, such as, lipid encapsulation and prodrug strategies. Lipid encapsulation strategies take advantage of the observation that tumor blood vessels are relatively permeable compared to vessels in normal healthy tissue. Thus lipid vesicle-based delivery systems can be designed that can escape into tumor tissue through leaky vessels but not into normal healthy tissue (Harashima et al., 1999). Prodrug strategies make use of tumor-selective enzymes to convert harmless precursor drug molecules to the active component specifically at the tumor site. Examples of the latter approach include capecitabine (Reigner et al., 2001), etoposide phosphate (Budman et al., 1994), and peptide-drug conjugates with cleavage sites for tumor-specific proteases.

These agents are typically less systemically toxic than the parent drug. However, the challenge with these approaches is to ensure that the mechanism that allows for conversion to the active agent at the tumor site is sufficiently efficient to permit effective generation of active drug. Otherwise, improved safety will come about at the expense of reduced efficacy, with little overall gain in therapeutic margin.

12.9 Prospects for Novel Therapies

Over the past 20–30 years, our understanding of the basic molecular mechanisms that contribute to human cancer has dramatically increased. This understanding has led to the emergence of a host of potential novel therapeutic targets. The hope is that agents directed to these targets will have improved efficacy and reduced toxicity. So far, the fulfillment of this promise has been limited: Novel agents have tended to have limited efficacy, at best, and often exhibit substantial if different toxicities. Potential reasons that contribute to this situation are probably the multifactorial and heterogeneous nature of tumors, making it likely that agents targeting multiple mechanisms may have to be delivered in combination to achieve broad therapeutic efficacy. The exact combination necessary may vary with the target tumor and with the tumor stage. In addition, as stated earlier, even some of these newer targets are not cancer cell specific, so that inhibition may carry some mechanism-based toxic potential. Finally, the novel agents tested are often not as selective for the intended target as desired. For example, selectivity for kinase targets is particularly difficult to achieve given the large number (> 600) of kinases and the fact that most small molecules that are effective kinase inhibitors are purine analogs that directly compete with ATP binding.

For these newer mechanisms, dosing to MTD may not be the appropriate approach to deliver these agents effectively (see Chapter 14). PK methods will need to be developed to determine when the agent has been dosed to affect the target maximally. If maximally effective drug levels can be achieved (with respect to target modulation), without unacceptable toxicities, then it will no longer make sense to dose these drugs based on MTD but rather to dose them to levels that efficacious in regard to mechanism. If this can be achieved, these agents could be combined in rational combinations to achieve antitumor efficacy, to dose earlier stage patients, and to explore chemopreventive therapies. This combinatorial approach will depend on emergence of therapies with nonoverlapping toxicities.

An improved understanding of the basic biology of cellular transformation and tumor formation has permitted the entry into clinical testing of a variety of novel agents with specific molecular targets in major oncogenic pathways. These include agents targeted to *ras* and *rho* GTPases (farnesyl transferase inhibitors), the epidermal growth factor (EGF) receptor (Iressa, Tarceva, and Erbitux), the *abl* tyrosine kinase (Gleevec), to specific cell cycle regulatory kinases (flavopiridol), and to kinases in various signal transduction pathways that are important for transformation and survival. Novel therapies have also targeted angiogenesis, the process by which new tissue, including tumor tissue, generates a blood supply to allow tissue expansion above a certain size (e.g., marimistat, batimistat, angiostatin, thalidomide).

Although many of these therapies have had specific molecular targets, for the most part they have been tested preclinically and clinically using the principles developed for less specific cytotoxic agents. Very few studies have determined tumor drug levels that are required for target inhibition and related these to any observed antitumor efficacy or toxicity. Part of the problem relates to the challenges posed by some of the targets themselves, which may be hard to monitor by PD measurements (see Chapter 12) and the difficulty in generating and characterizing inhibitory agents that are truly specific for the intended target. For example Gleevec, an oral agent targeted to the *abl* kinase, has inhibitory activity on additional receptor tyrosine kinases in addition to the primary target. As a result, efficacy and toxicity readouts may reflect multiple activities of the tested agent. Improving the selectivity of molecules like Gleevec for their target kinase, relative to off-target kinases, may reduce the liabilities associated with administration without impairing the efficacy of the agent.

The practical goal for those in search of new therapies is perhaps appropriately driven by the need to discover and develop a novel effective agent, rather than to test rigorously the merit of targeting a specific pathway. However, the challenge for the future remains to determine whether molecular-targeted drugs, either alone in combination, when delivered appropriately to mechanism, will improve the therapeutic options available to cancer patients at all disease stages.

As expected, the toxicities reported for some of the existing novel agents differ somewhat from those associated with conventional therapy, but, to date, are still significant. Iressa, an EGF receptor–tyrosine kinase inhibitor with oral activity, has been reported to produce an acne-like rash and diarrhea that constitute the dose-limiting toxicity. Side effects associated with Gleevec include nausea, fluid retention, muscle cramps, diarrhea, muscle and bone pain, skin rash, headache, fatigue, joint pain, and shortness of breath. Serious liver problems have been observed in some patients.

In addition, there are some cases in which drug–drug interactions may influence use of the agent. Gleevec, for example, has drug–drug interactions with acetaminophen and oral contraceptives. It is also metabolized by the same P450 as the antithrombotic agent Coumadin. Some of these interactions may influence the use of this agent, particularly in an older patient population.

12.10 Unconventional Therapies: Antisense, Gene Therapy, Immunomodulation

Because cancer can be such a devastating and life-threatening disease, it represents one arena in which it is ethical – in terms of risk–benefit considerations – for the patient to test some of the more novel and potentially risky forms of therapy. Included in this category is antisense therapy and gene therapy, both of which possess some potential for mutagenicity. Disseminated intravascular coagulation has also plagued early gene therapy trials. Because these agents are very new on the scene, the guidelines for preclinical and clinical testing of these agents are still evolving and will likely require different approaches to classical agents, cytotoxics, or otherwise.

Immunomodulatory therapy in which the general immune system function is enhanced, or tumor-associated antigens are rendered immunogenic (tumor vaccines) are also under development as novel anticancer strategies. Careful management of generalized or local immune-mediated toxicity will likely be required for the safe administration of such agents.

12.11 Combination Therapy and Its Implications

Among the challenges faced in the treatment of cancer are the heterogeneity of most human cancers and the genetic instability of the cancer cell. This genetic instability increases the likelihood that mutations that license drug resistance will occur; such mutations can be a major contributor to tumor heterogeneity. Probably as a result, human tumors rarely respond strongly to a single agent, and combination therapy is used commonly in the clinical setting. An important issue to be considered with any novel agent is how well it will interface with preexisting therapies. Not only does this requires a consideration of mechanism of action issues also, but in a setting in which established drugs are frequently dosed to the MTD, toxicity becomes a critical issue in combination therapy. Agents can be used most effectively in combination if they lack overlapping side effects. Toxicity considerations also play an important role in determining which anticancer agents can be used together. For example the potential cardiotoxicity reported for Herceptin has limited its use in combination with anthracycline, for which cardiotoxicity is also a significant side effect.

Drug–drug interactions may also be important considerations for drug PK. The mechanism of clearance of a new agent needs to be evaluated in the context of its potential interference with clearance pathways for co-administered drugs. An agent showing inhibition of P450 enzymes, for example, may alter the PK of another drug metabolized by that same P450 enzyme. These possible drug–drug interactions need to be evaluated carefully. This is important for any new therapeutic, but especially for cancer therapeutic in which the agents may be significantly toxic and will almost certainly be used in combination with other therapies. The fact that the patient population to be treated tends to be older and receiving medication for accompanying co-morbid conditions, the risks from toxicity that emerge from drug combinations are likely considerably higher than for other therapeutic settings.

12.12 Supportive Care

The significant toxic side effects associated with standard chemotherapy themselves provide opportunity for therapeutic intervention. Hematopoietic side effects have been counteracted by the administration of hematopoietic growth factors (erythropoietin, G-CSF, GM-CSF, etc.). Gastrointestinal side effects have been treated symptomatically with the use of antiemetic and antinausea agents. Of course, the strategy for testing these supportive agents is significantly different from that used to test therapeutic agents and the tolerance for toxic side effects is much less.

12.13 Summary

Historically, cancer therapy has been dominated by cytotoxic agents, with potentially multiple mechanisms of action, significant systemic toxicity, and few defined enzyme targets. The therapeutic efficacy of these agents is limited by their systemic toxicity; and as result, they have been dosed clinically to the MTD. Interpatient variations in PK have complicated safe dosing, and more recent studies have attempted to guide dosing based on the maximum exposure that can be tolerated without unacceptable toxicity. This approach may allow individual patients to derive maximum benefit. The improved understanding of cellular transformation and survival pathways has provided us with the potential for the discovery of newer more sophisticated therapies more specifically targeted to the cancer cell, perhaps showing efficacy with much lower toxic penalty. If mechanism-based agents turn out to be safer for the patient, and can be dosed to maximal target efficacy, then we may see the dawn of a new era in which PK considerations will take on an increasingly important role, using PK to guide dosing based on efficacy considerations (as is the normal practice in other therapeutic disciplines). These agents may provide the potential to treat earlier disease and may also be useful in a chemopreventive mode. The heterogeneity of cancer as a disease and the genetic instability of tumor cells may mean that multiple targeted agents will be needed in combination to treat any given patient. Therefore, lack of overlapping toxicity will be an important factor in guiding which agents can be used in combination therapy.

Although it is an exciting time in oncology, with many new mechanisms emerging for clinical testing, it appears that to date many of these newer therapies have not so far lived up to their early promise. One explanation for this observation may be that imperfect agents have been tested inappropriately. Inhibitors designed against rational targets have ultimately been tested using the approaches developed historically for broadly cytotoxic agents. In the future, it will be important in testing any new mechanism to do the following:

- Identify an agent that is highly selective for the desired mechanism.
- Determine the drug exposure (concentration, time of exposure) that is necessary to target that mechanism in the target tissue, usually the tumor.
- Determine whether the optimal drug levels can be delivered safely.
- Determine if at the achievable dose there is any therapeutic benefit, alone or in combination with other agents.

This logical strategy for improving selectivity, tumor targeting, and accurately tailoring pharmacokinetics to efficacy will permit a more effective exploration of the potential of novel target-based anticancer therapies.

References

Akinaga, S., Sugiyama, K., and Akiyama, T. UCN-01 (7-hydroxystaurosporine) and other indolocarbazole compounds: a new generation of anti-cancer agents for the new century? *Anti-Cancer Drug Design* **15,** 43–52 (2000).

Ayash, L. J., Wright, J. E., Tretyakov, G., et al. Cyclophosphamide pharmacokinetics; correlation with cardiac toxicity and tumor response. *J. Clin. Oncol.* **10,** 995–1000 (1992).

Budman, D. R., Igwemezie, L., Rauls, S. et al. Pharmacodynamic finding with etoposide phosphate (BMY 4081) a water soluble prodrug [Abstract]. *Proc. Am. Soc. Clin. Oncol.* **13,** 146 (1994).

Buzdar, A. U., and Horrobagyi, G. N. Tamoxifen and toremifene in breast cancer: Comparison of safety and efficacy. *J. Clin. Oncol.* **16,** 348–353 (1998).

Buzdar, A. U., Robertson, J. F. R., Elermann, W., and Nabholtz, J.-M. An overview of the pharmacology and pharmokinetics of the newer generation aromatase inhibitors Anastrozole, Letrozole, and Exemestane. *Cancer* **95,** 2006–2016 (2002).

Clarke, S. J., and Rivory, L. P. Clinical pharmacokinetics of Docetaxel. *Clin. Pharmacokinetics* **36,** 99–114 (1999).

Danesi, R., Fogli, S., Gennari, A., et al. Pharmacokinetic-pharmacodynamic relationships of the anthracycline anticancer drugs. *Clin. Pharmacokinetics* **41,** 431–444 (2002).

Desoize, B., Mareschal, F., and Cattan, A. Clinical pharmacokinetics of etoposide during 120 hours continuous infusions in solid tumors. *Br. J. Cancer* **62,** 840–841 (1990).

Eisnehauer, E. A., and Vermorken, J. B. The taxoids, comparative clinical pharmacology and therapeutic potential. *Drugs* **55,** 5–30 (1998).

Escande, D. Pharmacogenetics of cardiac K^+channels. *Eur. J. Pharmacol.* **410,** 281–287 (2000).

Etienne, M. C., Lagrange, J. L., Dassonville, O. et al. Population study of dihydropyrimidine dehydrogenase in cancer patients. *J. Clin. Oncol.* **12,** 2248–2253 (1994).

Harashima, H., Iida, S., Urakami, Y., et al. Optimization of anti-tumor effect for liposomally encapsulated doxorubicin based on simulations by pharmacokinetic-pharmacodynamic modeling. *J. Control. Release* **61,** 93–106 (1999).

Huizing, M. T., Giaccone, G., Van Warmerdam, L. J. C., et al. Pharmacokinetics of paclitaxel and carboplatin in a dose escalating and sequencing study in patients with non-small-cell lung cancer. *J. Cliin. Oncol.* **15,** 317–329 (1997).

Ingelman-Sundberg, M. Polymorphism of cytochrome P450 and xenobiotic toxicity. *Toxicology* **27,** 447–452 (2002).

Johnson, S. A. Clinical pharmacokinetics of nucleoside analogues. *Clin. Pharmacokinetics* **39,** 5–26 (2000).

Levi, F., Metzger, G., Massari, C., and Milano, G. Oxaliplatin. *Clin. Pharmacokinetics* **38,** 1–21 (2000).

Lichtman, S. M., Ratain, M. J., Van Echo, D. A., et al. Phase I trial of granulocyte-macrophage-colony-stimulating factor plus high dose cyclophosphamide given every 2 weeks, a cancer leukemia group B study. *J. Natl. Cancer Inst.* **85,** 1319–1326 (1993).

Mathijssen, R. H. J., van Alphen, R. J., Verweij, J., et al. Clinical pharmacokinetics and metabolism of irinotecan (CPT-11). *Clin. Cancer Res.* **7,** 2182–2194 (2001).

Milano, G., and McLeod, H. L. Can dihydropyrimidine dehydrogenase impact 5-fluorouracil-based treatment? *Eur. J. Cancer* **36,** 37–42 (2000).

O'Dwyer, P. J., Stevenson, J. P., and Johnson, S. W. Clinical pharmacokinetics and administration of established platinum drugs. *Drugs* **59(suppl 4),** 19–27 (2000).

Reigner, B. M, Blesch, K., and Weidekamm, E. Clinical pharmacokinetics of capecitabine. *Clin. Pharmacokinetics* **40,** 85–104 (2001).

Rowland, M., and Tozer, T. N. *Clinical Pharmacokinetics – Concepts and Applications.* 3rd ed. Philadelphia, Lee & Febiger (1995).

Slamon, D. J., Leyland-Jones, B., Shak, S., et al. Use of chemotherapy plus a monoclonal antibody against HER2 for metastatic breast cancer that overexpresses HER2. *New Engl. J. Med.* **344,** 783–792 (2001).

Taras, T. L., Wurz, G. T., Linares, G. R., and DeGregorio, M. W. Clinical pharmacokinetics of Toremifene. *Clin. Pharmacol.* **39,** 327–334 (2000).

Vaishampayan, U., Parchment, R. E., Jasti, B. R., and Hussain M. Taxanes: An overview of the pharmacokinetics and pharmacodynamics. *Urology* **54(suppl 6A),** 22–29 (1999).

Van den Bongard, H. J. G. D., Mathot, R. A. A., Beijneen, J. H., and Schellens, H. M. Pharmacokinetically guided administration of chemotherapeutic agents. *Clin. Pharmacokinetics* **39,** 345–367 (2000).

CHAPTER 13

Clinical Development Issues

Steven D. Averbuch, Michael K. Wolf, Basil F. El-Rayes, and Patricia M. LoRusso

Molecular Cancer Therapeutics: Strategies for Drug Discovery and Development, edited by George C. Prendergast
ISBN 0-471-43202-4

In recent years, scientific insights into the molecular events central to the initiation and progression of human cancer have guided the search for novel anticancer agents. Innovative approaches to drug discovery have resulted in a move from identifying cytotoxic agents to molecular-targeted agents. Such agents directed at tumor cell targets and/or the stromal environment include those interfering with genetic or protein regulation of proliferation, survival signals, angiogenesis, invasion, and metastasis (Table 13.1). While the promise of this paradigm shift is the discovery and development of more selective, effective, and less toxic therapies, current anticancer clinical drug development strategies and standards are still largely based on historical precedent for cytotoxic agents. Initial clinical trials of these novel agents have indicated that the steps and end points routinely applied in the evaluation of traditional cytotoxic drugs are not necessarily appropriate for the evaluation of molecular targeted agents (Dancey and Freidlin, 2003; Gasparini and Gion, 2000; Gelmon et al., 1999; Korn et al., 2001).

The clinical development of anticancer agents, involving sufficient evidence for an efficacy and safety profile and leading to market approval by regulatory authorities, has traditionally been conducted through a series of sequential clinical trials classified as Phase I, II, III, and IV. Observations throughout these clinical investigations enable decisions regarding proof of principle, dose and schedule, registration strategy, overall pharmacological profile, and therapeutic index. Each phase typically has distinct objectives and involves increasingly larger populations of patients to achieve these objectives (Table 13.2). Phase I trials provide the first human exposure to an investigational agent with the principal aims of identifying toxicities, pharmacokinetic (PK) profile, and ascertaining dose(s) for Phase II trials.

Table 13.1 Selected Classes of Targeted Anticancer Agents

Target	Class
Estrogen receptor	Receptor antagonists: Tamoxifen Inhibitors of estrogen synthesis: LHRH agonists, aromatase inhibitors
EGFR	Monoclonal antibodies: Cetuximab Tyrosine kinase inhibitors: Gefitinib, Erlotinib
VEGFR	Monoclonal antibodies: Bevacizumab Tyrosine kinase inhibitors: Gefitinib
bcr-abl	Tyrosine kinase inhibitors: Imitanib
c-*kit*	Tyrosine kinase inhibitors: Imitanib
Cyclooxygenase pathway	NSAIDs: sulindac Cyclooxygenase 2 selective inhibitors: celecoxib
ras/rho signaling pathways	Farnesyl transferase inhibitors
Cell cycle regulation	Cyclin-dependent kinase inhibitors: flavopiridol

EGFR, epidermal growth factor receptor; *VEGFR,* vascular endothelial growth factor receptor; *LHRH,* luteinizing hormone releasing hormone.

Table 13.2 Summary of Phase I–III Clinical Trials

Phase and Purpose	Number of Patients	Trial Duration	Typical Assessments
Phase I: safety and dosing	20–100	Several months	PK, PD, safety, objective response
Phase II: safety and efficacy	Up to hundreds	Several months to 2 years	Objective response safety, survival time to progression quality of life
Phase III: controlled safety and efficacy	Several hundreds to several thousands	1–4 years	Safety, survival time to progression Quality of life Objective response

The dose selected for Phase II trials is commonly that at or slightly below the maximum tolerated dose (MTD) established in Phase I trials. Phase II trials are designed to demonstrate or reject a hypothesis regarding a prospectively defined level of evidence for anticancer activity in patients with specific cancer types. Historically, this evidence has been based on measurable reduction of tumor size or objective response rates as defined by standard criteria (Therasse et al., 2000). Phase III trials are designed to satisfy a specific efficacy or clinical outcome hypothesis by a randomized, controlled comparison of the new drug with established standard treatment. The traditional and widely accepted end points for Phase III trials are survival, time to progression, or quality of life. Often involving hundreds to thousands of patients, Phase III trials provide a great deal of information regarding safety and other aspects of an agent's profile that will define prescribing information following market approval.

13.1 Preclinical Development

Before the initiation of human trials of novel anticancer agents, it is necessary to perform preclinical trials to evaluate the agent's toxic and pharmacologic effects in vitro and in vivo. The U.S. Food and Drug Administration (FDA) requires manufacturers to develop a pharmacologic profile of the agent, determine its acute toxicity in at least two species of animals, and conduct short-term toxicity studies (which can range from 2 weeks to 3 months) (FDA, 2003). The use of animal models to identify initial safe doses for use in Phase I clinical trials has been expanded to cover dose escalation during the trials. Assuming that drug efficacy and toxicity are related to drug exposure, the area under the dose–time curve (AUC) in mice at the LD_{10} dose (the dose that is fatal in 10% of the animals) can be used to guide dose escalation. For example, if the AUC of the initial dose in humans is significantly lower than the AUC of the LD_{10} dose in mice, the dose can be escalated more quickly than it would be normally using standard modified Fibonacci increments; the speed of escalation depends on the therapeutic index of the drug and the difference in the AUCs seen in the animal models and the human trials (Curt, 1994).

13.2 Phase I Development

In addition to establishing an initial safety profile and defining MTD, a main objective of Phase I studies is to select appropriate doses for subsequent trials. With cytotoxic agents, antitumor activity and toxic effects are generally seen within a similar dose range, because the nonselective mechanism of action of cytotoxic agents works similarly on replicating tumor and normal tissue. This has led to the common practice of adopting dose-limiting toxicity as a surrogate end point for cytotoxic agent antitumor activity in Phase I–II trials. However, for biologically targeted agents that are more selective against a cellular or host target critical to the cancer, it would be expected that the maximal biologic effect may occur at doses lower than the MTD, as determined in conventional dose-escalation Phase I trials (Rowinsky, 2000). Therefore, it is important to define an optimal biologic dose (OBD) or dose range – that is, the doses that provide maximal biologic effect without causing dose-limiting toxicity in these early clinical trials (Fig. 13.1). The advantage of defining the OBD as an end point in the Phase I setting include identification of the active dose range for Phase II and III trials and demonstration of the intended biologic effect in vivo for providing higher confidence in taking the agent through further development. Failure to determine the biologic activity of a targeted agent in the Phase I setting is likely to confound interpretation of the results of subsequent trials.

Several targeted agents have already been evaluated in Phase I trials using the OBD as a primary end point. Different trial designs aimed at evaluating the biologically active dose have used other end points, including tissue-based assays, serum surrogate markers, serum drug levels, and functional imaging.

13.2.1 TISSUE-BASED ASSAYS

The most direct way of determining biologic activity is to assay the effects on pretreatment and posttreatment biopsies from the tumor, however, such designs are limited by the availability of tissue, the ethical considerations of subjecting patients to repeated invasive biopsy procedures, and the availability of robust validated assays. Recently, attempts to use tissue-based assays have been addressed by enrolling patients with easily accessible tumors, such as head and neck or subcutaneous lesions, in Phase I trials (Hidalgo et al., 2001). This approach has limitations dictated by the selected tumors and in practical terms, repeated biopsies of tumors has proven to be difficult.

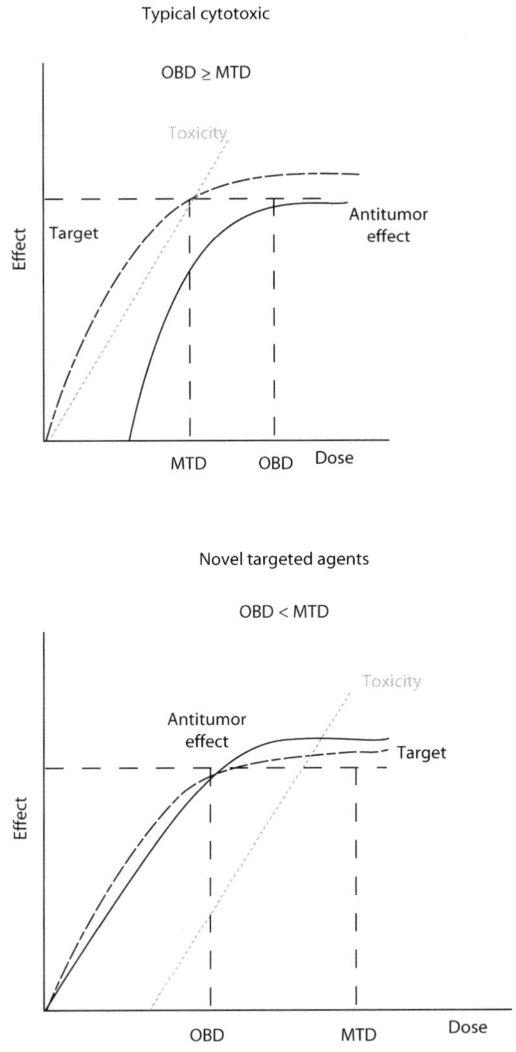

Figure 13.1 Idealized curves reflecting molecular target effects, antitumor effects, and toxicity as functions of dose for a typical cytotoxic agent and a hypothetical target-based antiproliferative agent. [Reprinted from Rowinsky (2000). Permission required].

An alternative approach to tumor tissue-based assays is to make use of surrogate normal tissue, such as peripheral blood or bone marrow cells or skin, to evaluate the biologically active dose of the targeted agent. An example of using normal tissue to establish dose-related inhibition of the target in human tissue comes from two Phase I trials of gefitinib (Iressa, ZD1839), an orally active small molecule inhibitor of the epidermal growth factor receptor (EGFR) tyrosine kinase. As the EGFR is present and important for skin homeostasis (Albanell et al., 2002), the biologic activity of gefitinib was assessed using markers of pharmacodynamic (PD) effects in punch skin biopsies taken from consenting patients before and during gefitinib therapy.

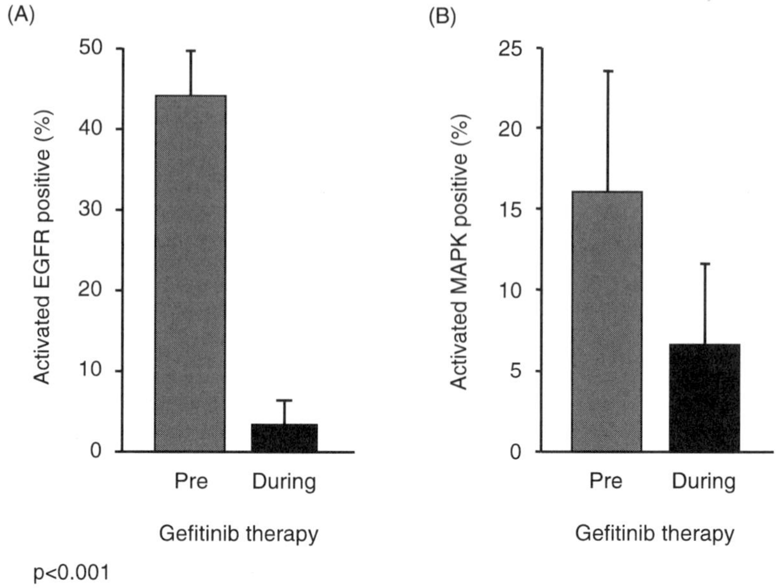

Figure 13.2 Activation of EGFR and MAPK before and after gefitinib treatment. [Reprinted from Albanell et al. (2002). Permission required].

These markers included activated molecules of the EGFR signal transduction pathway in addition to indicators of downstream cellular effects, such as Ki67 proliferative index, p27, and apoptotic index terminal deoxynucleotide transferase-mediated dUTP nick end labeling (TUNEL), which were all evaluated using validated semiquantitative immunohistochemical methods (Albanell et al., 2002) (Fig. 13.2). Skin biopsy studies showed that $\geq$ 150 mg/day gefitinib abolished EGFR autophosphorylation and significantly reduced mitogen activated protein kinase (MAPK) activation, indicating that gefitinib was exerting considerable biologic effect at the lowest dose studied (Albanell et al., 2002). Use of a surrogate normal tissue assumes that the pharmacologic effect of the targeted agent is similar in normal and malignant tissue and, more important, that the relative significance of these pathways and the degree of inhibition required to predict a meaningful antitumor effect has been defined. These relationships need definition in preclinical models, when possible, and the concordance of effect between normal and malignant tissue is needed in the early clinical trials.

13.2.2 SURROGATE MARKERS AND IMAGING

The use of serum surrogate markers has been incorporated into Phase I design with the assumption that the surrogate marker predicts the biologic effects of the drug on the tumor. In a Phase I study of bevacizumab, a vascular endothelial growth factor (VEGF) monoclonal antibody, serum VEGF levels were assayed. A bevacizumab dose of $\geq$ 0.3 mg/kg provided linear kinetics that was sufficient to achieve undetectable levels of free-serum VEGF (Gordon et al., 2003). However, this design did not address the effects of the antiangiogenic agent on the tumor. The failure to define the optimal biologically active dose in the Phase I trials with bevacizumab has resulted in a series of Phase II and III trials using different doses and with contradictory results. The Phase III trial in renal-cell cancer demonstrated that bevacizumab at the higher dose of 10 mg/kg was more active than the 3 mg/kg dose level (Yang et al., 2002). However, in the randomized Phase II trial conducted in colorectal cancer, the 5 mg/kg dose of bevacizumab was superior to the higher dose of 10 mg/kg (Kabbinavar et al., 2003). Trials confirming the predictive value of serum VEGF levels on angiogenesis in the tumor would have been helpful in deciding the effect of the low, intermediate, and high doses of bevacizumab on tumor angiogenesis and consequently the biologic dose for the Phase II and III trial design. Therefore, before using this design, it would be helpful to prove that the surrogate marker is predictive of the drug's biologic effects on the tumor. Another approach to evaluate biologic activity is the use of functional imaging. Determination of tumor blood flow by PET or MRI has been used in Phase I trials of antiangiogenic and vascular targeted agents (Galbraith et al., 2003; Herbst et al., 2002b).

13.2.3 PHARMACOKINETIC CRITERIA

Several examples that have successfully used pharmacokinetic parameters to guide dose selection include, imatinib (Gleevec, STI571), a small molecule tyrosine kinase inhibitor that inhibits the *bcr-abl* fusion protein involved in chronic myeloid leukemia (Deininger et al., 1997), and cetuximab (Erbitux, IMC-C225), a humanized monoclonal antibody that targets the EGFR (Baselga et al., 2000). The recommended dose of 400 mg/day imatinib was determined by the PK profile and the in vitro and in vivo inhibition of *bcr-abl* signaling (Druker et al., 2001). The selected dose of 200 mg/m^2 cetuximab was the lowest dose that saturates the antibody systemic clearance rate (Baselga et al., 2000). In both of these Phase I clinical trial programs, the selected dose for Phase II and III trials was significantly lower than the MTD. The PK end point in these trials is based on the supposition that the concentration needed to saturate the receptor in the preclinical models is predictive of the concentration required to saturate the tumor receptors in vivo. Tumor biopsies confirming this assumption were not performed before Phase II and III trials.

13.2.4 TOXICITY EVALUATION

In addition to determining a biologically active dose, Phase I trials of targeted agents evaluate the toxicity profile. Since these agents have diverse mechanisms of action, the toxicity profiles differ considerably from conventional cytotoxic agents. Phase I trials should be designed to evaluate these unique toxicities and determine the patient characteristics, such as polymorphisms for drug metabolism, that predispose to these toxicities. Defining the patient population at risk for side effects allows early interventions aimed at preventing these detrimental toxicities in the Phase II and III trials, including measures to select and exclude patients for whom risks may outweigh benefit in these phases of clinical developments (Nagasubramanian et al., 2003; Somer et al., 2003).

13.2.5 THE GEFITINIB EXAMPLE

The initial trials used in the evaluation of gefitinib illustrate how alternative end points may be used in the early clinical development of biologically targeted agents. As gefitinib had a mild and predictable mechanism-based toxicity profile in preclinical models, it was possible to perform Phase I trials in healthy volunteers. These included a rising single-dose study, a food-interaction study, a single radiolabeled metabolism and disposition study, a multiple rising-dose study, and an itraconazole interaction study (Laight et al., 2002; Swaisland et al., 2001, 2002). The safety of gefitinib in cancer patients was initially evaluated in two Phase I clinical trials in which oral doses of 50–925 mg/day were administered for 14 days followed by 14 days of observation (Nakagawa et al., 2003; Ranson et al., 2002). In these trials the MTD was 700 mg/day. At doses $\geq$ 100 mg/day, mild reversible diarrhea and acneiform rash were observed with dose-related increased incidence and severity up to the MTD (Fox et al., 2002). In contrast to the typical experience with conventional cytotoxic agents, in which objective responses are usually observed near the MTD, objective responses were observed across the doses investigated $\geq$ 225 mg/day, and stable disease was seen at doses $\geq$ 50 mg/day (Nakagawa et al., 2003; Ranson et al., 2002). These two trials showed that it was feasible to administer gefitinib for 14 days, followed by 14 days of observation.

To determine a safe, tolerated dose range for gefitinib given orally on a continuous daily schedule and to further define the OBD of gefitinib, two identical multicenter Phase I PD trials were then performed in patients with five tumor types known to express EGFR (non-small cell lung cancer [NSCLC], head and neck, ovarian, colorectal, or prostate cancer) (Baselga et al., 2002; Herbst et al., 2002a). Secondary objectives were to determine the PK profile, to investigate the feasibility and sensitivity of the Functional Assessment of Cancer Therapy (FACT) questionnaire and the seven-item Lung Cancer Subscale (LCS) of FACT in assessing improvements in quality of life- and disease-related symptoms, respectively (Cella et al., 2002), and to select tumors for Phase II and III studies. Dose escalation proceeded until the MTD (800 mg/day) was determined. As with the trials in which gefitinib was given in a 14 days on-drug, 14 days off-drug schedule, common adverse events (AEs) were mild dose-related skin toxicity and diarrhea. As discussed previously, PK and PD assessments demonstrated that at doses ≥ 150 mg/day, biologically relevant plasma concentrations were maintained and EGFR was maximally inhibited in normal skin. In both trials, disease control (response plus stable disease) was observed over the dose range studied (150–1000 mg/day).

These gefitinib trials integrated multiple efficacy end points and demonstrated that indirect surrogate markers of antitumor activity may include a meaningful lack of tumor progression (disease control) combined with improvements in quality of life and disease-related symptoms. Both the LCS and FACT questionnaires were found to be feasible and sensitive tools with which to assess improvements in these areas. Patients with NSCLC who had stable disease for ≥ 6 months also had improvements or stabilization in disease-related symptoms (LCS scores), whereas those patients with disease progression had worsened LCS scores (LoRusso et al., 2003). Patients with NSCLC, head and neck cancer, and colorectal cancer had no significant deterioration in median quality-of-life scores. These trials demonstrated the utility of symptom and quality-of-life end points in early clinical trials of novel, targeted, anticancer agents (Patrick-Miller 2003).

13.3 Phase II Development

13.3.1 END POINTS FOR PHASE II TRIALS

Phase II trials are designed to estimate the effectiveness of new agents. The most commonly used end point in this setting is response rate. A distinguishing feature of many biologically targeted agents, in contrast to cytotoxic agents, is their ability to inhibit cellular growth regulatory pathways, resulting in cytostatic rather than cytocidal effects on tumor cells in vitro. Clinically, this antitumor effect could be manifest as an inhibition of tumor growth rather than tumor shrinkage so that the traditional surrogate end point of objective response could underestimate the potential for clinical benefit. Possible alternative surrogate end points include reduction in rate of tumor growth (difficult to measure sensitively), time to progression, symptom improvement, quality of life, reduction in serum tumor markers (e.g., prostate-specific antigen, carcinoembryonic antigen, CA-125), changes in mechanism-related PD markers, or changes observed by functional imaging, such as physiologic changes as measured by FDG-fluorodeoxyglucose PET scan or change in tumor vasculature for antiangiogenic agents (Bubley et al., 1999; Eisenhauer, 1998; Galbraith et al., 2003; Herbst et al., 2002b; Stadler and Ratain, 2000). However, no evidence exists as yet to prove that these suggested end points could predict effectiveness in subsequent randomized trials; therefore, novel end points for Phase II studies should undergo evaluation to confirm their clinical usefulness before routine incorporation into clinical trial design.

Targeted agents that were considered to be cytostatic with relatively little potential for producing responses, such as gefitinib (Fukuoka et al., 2003; Kris et al., 2003) and bevacizumab (Kabbinavar et al., 2003), did produce objective responses in Phase II trials. Although the assessment of alternative end points for Phase II trials could be valuable, for the time being response rates continue to be the most widely accepted end point in screening the effectiveness of targeted agents. However, for a targeted agent that has achieved a pharmacodynamic proof of concept in Phase I, the absence of objective tumor regression in early clinical trials may or may not be predictive for ultimate effectiveness. Therefore, conventional empiric Phase II testing using single-arm Phase II designs with traditional end points (such as response rate) are likely to have limited value. Alternative Phase II designs need to be considered and employed for molecular targeted agents.

13.3.2 TRIAL DESIGNS TO EVALUATE CYTOSTATIC EFFECTS OF MOLECULAR TARGETED AGENTS

A number of alternative trials designs to evaluate targeted agents in Phase II trials have been proposed (Dancey and Freidlin, 2003; Estey and Thall, 2003; Fox et al., 2002; Gasparini and Gion, 2000; Gelmon et al., 1999; Korn et al., 2001; Rowinsky, 2000; Stadler and Ratain, 2000). However, most of these have not been fully applied or validated in practice.

Eisenhauer (1998) proposed that using a multinomial Phase II stopping rule that incorporates both objective response and early progression rates may be particularly useful in assessing agents that inhibit tumor growth. This approach has been formally validated only retrospectively for cytotoxic agents, so it is unknown whether or not this design would add utility in the Phase II evaluation of targeted agents (Dent et al., 2001).

Randomized Phase II designs incorporating biologically and clinically relevant end points are likely to be more informative and useful for critical decision making with higher confidence (Estey and Thall, 2003). Phase IIa dose randomization using surrogate tissue or tumor-based assays in a preoperative setting has been proposed to demonstrate target modulation that may be sufficient to support Phase III clinical trials (Stadler and Ratain, 2000).

For patients with advanced disease, the following proposed design and analysis demonstrates how multiple surrogate efficacy end points could be integrated to help predict for meaningful antitumor outcomes to select the dose for Phase III studies: Following determination of the MTD from dose escalation, four biologically active doses that are well tolerated can be selected for further investigation. Patients are then randomized across the doses and several potentially relevant measurements are obtained. These data would be collated for each individual and a blinded panel of clinical experts would be asked to dichotomize the data for each patient as either yes or no evidence for a targeted biologic effect (TBE) (Fig. 13.3A). Based on these assessments, a regression analysis for dose response would be constructed from which Phase III dose selection may be derived (Fig. 13.3B). Additional evaluation

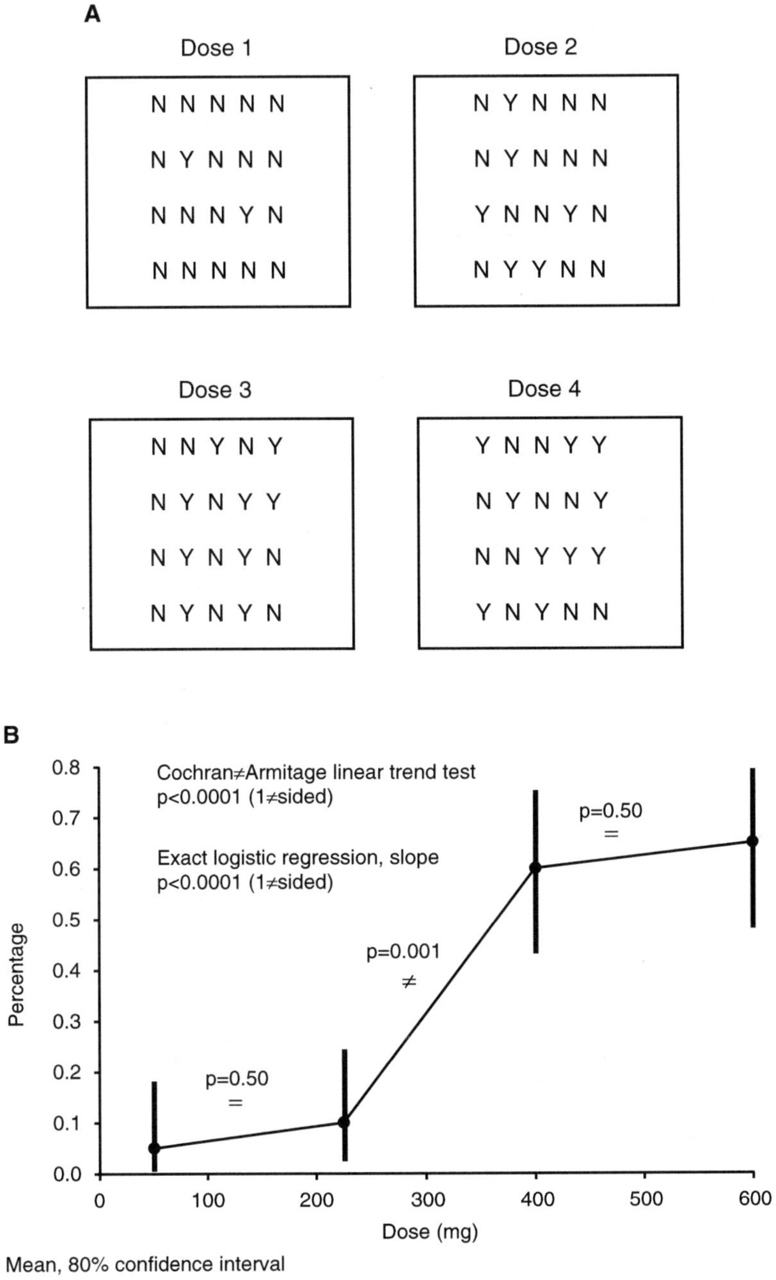

Figure 13.3 **A,** Assessment of TBE by dose ($n = 20$/dose level). **B,** TBE rate by dose. An ideal case in which the TBE rate at the low end is distinct from the high end. A closed test procedure indicates that the 150 and 225 mg levels are the same, but that 400 mg has a higher rate than 225 mg, and 400 mg, in turn, has a lower rate than 600 mg. In this case, 400 mg and 600 mg should be selected.

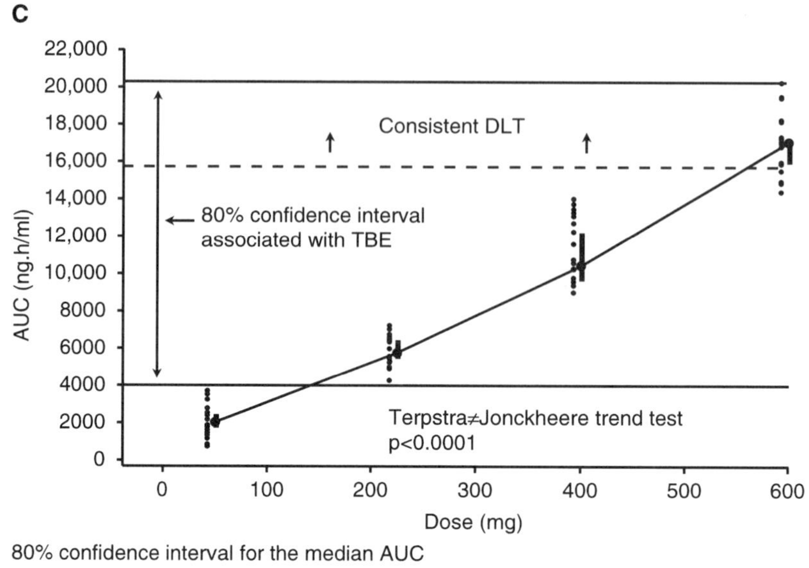

Figure 13.3 (*Continued*) **C,** AUC versus dose. This scenario assumes that concentration is proportional with dose, with a common standard deviation at each dose level. The Terpstra-Jonckheere trend test can be used to confirm the overall trend and to evaluate whether the adjacent distributions overlap (in this case, they do not). An imaginary confidence interval is shown for the AUC distribution associated with target effect. This can be used to estimate the probability of achieving an active concentration at each dose level.

of TBE as a function of plasma concentrations would be used to establish a PK–PD relationship (Fig. 13.3C).

Additional suggested trial designs evaluating the effect of targeted agents on tumor growth inhibition involves a randomized discontinuation design after a short course of the investigational agent or the randomization of the targeted agent versus placebo after maximal cytoreduction, such as after definitive chemoradiotherapy or surgery. Trials to evaluate gefitinib as maintenance therapy in NSCLC after chemoradiotherapy in stage III disease or after surgery for stage I to II disease are ongoing (Sridhar et al., 2003).

Clinical trial designs that demonstrate the ability of targeted agents to sensitize resistant cancer cells to the effects of chemotherapy have been proposed. One such design is based on adding a targeted agent to a chemotherapeutic regimen at time of disease progression. The rationale for such a design is that progression defines the emergence of cancer cells resistant to the chemotherapeutic agent. The demonstration of responses at that time would confirm the potential of targeted agents to modulate the mechanisms of resistance. The results of such a trial with cetuximab and irinotecan in advanced colorectal cancer have been reported (Saltz et al., 2002). This trial demonstrates the potential for cetuximab to modulate chemoresistance to irinotecan. The criticisms for this design are the absence of a standard definition of chemoresistance, disease progression, and the correlation of the two phenomena. Randomized Phase II trials of a chemotherapeutic agent with a targeted drug versus the targeted drug at time of disease progression appears to be a useful model for screening the ability of these agents to modulate chemoresistance. Such designs would require a more stringent definition of progression to ensure the similarity of the patient populations in different studies. The addition of biologic assays evaluating pathways of apoptosis and chemoresistance to such designs would help in confirming the observed clinical effects.

13.3.3 DURATION OF THERAPY

The persistence of expression of a molecular target after disease progression raises the possibility of benefit from continued therapy with the targeted agents. In the pivotal Phase III study of trastuzumab in advanced breast cancer, trastuzumab was continued beyond progression while the cytotoxic agents were changed (Slamon et al., 2001). The role of continuing targeted agents after disease progression remains undefined, and clinical trials randomizing patients to continuation versus discontinuation of these agents at progression need to be conducted.

13.3.4 PREDICTORS OF RESPONSE

Since targeted agents are designed to inhibit specific molecular pathways, attempts at defining predictors of response have been incorporated into study designs. The advantages of defining predictors of response include preventing the exposure of patients to potentially harmful and ineffective agents, increasing the effectiveness of therapy through selecting a group of patients with a higher likelihood of response, allowing smaller Phase II and III trials, and the identification of a patient population that requires a different strategy for therapy.

The success of trials aimed at defining predictors of response has been mixed. For patients with breast cancer, response to therapy with trastuzumab correlates with the level of *her-2/neu* expression (Fornier et al., 2002). In contrast, Saltz et al. (2001) found no association between EGFR expression by immunohistochemistry in colorectal cancer and response to cetuximab. Similarly, Bailey et al. (2003) did not find a correlation of EGFR expression with clinical outcome following gefitinib treatment in patients with advanced NSCLC using a rigorous assessment based on a validated immunohistochemical assay. The unifying theme from these trials is that evaluation of the expression of a target might not be a sufficient predictor of response. The level of expression of estrogen receptor in breast cancer does not predict the likelihood of response to tamoxifen, whereas co-expression of the estrogen receptor and progestin receptor has the highest predictive value for response. Progestin receptor expression is under the transcriptional control of the estrogen receptor; therefore, expression of the progestin receptor is evidence of the functionality of the estrogen receptor (Osborne, 1998). Similarly, in the imatinib trials for gastrointestinal stromal tumors, the presence of a functional mutation of the c-*kit* growth factor was the predictor of response (Heinrich et al., 2002). Therefore, functional assays offer a better chance of determining biologic predictors of response so that developing and validating such assays in preclinical models will facilitate drug development in the Phase II setting.

Another concern with current trial designs is the availability of several different assays measuring the target of interest and the absence of standardization of these assays. Incorporation of the different assays into clinical trials will help in defining their relative predictive value. For example, *her-2-neu* can be measured by immunohistochemistry or fluorescence in situ hybridization (FISH). Clinical trials have demonstrated the superiority of FISH-based assays as predictors of response versus the immunohistochemistry-based assays (Vogel et al., 2001). The centralization of tissue processing and evaluation of molecular end points to a few national centers will standardize the assays and consequently enable general conclusions to be derived from the results of several trials.

Finally, there is recent interest in the potential for molecular profiling to define patient and biological parameters that may predict for clinical outcome following targeted therapy. Retrospective analyses have suggested that genomic patterns or signatures may be strong independent factors predictive of survival outcomes for patients with newly diagnosed cancer (Beer et al., 2002; van de Vijver et al., 2002). More recently, preliminary studies by Pusztai et al. (2003) using DNA and Natale et al. (2003) using RNA have demonstrated that genomic signatures may be predictive for chemotherapy outcomes in breast cancer and gefitinib outcomes in lung cancer, respectively. Should larger, prospective trials confirm these results, molecular profiling has vast potential to select patients based on their individual susceptibility to the therapeutic benefit from treatment with a specific targeted agent. This would provide the opportunity for greater efficiency by designing smaller trials for with sufficient power to detect meaningful outcomes (Rothenberg et al., 2003).

13.3.5 THE GEFITINIB EXAMPLE

As described above, Phase I PD trials suggested that gefitinib had maximal biologic and clinical activity over the dose range studied (150–1000 mg/day). In addition, PK–PD studies indicated that plasma levels of gefitinib over this dose range were sufficient for effective EGFR inhibition. These data enabled two dose levels that were well tolerated and gave the widest possible separation of doses and systemic exposures to be selected for Phase II and III studies; these were 250 and 500 mg/day. The former was above the lowest dose shown to produce biologic and antitumor activity, thereby ensuring adequate gefitinib drug exposure and minimizing the chance of any individual having subtherapeutic exposure as a result of interpatient PK variability. The latter was the highest dose tolerated by most patients on a chronic daily dosing schedule, and it provided maximal exposure. Both doses were significantly below the MTD (approximately one third and two thirds of the MTD, respectively), unlike conventional dose selection for chemotherapy agents.

Two large, dose-randomized, double-blind, parallel-group, multicenter Phase II trials (IDEAL 1 and 2) independently evaluated the activity of 250 and 500 mg/day gefitinib in 425 patients with advanced NSCLC (Fukuoka et al., 2003; Kris et al., 2002). These trials allowed a more detailed evaluation of the OBD using the doses selected from the Phase I trials, and they included symptom improvement as an additional endpoint.

In both trials, fewer and less severe AEs were observed using 250 mg/day compared with 500 mg/day, while no differences in efficacy end points (response rate, disease control rate, overall survival, and symptom improvement) were seen between the two doses (Table 13.3). Response rates ranged from 9 to 19%; overall, approximately 40% of patients experienced disease control and symptom improvement. The results from these two double-blind dose-randomized phase II trials confirmed the proposition that selection of an OBD below the MTD is the correct developmental strategy for molecularly targeted agents.

13.4 Phase III Development

The demonstration of efficacy and safety of a novel targeted agent in the Phase III setting is challenged by the practical difficulty of randomizing patients to a no treatment or placebo arm. For most advanced cancers, cytotoxic chemotherapy is established as the standard of care based on modest improvements in survival compared to best supportive care. Therefore, a no treatment arm is generally considered not in patients' best interest. Randomized comparisons between standard therapy and the targeted agent have been criticized based on the difficulty of interpreting results in the presence of an active control (Temple and Ellenberg, 2000). However, based on the strength of the Phase II data in chronic myelogenous leukemia, this randomized Phase III

Table 13.3 IDEAL 1 and 2 Safety and Efficacy of Gefitinib in Advanced NSCLC[a]

	IDEAL 1		IDEAL 2	
	250 mg/day (n = 103)	500 mg/day (n = 106)	250 mg/day (n = 102)	500 mg/day (n = 114)
Drug-related, grade 3/4 (AES %)	8.7	30.2	6.9	17.5
Response rate (%)	18.4	19.0	11.8	8.8
Disease control rate (%)	54.4	51.4	42.2	36.0
Overall survival (months)	7.6	8.1	6.5	5.9
Symptom improvement (%)	40.3	37.0	43.1	35.1
Patients with partial response experiencing symptom improvement (%)	69.2	85.7	100.0	90.0
Improved quality of life	23.9	21.9	34.3	22.8

[a] Data from Fukuoka et al. (2003) and Kris et al. (2003).

design was employed and proven to be successful in the case of imatinib (O'Brien et al., 2003).

Combining targeted agents with cytotoxic therapy is a rational approach in the development of these new agents. These combinations are based on the nonoverlapping toxicity with cytotoxic drugs and the preclinical work suggesting that targeted agents could modulate chemoresistance pathways and promote apoptosis in cancer cells. The positive effect of combining trastuzumab with chemotherapy in advanced breast cancer encouraged the rapid development of similar regimens in different disease models (Slamon et al., 2001). The results of subsequent Phase III trials with small molecule targeted agents such as matrix metalloproteinase inhibitors and EGFR tyrosine kinase inhibitors have been, in the majority, disappointing (Rothenberg et al., 2003). In contrast, combinations of targeted monoclonal antibodies, such as rituximab in non-Hodgkin lymphoma (Plosker and Figgitt, 2003) and cetuximab and bevascizumab in colorectal cancer, have provided improvements in survival.

Several possible interpretations for the observed negative results of combinations of cytotoxic therapy with small molecule targeted agents have been proposed. First, redundant activity on the same tumor cell populations between the targeted agents and chemotherapeutic drugs would limit the benefits of such combinations. In the case of gefitinib, the observed activity of this agent after failure of first- and second-line chemotherapy suggests at least partial non-cross-resistance (Fukuoka et al., 2003; Kris et al., 2003). Second, the effect of the targeted agent may be diluted when given to patients nonselectively relative to the biological target. In the trastuzumab study, only patients with high *her-2/neu* expression were enrolled in the study (Slamon et al., 2001). Given the poor response in patients whose tumors do not express *her-2/neu,* had all patients been enrolled in the trial regardless of *her-2/neu* expression, the outcome would not have been positive. Finally, targeted agents affect numerous aspects of cancer cell biology. For example, EGFR inhibitors affect cell cycle progression, induction of apoptosis, and angiogenesis. Therefore, the interaction of targeted agents with chemotherapy is complex and could be antagonistic. Predictive preclinical models evaluating the impact of modulation of cancer cell biology by targeted agents on chemosensitivity are needed. In particular, attention should given to the entire spectrum of outcomes in preclinical models rather than just the positive results in selected models. For example, targeted agent cytotoxic combination treatments in preclinical human xenograft models that demonstrate tumor- or model-dependent outcomes of no effect, antagonism, additivity, and synergy all need to be examined equally to understand the underlying mechanisms for each outcome and how they may be relevant to clinical outcomes.

The preliminary results of Phase II trials of combinations of targeted agents with cytotoxic chemotherapy suggest that these agents have the potential to modulate the side effects of chemotherapeutic drugs. For example, celecoxib, a cyclooxygenase 2 inhibitor, can decrease the incidence of hand–foot syndrome when combined with fluoropyrimidines (Lin et al., 2002), and arthralgias when combined with taxanes. However, trastuzumab increased the cardiotoxicity of anthracyclines (Behr et al., 2001). Careful monitoring and reporting of toxicities encountered in combination clinical trials should be performed.

In summary, Phase III trials are intended to conclusively determine the role of new therapeutic regimens in the care of cancer patients and as such should use the traditional end points of efficacy. Several recently completed Phase III trials on combinations of targeted agents and cytotoxic drugs were negative, which could have been due to choosing an inappropriate target, an unsuitable patient population, or an ineffective dose or schedule of drug administration. As a result of these negative trials, the development of several potentially active agents was discontinued (Rothenberg et al., 2003). Therefore, Phase III trials of targeted agents should be initiated only after adequate preclinical and Phase I and II trials are conducted and biologic end points are firmly established.

13.5 Issues for the Future

With the recent description of the human genome and additional scientific discoveries, there has been a vast increase in the number of potential molecularly targeted cancer drugs in development. In addition to the need to create new clinical development and regulatory approval paradigms, a significant issue for the pharmaceutical and biotechnology industries will be how to prioritize the many potential new drugs competing for the finite resources required to fully develop them. In particular, drug development is usually constrained by fixed costs and timing as well as a limited number of patients available for clinical trials. The increasing knowledge of tumor biology leading to selective targeting is likely to result in more orphan drugs for narrow clinical indications that will have significant medical practice and financial implications for practitioners, academia, industry, and government. Therefore, in the future, these forces will need to work together in unprecedented ways to secure the successful translation of molecular therapeutics into meaningful prevention and treatment of cancer.

References

Albanell, J., Rojo, F., Averbuch, S., et al. Pharmacodynamic studies of the epidermal growth factor receptor inhibitor ZD1839 in skin from cancer patients: Histopathologic and molecular consequences of receptor inhibition. *J. Clin. Oncol.* **20,** 110–124 (2002).

Bailey, R., Kris, M., Wolf, M., et al. Tumor EGFR membrane staining is not clinically relevant for predicting response in patients receiving gefitinib ('Iressa', ZD1839) monotherapy for pretreated advanced non-small-cell lung cancer: IDEAL 1 and 2 (Abstract). *Proc. Am. Assoc. Cancer Res.* **44,** 1362 (2003).

Baselga, J., Pfister, D., Cooper, M. R., et al. Phase I studies of anti-epidermal growth factor receptor chimeric antibody C225 alone and in combination with cisplatin. *J. Clin. Oncol.* **18,** 904–914 (2000).

Baselga, J., Rischin, D., Ranson, M., et al. Phase I safety, pharmacokinetic, and pharmacodynamic trial of ZD1839, a selective oral epidermal growth factor receptor tyrosine kinase inhibitor, in patients with five selected solid tumor types. *J. Clin. Oncol.* **20,** 4292–4302 (2002).

Beer, D. G., Kardia, S. L., Huang, C. C., et al. Gene-expression profiles predict survival of patients with lung adenocarcinomas. *Nature Med.* **8,** 816–824 (2002).

Behr, T. M., Behe, M., and Wormann, B. Trastuzumab and breast cancer. *N. Engl. J. Med.* **345,** 995–996 (2001).

Bubley, G. J., Carducci, M., Dahut, W., et al. Eligibility and response guidelines for phase II clinical trials in androgen-independent prostate cancer: Recommendations from the prostate-specific antigen working group. *J. Clin. Oncol.* **17,** 3461–3467 (1999).

Cella, D., Eton, D. T., Fairclough, D. L., et al. What is a clinically meaningful change on the Functional Assessment of Cancer Therapy-Lung (FACT-L) questionnaire? Results from the Eastern Cooperative Oncology Group (ECOG) study 5592. *J. Clin. Epidemiol.* **55,** 285–295 (2002).

Curt, G. A. The use of animal models in cancer drug discovery and development. *Stem Cells* **12,** 23–29 (1994).

Dancey, J. E., and Freidlin, B. Targeting epidermal growth factor receptor – Are we missing the mark? *Lancet* **362,** 62–64 (2003).

Deininger, M. W. N., Goldman, J. M., Lydon, N., and Melo, J. V. The tyrosine kinase inhibitor CGP57148B selectively inhibits the growth of BCR-ABL-positive cells. *Blood* **90,** 3691–3698 (1997).

Dent, S., Zee, B., Dancey, J., et al. Application of a new multinomial phase II stopping rule using response and early progression. *J. Clin. Oncol.* **19,** 785–791 (2001).

Druker, B. J., Talpaz, M., Resta, D. J., et al. Efficacy and safety of a specific inhibitor of the BCR-ABL tyrosine kinase in chronic myeloid leukemia. *N. Engl. J. Med.* **344,** 1031–1037 (2001).

Eisenhauer, E. A. Phase I and II trials of novel anti-cancer agents: Endpoints, efficacy and existentialism. *Ann. Oncol.* **9,** 1047–1052 (1998).

Estey, E. H., and Thall, P. F. New designs for phase 2 clinical trials. *Blood* **102,** 442–448 (2003).

FDA, Center for Drug Evaluation and Research. Drug development and review definitions. Available at *www.fda.gov/cder/about/smallbiz/definitions.htm* (2003).

Fornier, M., Risio, M., van Poznak, C., and Seidman, A. HER2 testing and correlation with efficacy of trastuzumab therapy. *Oncology* **16,** 1340–1352 (2002).

Fox, E., Curt, G. A., and Balis, F. M. Clinical trial design for target-based therapy. *Oncologist* **7,** 401–409 (2002).

Fukuoka, M., Yano, S., Giaccone, G., et al. Multi-institutional randomized phase II trial of gefitinib for previously treated patients with advanced non-small-cell lung cancer. *J. Clin. Oncol.* **21,** 2237–2246 (2003).

Gailbraith, S. M., Maxwell, R. J., Lodge, M. A., et al. Cobretastatin A4 phosphate has tumor antivascular activity in rat and man as demonstrated by dynamic magnetic resonance imaging. *J. Clin. Oncol.* **21,** 2831–2842 (2003).

Gasparini, G., and Gion, M. Molecular-targeted anticancer therapy: Challenges related to study design and choice of proper endpoints. *Cancer J.* **6,** 117–131 (2000).

Gelmon, K. A., Eisenhauer, E. A., Harris, A. L., et al. Anticancer agents targeting signaling molecules and cancer cell environment: challenges for drug development? *J. Natl. Cancer. Inst.* **91,** 1281–1287 (1999).

Gordon, M. S., Margolin, K., Talpaz, M., et al. Phase I safety and pharmacokinetic study of recombinant human anti-vascular endothelial growth factor in patients with advanced cancer. *J. Clin. Oncol.* **19,** 843–850 (2003).

Heinrich, M. C., Corless, C. L., Blanke, C., et al. KIT mutational status predicts clinical response to STI571 in patients with metastatic gastrointestinal stromal tumors (GISTs) [Abstract]. *Proc. Am. Soc. Clin. Oncol.* **21,** 2a (2002).

Herbst, R. S., Maddox, A. M., Rothenberg, M. L., et al. Selective oral epidermal growth factor receptor tyrosine kinase inhibitor ZD1839 is generally well-tolerated and has activity in non-small-cell lung cancer and other solid tumors: results of a phase I trial. *J. Clin. Oncol.* **20,** 3815–3825 (2002a).

Herbst, R. S., Mullani, N. A., Davis, D. W., et al. Development of biologic markers of response and assessment of antiangiogenic activity in a clinical trial of human recombinant endostatin. *J. Clin. Oncol.* **20,** 3804–3814 (2002b).

Hidalgo, M., Malik, S., Rowinsky, E., et al. Inhibition of the epidermal growth factor receptor (EGFR) by OSI-774, a specific EGFR inhibitor in malignant and normal tissues of cancer patients [Abstract]. *Proc. Am. Soc. Clin. Oncol.* **20,** 71a (2001).

Kabbinavar, F., Hurwitz, H. I., Fehrenbacher, L., et al. Phase II, randomized trial comparing bevacizumab plus fluorouracil (FU)/leucovorin (LV) with FU/LV alone in patients with metastatic colorectal cancer. *J. Clin. Oncol.* **21,** 60–65 (2003).

Korn, E. L., Arbuck, S. G., Pluda, J. M., et al. Clinical trial designs for cytostatic agents: Are new approaches needed? *J. Clin. Oncol.* **19,** 265–272 (2001).

Kris, M. G., Natale, R. B., Herbst, R. S., et al. Efficacy and safety of gefitinib (ZD1839, Iressa^{TM}), an inhibitor of the epidermal growth factor receptor tyrosine kinase, in symptomatic patients with advanced non-small cell lung cancer. *JAMA* **290,** 2149–2158 (2003).

Kris, M. G., Natale, R. B., Herbst, R. S., et al. A phase II trial of ZD1839 ('Iressa') in advanced non-small cell lung cancer (NSCLC) patients who had failed platinum- and docetaxel-based regimens (IDEAL 2) [Abstract]. *Proc. Am. Soc. Clin. Oncol.* **21,** 292a (2002).

Laight, A., Swaisland, H. C., Partridge, E. A., et al. Metabolism of [^{14}C]-ZD1839 ('Iressa') in healthy male volunteers [Abstract]. *Proc. Eur. Soc. Med. Oncol.* **13,** 27 (2002).

Lin, E. H., Morris, J., Chau, N. K., et al. Celecoxib attenuated capecitabine induced hand-and-foot syndrome (HFS) and diarrhea and improved time to tumor progression in metastatic colorectal cancer (MCRC) [Abstract]. *Proc. Am. Soc. Clin. Oncol.* **21,** 138b (2002).

LoRusso, P. M., Herbst, R. S., Rischin, D., et al. Improvements in quality of life and disease-related symptoms in phase I trials of the selective oral epidermal growth factor receptor tyrosine kinase inhibitor ZD1839 in non-small cell lung cancer and other solid tumors. *Clin. Cancer Res.* **9,** 2040–2048 (2003).

Nagasubramanian, R., Innocenti, F., and Ratain, M. J. Pharmacogenetics in cancer treatment. *Annu. Rev. Med.* **54,** 437–452 (2003).

Nakagawa, K., Tamura, T., Negoro, S., et al. Phase I pharmacokinetic trial of the selective oral epidermal growth factor receptor tyrosine kinase inhibitor gefitinib ('Iressa', ZD1839) in Japanese patients with solid malignant tumors. *Ann. Oncol.* **14,** 922–930 (2003).

Natale, R. B., Shak, S., Aronson, N., et al. Quantitative gene expression in non-small cell lung cancer from paraffin-embedded tissue specimens: Predicting response to gefitinib, and EGFR kinase inhibitor [Abstract]. *Proc. Am. Soc. Clin. Oncol.* **22,** 190 (2003).

O'Brien, S. G., Guilhot, F., Larson, R. A., et al. Imatinib compared with interferon and low-dose cytarabine for newly diagnosed chronic-phase chronic myeloid leukemia. *N. Engl. J. Med.* **348,** 994–1004 (2003).

Osborne, C. K. Tamoxifen in the treatment of breast cancer. *N. Engl. J. Med.* **339,** 1609–1618 (1998).

Patrick-Miller, L. J. Is there a role for the assessment of health-related quality of life in the clinical evaluation of novel cytostatic agents? Commentary re: P. M. LoRusso, Improvements in quality of life and disease-related symptoms in Phase I trials of the selective oral epidermal growth factor receptor tyrosine kinase inhibitor ZD1839 in non-small cell lung cancer and other solid tumors. *Clin. Cancer Res.* **9,** 2040–2048 (2003).

Plosker, G. L., and Figgitt, D. P. Rituximab: A review of its use in non-Hodgkins' lymphoma and chronic lymphocytic leukaemia. *Drugs* **63,** 803–843 (2003).

Pusztai, L., Ayers, M., Simmans, F. W., et al. Emerging science: Prospective validation of gene expression profiling-based prediction of complete pathologic response to neoadjuvant paclitaxel/FAC chemotherapy in breast cancer [Abstract]. *Proc. Am. Soc. Clin. Oncol.* **22,** 1 (2003).

Ranson, M., Hammond, L. A., Ferry, D., et al. ZD1839, a selective oral epidermal growth factor receptor-tyrosine kinase inhibitor, is well tolerated and active in patients with solid, malignant tumors: results of a phase I trial. *J. Clin. Oncol.* **20,** 2240–2250 (2002).

Rothenberg, M. L., Carbone, D. P., and Johnson, D. H. Improving the evaluation of new cancer treatments: challenges and opportunities. *Nature Rev. Cancer* **3,** 303–309 (2003).

Rowinsky, E. K. The pursuit of optimal outcomes in cancer therapy in a new age of rationally designed target-based anticancer agents. *Drugs* **60(suppl 1),** 1–14 (2000).

Saltz, L., Rubin, M., Hochster, H., et al. Cetuximb (IMC-C225) plus irinotecan (CPT-11) is active in CPT-11-refractory colorectal cancer (CRC) that expresses epidermal growth factor receptor (EGFR) [Abstract]. *Proc. Am. Soc. Clin. Oncol.* **20,** 3a (2001).

Saltz, L. B., Meropol, N. J., Loehrer, P. J., et al. Single agent IMC-C225 (Erbitux) has activity in CPT-11-refractory colorectal cancer (CRC) that express the epidermal growth factor receptor (EGFR) [Abstract]. *Proc. Am. Soc. Clin. Oncol.* **21,** 127a (2002).

Slamon, D. J., Leyland-Jones, B., Shak, S., et al. Use of chemotherapy plus a monoclonal antibody against HER2 for metastatic breast cancer that overexpresses HER2. *N. Engl. J. Med.* **344,** 783–792 (2001).

Somer, R. A., Langer, C. J., Millenson, M., et al. Restrictive eligibility criteria may limit applicability of new therapies in NSCLC [Abstract]. *Proc. Am. Soc. Clin. Oncol.* **22,** 668 (2003).

Sridhar, S. S., Seymour, L., and Sheperd, F. A. Inhibitors of epidermal-growth-factor receptors: A review of clinical research with a focus on non-small-cell lung cancer. *Lancet Oncol.* **4,** 397–406 (2003).

Stadler, W., and Ratain, M. Development of target-based antineoplastic agents. *Invest. N. Drugs* **18,** 7–16 (2000).

Swaisland, H., Laight, A., Stafford, L., et al. Pharmacokinetics and tolerability of the orally active selective epidermal growth factor receptor tyrosine kinase inhibitor ZD1839 in healthy volunteers. *Clin. Pharmacokinet.* **40,** 297–306 (2001).

Swaisland, H. C., Ranson, M., Smith, R., et al. Clinical drug interactions with ZD1839 ('Iressa'), a selective epidermal growth factor receptor tyrosine kinase inhibitor [Abstract]. *Proc. Eur. Soc. Med. Oncol.* 27 (2002).

Temple, R., and Ellenberg, S. S. Placebo-controlled trials and active-control trials in the evaluation of new treatments. Part 1: ethical and scientific issues. *Ann. Intern. Med.* **133,** 455–463 (2000).

Therasse, P., Arbuck, S. G., Eisenhauer, E. A., et al. New guidelines to evaluate the response to treatment in solid tumors. *J. Natl. Cancer Inst.* **92,** 205–216 (2000).

van de Vijver, M. J., He, Y. D., van't Veer, L. J., et al. A gene-expression signature as a predictor of survival in breast cancer. *N. Engl. J. Med.* **347,** 1999–2009 (2002).

Vogel, C. L., Cobleigh, M., Tripathy, D., et al. Superior outcomes with herceptin (trastuzumab) (H) in fluorescence in situ hybridization (FISH)-selected patients [Abstract]. *Proc. Am. Soc. Clin. Oncol.* **20,** 22a (2001).

Yang, J. C., Haworth, L., Steinberg, S. M., et al. A randomized double-blind placebo-controlled trial of bevacizumab (anti-VEGF antibody) demonstrating a prolongation in time to progression in patients with metastatic renal cancer [Abstract]. *Proc. Am. Soc. Clin. Oncol.* **21,** 5a (2002).

CHAPTER 14

Intellectual Property and Commercialization Issues in Drug Discovery

Lisa Gail Malseed

Molecular Cancer Therapeutics: Strategies for Drug Discovery and Development, edited by George C. Prendergast
ISBN 0-471-43202-4

Scientists conducting research at most institutions are required to be familiar with intellectual property issues. Legislative action created the push for patents at academic institutions and the growth of technology transfer from these institutes to pharmaceutical or biotechnology companies. As a result, institutions now require scientific employees to identify inventions, assist in the preparation and prosecution of patents, and automatically assign patent rights to the university, enabling the institution to out-license the commercial rights to inventions developed within the laboratories. Since the patent process initially requires secrecy and, eventually, a patent holder requires users to license the invention, the institutional emphasis on securing patent protection may impair an individual scientist's ability to build upon the progress of colleagues in science.

In the mid-1990s, some influential scientists (including members of the National Academy of Science) disputed whether the value gained from out-licensing (applied science supported by private money) outweighed the value lost (free public access to scientific achievement for the next generation) (Eisenberg, 1996). The discussion culminated in research scientists' anger over the license restrictions imposed by a commercial entity on its fundamental technology. After much public debate, the nation's largest institution, the National Institutes of Health (NIH), became the grantor of noncommercial research licenses with respect to these fundamental technologies, brokering a truce between scientists and commercial entities. Creating a "research purpose only" license and transfer agreement resolved the debate, and the existing system of invent, patent, and out-license technologies to commercial organizations remains the favored paradigm in modern drug development.

This chapter discusses the patenting process, the licensing process, and the issues around choosing the right development partner. This chapter offers a basic review of the patent statutory system, and the laboratory practices necessitated by it. This chapter also explores issues inherent in the system of licensing scientific developments, such as a discussion of the expanding definition of patentable invention, the appropriate grant (exclusive versus nonexclusive, field of use limits) necessary to develop technologies, and the issues that still surround granting research licenses to nonprofit organizations, while requiring commercial interests to pay for the same privilege. Finally, the chapter discusses financial issues along the development time line.

14.1 Intellectual Property

Intellectual property is defined as intangible personal property afforded legal protection that allows the owner to prevent third parties from gaining unapproved commercial advantage. The protection is statutory and includes patent, trademark, service mark and copyright. Congress enacted U.S. patent laws under its constitutional grant of authority to protect the discoveries of inventors. U.S. Constitution, Article 1, §8 reads, in part: "The Congress shall have power to lay and collect taxes, duties, imposts and excises, to pay the debts and provide for the common defense and general welfare of the United

States; but all duties, imposts and excises shall be uniform throughout the United States; . . . To promote the progress of science and useful arts, by securing for limited times to authors and inventors the exclusive right to their respective writings and discoveries."

Patent statutory law is found in Title 35 of the United States Code and Title 37 of the Code of Federal Regulations. A patent can be obtained for a process that produces a useful, concrete, and tangible result; a machine; an article of manufacturing; a composition of matter or genetically altered life form; or an improvement to anything falling into the listed categories (35 USC Section). In 1980, the U.S. Supreme Court held in the case of *Diamond v. Chakrabarty,* 447 U.S. 303 (1980), that a living, genetically altered organism may qualify for patent protection as a new manufacture or composition of matter under Section 101 of the U.S. Patent Code. Characterizing Chakrabarty's invention as "a new bacterium with markedly different characteristics from any found in nature" and "not nature's handiwork, but his own," the Court indicated that Congress intended the patent laws to cover "anything under the sun that is made by man." A good general discussion of the patent process can be found on the Internet on www.nolo.com as well as on several law firm sites (such as the Web site of Litman Law Offices, Ltd.: www.litmanlaw.com). These and many other similar sites can be helpful, but each is a restatement of the statutory requirements as set forth in 35 USC.

Provided certain statutory definitions are met, a patent may also be obtained on a design or a plant. A design patent protects the decorative aspects of an invention when the essence of an invention is its appearance, or if the final version has a unique look (35 USC). A plant patent is granted for the asexual reproduction of a plant with distinctive traits (35 USC). Once issued, a patent allows the owner to exclude others from making, using, offering for sale, selling, or importing the invention claimed into the relevant jurisdiction. Patent protection for applications filed after 1995 extends for 20 years from the date of filing of the regular patent application. (The filing of a Provisional Patent Application does not begin the 20-year period.)

Trademarks (for goods) and service marks (for services) use words, symbols, numbers, slogans, designs, packaging, and even sounds, to identify the source of the products (branding) (15 USC 1051–1127). Copyright applies to the ownership of tangible authored works (books, movies, music) (17 USC 102). All of these protections are offered to owners who are willing to publicly reveal the details of their property. Trade secret status covered by state law is afforded to information owned by an individual, organization, and/or corporation that is kept secret from third parties (most notably, the recipe for Coca-Cola). Until the late 1970s, intellectual property protection of scientific research results was inconsistent based on the cost to prosecute as well as the climate of free access.

The protection of intellectual property arising from scientific research conducted using federal funding was altered in 1980 when the Congress enacted two laws: the Bayh-Dole Act (Pub. Law No. 96-517, 6(a), 94 Stat. 3015, 3019-27 [1980]) and the Stevenson-Wydler Technology Innovation Act (Pub. Law No. 96-480, 94 Stat. 2311 [1980]). Until 1980, all scientific information arising from research conducted using federal funding was required to be released in the public domain (National Research Council, 1996), to allow all researchers access to the advancements. Academic scientists and institutions concerned themselves with publications and the prestige of the journal selecting the article, often referred to as the publish or perish rule. However, with the passage of the Bayh-Dole Act and the Stevenson-Wydler Technology Innovation Act, government contractors, small businesses, and nonprofit organizations were allowed to retain certain patent rights in government-sponsored research and permitted to transfer the technology to third parties (Eisenberg, 1996). Eisenberg (1996) described the intention of the legislators and the almost immediate impact of the new legislation:

> The stated intent of Bayh-Dole was to ensure that the patented results of federally funded research would be broadly and rapidly available for all scientific investigation. Bayh-Dole effectively shifted federal policy from a position of putting the results of government-sponsored research directly into the public domain for use by all, to a pro-patent position that stressed the need for exclusive rights as an incentive for industry to undertake the costly investment necessary to bring new products to market. The policy was based on a belief that private entities, given the incentives of the patent system, would do a better job of commercializing inventions than federal agencies. The Act for the first time established a largely uniform government-wide policy on the treatment of inventions made during federally supported R and D.

Stevenson-Wydler is the basic federal technology law. A principal policy established by that act is that agencies should ensure the full use of the results of the nation's federal investment in research and development (R&D). Another is that the law requires federal laboratories to take an active role in the transfer of federally owned or originated technology to both state and local governments and to the private sector. Stevenson-Wydler required agencies to establish offices of research and technology applications at their federal laboratories and to devote a percentage of their R&D budgets to technology transfer (Eisenberg, 1996).

In effect, the first law said the inventor could own what was patented despite having received research funding from public monies and the second law said what is patented ought to be licensed and developed using private dollars. The new legislation is often cited as the cause for the increase in utility patent applications in the U.S. Patent and Trademark Office (USPTO). In 1979, 100,494 utility patent applications were filed with the USPTO; in 1989, 152,750 utility patent applications were filed; and by 2000, 295,926 patents were being filed annually (www.uspto.gov).

The legislation's other effect was the growth of the biotechnology company, a privately financed organization that in-licensed technologies developed at institutions to speed development from invention to product. Biotechnology companies were intended to take benchtop technologies out of the laboratory and develop the technologies quickly and cost effectively to become more interesting to established pharmaceutical companies or the medical professionals. The new legislation also required academic institutions accepting federal funding to establish technology transfer offices whose sole responsibility is securing private dollars to commercially develop federally funded inventions. Academic institutions became licensing houses, and scientists were required to have familiarity with intellectual property concepts.

The drug development process often involves the creation of intellectual property. For the academic investigator, applications scientist and the clinical development product manager, recognizing inventions and taking the steps necessary to obtain utility patent protection is essential. Therefore, the following section deals in detail with the process a laboratory may use successfully to obtain utility patent protection in the United States for its inventions, with a brief discussion of foreign jurisdiction application. These same processes may be used to identify and protect intellectual property with plant or design patents, trade or service marks, copyrights and, in some appropriate instances, trade secrets.

14.2 Laboratory Practices

Research and drug development laboratories plan and conduct experiments to corroborate or disprove theories about the use, application, and composition of compounds. They also study and discern the important mechanisms within living organisms. These experiments yield results that often lead to the continuation or alteration of the laboratories research program. The experiments and results are kept in laboratory notebooks, assigned to individuals, and they contain preprinted consecutively numbered pages with execution and witness signature spaces. (Laboratory notebooks and procedures are available commercially; see also *Federal Register* 21 CFR 58.) Laboratory notebooks have standardized formats (e.g., see www.snco.com). Current debate in this area rages around the issue of electronic laboratory notebooks (e.g., www.gensys.com) and the sufficiency of data security and signature authenticity. The debate has become so heated that the American Bar Association has issued an opinion on digital signatures (see www.abanet.org/scitech/ec/isc/dsgfree.html). The important aspects of notebooks as evidence of who, when, and what happened, at least in hard copy, is provided within bound, prenumbered pages that are completed and signed daily by the same laboratory person and witnessed by knowledgeable scientists. Notebooks that include pages that are unsigned, that present data that have been altered, or that are missing pages prevent the notebook from supporting the date and/or inventor of an invention.

Eventually, the results written in the laboratory notebooks lead to knowledge, which the laboratory considers valuable to others, and a journal article is written and submitted for publication. In some institutions, presubmission review is not required. However, given that the invention will not receive patent protection in most non-U.S. jurisdictions if a public disclosure (publication of an article, presentation to a peer review group) is made before the filing of the patent application, review by the administration of the institution is most often required before manuscript submission to a journal or abstract submission to a meeting.

Normally, it is at this point that an institutional reviewer first learns of the research result. During the review, determinations are made whether the results constitute intellectual property and, if so, whether an invention has been made. Better organizations hold quarterly reviews during which laboratory notebooks are reviewed and research results are discussed in a group setting with trained patent personnel in attendance. In either of these processes, the institutional administrators may request that the inventing scientist(s) complete an invention disclosure report, if the scientist has not already completed one. The invention disclosure report is a document written by a scientist for a patent lawyer's use in determining whether patent protection should be sought for the described invention. It is a confidential document provided the invention disclosure report is not made public or discussed with third parties who are not under obligations of confidentiality with the laboratory, institution, and/or company.

Invention disclosure reports vary in format, but generally contain a description of the invention, a thorough discussion of the status of the research field in the current publications and patents, the preferred practice of the invention, relevant research results and other data, and – for U.S. filings – all public disclosures of the invention (Silverman, 1994; see also www.ott.rice.edu/Policies/DisFormInvention.cfm and www.maineandasmus.com/disclos.htm). The authoring scientist should discuss potential practices of the invention and should consider alternative approaches to solving the problem addressed by the invention. The invention disclosure is then reviewed to determine if the described invention may be appropriate subject matter for a patent. It is at this point that the determination of whether other intellectual property protection (copyright, trademark, etc.) should be considered for the discovery.

To qualify for a utility patent, the invention must be a process, machine, article of manufacturing, composition of matter, and/or improvement, and the invention must have utility, be novel and be nonobvious. New drugs or composition of matter that are theoretically unsupportable and have no proof of efficacy will not support a patent. Unfortunately, ground-breaking scientific advancement can be difficult to patent until efficacy is established, whereas the USPTO tends to accept the slightest improvement to a known drug.

The invention or some constituent part of the invention must be different from prior inventions in some meaningful way to support the requirement that it be novel (www.becentral.com). To be the subject matter of a patent, an invention cannot have been published before the application date, or, in the United States, more than 1 year before that date. The patent examiner will investigate all prior inventions in the field of the current application (known as *prior art*) as well as past and contemporary publications to determine novelty. In the United States, an invention is not novel if it was described in a published document or put to public use more than 1 year before the filing date of the patent application (35 USC 102(b)). Foreign jurisdiction examiners will disallow patent issuance for lacking novelty if the publication or public use or sale occurred before (even the day before) the filing of the foreign application.

Institutions and scientists should understand that public disclosure or use or sale of the invention prior to 1 year before filing a patent application effectively prevents securing patent protection within the United States. In this context, public disclosure must be a printed (tangible form) publication with an enabling disclosure that is accessible to the public (see *MIT v. AB Forcia,* 74 Fed.2d. 1104). As an illustration, a lecture given to 50 people describing the invention with a printed publication handed to 6 was sufficient to constitute a "public disclosure" (*MIT v. AB Forcia,* 74 Fed.2d. 1104).

An invention may be used for experimentation without being considered for sale. The definition of what *for sale* means and when the experimental use exception is overridden may be found in a recent U.S. Supreme Court case (*Pfaff v. Wells,* 525 US 55). A sale or public use of the invention occurs if what was offered or sold was in fact something in the scope of the claims, regardless of whether the inventor or seller knew that it did (*Abbot v. Geneva,* 51 PQ2d 1307, 182 Fed.3rd 1315).

Finally, an invention must be nonobvious. An invention is considered nonobvious if someone who is skilled in the particular field of the invention would view it as an unexpected or surprising development (35 USC 103 and 35 USC 112). Obviousness raises particular difficulty for institutions and scientists who tend to see developments as natural progression from earlier work. Research programs grow out of discoveries of other laboratories or even prophetic statements made at the end of publications. That a result arises from earlier discoveries does not condemn that result from being suitable for patenting. The standard is only that some cleverness or original thought be required in the research leading to the result.

Once the invention disclosure report has been received by patent counsel, the process of application development begins. Patent law, which is statutory law, provides a list of required documents and sections for utility patents (35 USC §111). The most challenging of these sections may be the specification and the claims as described in §112 of 35 USC:

> The specification shall contain a written description of the invention, and of the manner and process of making and using it, in such full, clear, concise, and exact terms as to enable any person skilled in the art to which it pertains, or with which it is most nearly connected, to make and use the same, and shall set forth the best mode contemplated by the inventor of carrying out his invention. The specification shall conclude with one or more claims particularly pointing out and distinctly claiming the subject matter which the applicant regards as his invention.

A claim may be written in independent or, if the nature of the case admits, in dependent or multiple dependent form.

> Subject to the following paragraph, a claim in dependent form shall contain a reference to a claim previously set forth and then specify a further limitation of the subject matter claimed. A claim in dependent form shall be construed to incorporate by reference all the limitations of the claim to which it refers.

> A claim in multiple dependent form shall contain a reference, in the alternative only, to more than one claim previously set forth and then specify a further limitation of the subject matter claimed. A multiple dependent claim shall not serve as a basis for any other multiple dependent claim. A multiple dependent claim shall be construed to incorporate by reference all the limitations of the particular claim in relation to which it is being considered.
>
> An element in a claim for a combination may be expressed as a means or step for performing a specified function without the recital of structure, material, or acts in support thereof, and such claim shall be construed to cover the corresponding structure, material, or acts described in the specification and equivalents thereof.

When the scientist, laboratory, and institution are determining the existence of an invention, they should also consider the funding source of the research versus the commercial value (or potential value) of the invention. Currently, federally funded programs that lead to inventions have a requirement of patenting "to pursue commercialization in the private sector." The cost of having a patent firm file a utility patent application varies widely ($5,000–25,000 per application). Given the technical nature of the patent law as well as the expertise required to understand the novelty, obviousness, prior art and the construction of claims, these amounts are fair if the invention has a commercial value. Determining the commercial value of a patentable invention that is merely a potential product can be difficult especially if the funded program does not allow sufficient funds for the initial patent filing and early prosecution. The question of who will finance the early protection of new inventions has kept scientists, laboratory managers, and institutional administrators in disagreement, especially when reduction to practice is ongoing.

In part to allow scientists to continue to perfect and understand their discoveries, in June 1995, Congress enabled inventors to file a provisional patent application (PPA) on inventions. Filing a PPA allows an inventor to claim patent-pending status for the invention but involves only a small fraction of the work and cost of a regular patent application. A PPA may be filed with a cover sheet, a small fee, and a detailed description of the invention, which can be a publication manuscript or technical paper detailing the invention (35 USC 111(b) 1–3). Informal drawings may be required if they are required to understand the invention. A PPA does not require an abstract, patent claims, or prior art review.

The benefit of filing a PPA is that if a regular application is filed within a year of the PPA application and the claims rely on the information disclosed within the PPA, the regular application is afforded the filing date of the PPA (35 USC 119). The regular application may include any new matter (technical information about the invention) that was not in the PPA, but the claims based on that new information may not rely on the PPA's filing date. Failure to file a regular application within 1 year will result in the loss of the PPA filing date.

In addition to an early filing date and the right to claim patent-pending status for the invention, filing a PPA can provide another important advantage.

The PPA filing date does not affect when the patent on the invention will expire. The expiration date is still 20 years from the date of the regular patent application. So a PPA has the practical effect of delaying examination of the regular patent application and extending – up to 1 year – the patent's expiration date.

Foreign jurisdictions vary in the procedures required to protect intellectual property. However, most require that the inventor of an invention maintain secrecy until the patent application has been filed (unlike the United States, wherein an inventor has 1 year from disclosure during which to file for protection). Foreign jurisdictions have particular technical requirements (language, length, type, and number of claims) that differ from the United States as well as from each other. When filing for patent protection, it is crucial to determine whether the invention might have value worldwide, and, if so, to take the steps necessary to secure patent protection in appropriate foreign jurisdictions. A good method is to file for U.S. and foreign protection before any disclosures by using the PTO PCT (Patent Convention Treaty) application process. The PCT application allows the patent application to be "filed" in every participating jurisdiction (only Japan, China, and India do not currently participate of the commercially significant countries) by the completing of a form, payment of fees, and attachment of the U.S. application. The filer then has up to 30 months to select, conform the application to, and pay the individual fees of other countries.

The end result of all the records, the discussions, the planning and the drafting is a patent application that describes the invention seeking protection. This application and subsequent patent then becomes a powerful tool for the further development and commercialization of the invention.

14.3 Ownership of Intellectual Property

Ownership of a patent gives the patent owner the right to exclude others from making, using, offering for sale, selling, or importing into the United States the invention claimed in the patent (35 USC 154(a)(1)). The ownership of the patent (or the application for the patent) initially vests in the named inventors of the invention of the patent (see *Beech Aircraft Corp. v. EDO Corp.*, 990 F.2d 1237, 1248, 26 USPQ2d 1572, 1582 (Fed. Cir. 1993)). The patent (or patent application) is then assignable by an instrument in writing, and the assignment of the patent, or patent application, transfers to the assignee(s) an alienable (transferable) ownership interest in the patent or application (35 USC 261). In 37 CFR 3.1, assignment of patent rights is defined as "a transfer by a party of all or part of its right, title and interest in a patent or patent application." Thus the assignment process is the mechanism whereby an inventor can transfer some or all of the ownership interest to another party.

In both industry and academia, institutions require employees (potential inventors) to sign agreements (called intellectual property assignments and sometimes contained within employment contracts), which require that all inventions conceived of by employees be assigned to the institution automatically on invention. These agreements often require the inventor (even if employment has been terminated) to assist the institution in drafting and prosecuting the patent. Many institutions and virtually all companies do not provide additional compensation to the inventor for this achievement or other postinvention work. The rationale seems to be that this activity is within the scope of employment for which the employee is already compensated.

14.4 Commercialization of the Patent

A generation ago, the prevailing wisdom was that the best way to ensure full use of publicly sponsored research results for the public good was to make them freely available to the public. Today, federal policy reflects the opposite assumption. The current belief is that if research results are made widely available to anyone who wants them, they will languish in government and university archives, and institutions will be unable to generate commercial interest to pick up where the government leaves off by encouraging other to use the results to develop commercial products (National Research Council, 1996). To make government-sponsored research discoveries attractive candidates for commercial development, institutions performing the research are motivated or required to obtain patents and to offer licenses to the private sector.

As compared to assignment of patent rights, the licensing of a patent transfers a bundle of rights that is less than the entire ownership interest, for example, rights that may be limited as to time, geographical area, or field of use. (Unlike the statutory protection of assignment (35 USC 261), patent licenses are covered by contract law, by which patents are treated as personal property.) A patent license is, in effect, a contractual agreement that the patent owner will not sue the licensee for patent infringement if the licensee makes, uses, offers for sale, sells, or imports the claimed invention, as long as the licensee compensates the patent owner and otherwise fulfills its obligations and operates within the bounds delineated by the license agreement.

It is at this point that the three groups most interested in the invention (scientist-inventor, assignee-institution, and company) may have interests that are in conflict. It is essential, therefore, to understand each bundle of interests to maximize the areas of convergence and minimize the disagreements. The following is a process that may be followed.

14.5 Protecting the Protected

A company often expresses interest in an institution's patent and patent-pending portfolio, requiring confidential information exchanges between the parties. Moreover, as research laboratories increasingly require private funding to continue interesting research and/or for reduction to practice of invention, disclosing confidential information without harming future protection of intellectual property is important. As discussed above, an invention disclosure document is a confidential document between inventor, assignee (normally institution or company), and the patent attorney. All information contained within the document is treated as confidential provided no information is made public. However, a third party sought out for funding or interested in published results of the laboratory may express interest in reviewing applicable invention disclosures or the subsequent descriptive documents (PPA, patent application, draft licensing agreements).

The mechanism for such disclosures is called a nondisclosure agreement or a confidentiality agreement. As with all legal documents, these agreements vary in format but most include (1) a definition of what constitutes confidential information, (2) the receiving party's required acts to keep the information from becoming public knowledge (known generally as the receiving party's obligations), and (3) the time frame during which the confidential information is to be treated by both parties as confidential. Disclosures of inventions, whether written or oral, include information generally known by all as well as information that is known only to the inventor and others who are bound to keep it secret. It is this second type of information that the discloser wishes to require the receiving party to keep confidential. The receiving party is often required to keep the disclosure secret from everyone not specifically included in the agreement and, for all included, most are required to be subject to an obligation of confidentiality with respect to other third parties. The length of time during which the receiving party must continue to treat the disclosed information as confidential is subject to negotiations. With publication of European patent applications (typically 18 months after submission of a foreign patent application), these time limits have begun to standardize at 3 years, although some agreements' confidentiality terms are 1 year and others require as many as 10 years.

14.6 The Three-Sided Talk: Focus on the Invention

Once the confidentiality agreement is signed, the interested party may review all information that the scientist and institution agree to release. If the invention is interesting to the organization, negotiations begin between the scientist, institution, and company. This negotiation often highlights the different interests of the parties. The inventing scientist often prioritizes funding for additional research in the laboratory, ensuring the continued progress of the research. The institution supports some research funding but may be more interested in recouping costs borne by the institution in the patent process as well as securing dependable future milestone and royalty payments necessary to ensure the continuation of the institution. Business development within the company attempts to keep the immediate and midterm costs and on-going support obligations minimal as the company agrees to undertake the often long-term responsibility for developing the invention into a commercial product.

Taken individually, each group's stated goal provides a mechanism to maintain the existence of that group. A scientific laboratory needs capital to perform research. An institution needs tremendous amounts of outside financial support to maintain the excellence of the staff, facility, and research. A biotechnology firm must raise capital in the equity markets based on its ability to get a product to market. Each group needs something from the licensing transaction, but negotiating through the differences often results in relational fractures. Although some disharmony is unavoidable, careful consideration of the invention itself often leads to an equitable result.

Consideration should be given to the history of the invention. Questions should be raised about the status of the current field of research/product development/therapy; the importance of this development to that field; the amount, reliability, repeatability, and efficacy of the existing data; effectiveness in one or more systems; and the current applicability of the invention for biotechnology and/or pharmaceutical research and development. Bottom-line questions must be posed – What is the status of the invention today? What it may become is important, but knowing and discussing what it is today is a critical, if overlooked, part of licensing negotiations.

Ownership of the invention must be discussed. Protected intellectual property has had increasing value to institutions, which always need to find dollars to continue their growth. Even not-for-profit and charitable organizations that support basic research have recognized the important financial aspects of intellectual property. Grants from these organizations often include discussion of ownership of the resulting intellectual property. Thus companies can be negotiating for rights to property owned by more than one organization making the negotiation process long and tedious. Moreover, the need for additional technologies must be considered as the cost to license that property impacts the value of the new technology.

Once the status of the invention is thoroughly understood, the parties need to discuss the potential of the invention. Suppose that the invention is a cancer model, which mimics certain aspects of a disease state. The parties might discuss whether the hypothetical disease model may be used in studies using existing drugs, or is it more suitable as a targeting screen for potential drugs, or is the model's future value in the research results which arise from the use of the model as scientists attempt to modulate regulation leading to a better understanding of the function or the cascade events in cancer. All of these potential applications are valuable, but their current monetary value is substantially different. Almost any businessperson could make a profit with an acceptable drug efficacy-testing model. A number of commercial entities would be interested in using a drug target model. Functional studies are the basics of research and, therefore, a model would be interesting to all researchers, academic or industrial.

Getting back to the licensing table, however, these three invention potentials would lead to different deals with different partners. The drug screen might be licensed immediately for cash, and, therefore, the scientist and institution have, in effect, a product. These licenses include research funding only to the extent that the laboratory may continue to produce similar products. These licenses tend to have bigger upfront payments to the institution (and agreements that some portion be given to the inventing laboratory), few if any milestones and minimum annual royalties in lieu of or in addition to annual licensing fees.

In the targeting model, the biotechnology firm must continue research at its site (or at the university) to determine whether in fact targeting can be achieved. Appropriate commercial partners ought to have targeting expertise as well as libraries of proprietary compounds. These agreements tend to focus more on consulting support from the inventor with fees paid to the scientist and/or the laboratory, and tend to give the institution less up front, low license maintenance fees, increasingly large milestones and royalties.

Finally, the function model has the least immediate cash value despite perhaps having a greater scientific value. It may be a good candidate for the nonexclusive license, with annual licensing fees, milestones, and cash payments around future related products arising from and/or sublicenses by the biotechnology company.

It is important for commercial concerns to know the actual costs spent by the university on legal and administrative costs incurred by the university. Most research has been conducted under federal or industrial financing, so the institution has been paid for their piece of that in the form of overhead. Up front payment to institutions for "promising" or "potential" technologies should be limited to reimbursement for actual costs (and those costs should be supported by financial documentation).

Addressing and balancing these competing desires takes time, but eventually, an agreement will be written. At times, a large organization is able to secure the assignment of the invention (this often requires a substantial one-time payment with future royalties) but usually the agreement constructed is in the form of a license. The license allows the institution to reclaim the commercial rights if the licensing party does not adequately perform its stated obligations as well as allowing the institution to maintain the asset of the patent.

14.7 Licensing the Invention

A license is written permission to use an invention. A license may be exclusive (if only one group is licensed to develop the invention) or nonexclusive (if a number of groups are licensed to use and/or develop it). The license may be for the duration of the patent or for a shorter period of time. The licensee itself may have the right to sublicense other companies to develop, manufacture, or distribute the product arising from the invention. The institution normally retains the right to continue research on the invention (and the licensee may have continuing obligations to support that research in exchange for license to additional intellectual property arising from the research) and may have the right to regain the commercial rights to the invention if the organization does not spend sufficient time, effort, and money to create, manufacture, and distribute a royalty-bearing product. These terms are included in the licensing agreement.

Some inventors start new companies to develop and market their patented inventions. Some academic institutions have created incubator space within their campus laboratories or off campus to allow their scientists to oversee the initial development and to allow the institution to own the start-up organization. Local governmental agencies, local industry and local venture capital firms often subsidize these incubators.

Institutions have allowed outside organizations to fully fund research within a named scientist's laboratory in exchange for ownership of all arising intellectual property. These arrangements shift the burden of pursuing protection to the third party while giving the institution short-term gain (often in the form of a percentage of the research contract called *overhead*), midterm gain (often in the form of milestone payments for development goals achieved), and long-term gain (often in the form of royalties or a percentage of the licensee company). However, once funded, the scientists within the laboratory often find their research dictated by the funding organization. Given the growth in the value of inventions and other intellectual property arising from scientific research, issues have arisen in the institutions and the drug-development industry concerning the proliferation of patents covering fundamental scientific research and its effect on scientific progress, the real value in ownership of rights in inventions that are years away from being commercial products and decades away from being patient therapies, and the viability of venture funded companies whose sole property is a patent estate covering "potential products."

Many cancer drugs are developed within the research laboratories of pharmaceutical companies. Scientists compete for the limited funding of the companies. If selected for development, pharmaceuticals support the drug's development internally or, at times, in partnership with affiliated companies. When a potential therapy is developed within and patented by a pharmaceutical, but is not selected for development, several outcomes may occur. First, the pharmaceutical's business development group may attempt to out-license the therapy to another pharmaceutical or biotechnology company. Second, the company may be persuaded to allow the inventing scientist to pursue external funding, providing that the pharmaceutical be compensated in the event of monetary success. Third, as the owner, the pharmaceutical may simply shelve the invention without further development. In the event the company chooses not to continue to pursue the invention, the company is obligated to allow the inventor to pursue the invention. However, patent prosecution costs tend to be high, especially when foreign jurisdiction work is necessary. Currently, pharmaceutical companies are in the process of merging and are reviewing two sets of research programs. This internal glut of research and development choices means that potential therapies may not be developed unless the scientific team is willing to work with business development to find more compatible laboratories.

14.8 Commercial Discussions

Eisenberg (1996) discussed the problem arising from the growing need of institutions to find nonfederal funding and the change in the statutory law surrounding the intellectual property arising from federally funded research. She noted that from the late 1970s, the explosion of commercial interest in research science, and the attendant emergence of commercial biotechnology companies, has amplified the importance of intellectual property in the biomedical sciences. Many biotechnology firms have found a market niche somewhere between the fundamental research that typifies the work of university and government laboratories and the end product development that occurs in more established commercial firms. To survive financially in this niche, biotechnology firms need intellectual property rights in discoveries that arise considerably upstream from commercial product markets. This creates pressure to patent discoveries that are closer to the work of research scientists than to ultimate consumer products (Eisenberg, 1996). Since the "Stanford mouse," patents have been issued covering numerous basic research procedures, animal models, experimental cell lines, reagents, and techniques. The issued patent may be used to prevent researchers from exploiting the development until a license can be secured. Thus a scientist may not be able to use a reagent until his or her technology transfer office has come to agreement with the technology transfer office of another institution. Although this time lag may be an annoyance, there is not critically collected evidence to establish that the wheels of research are being harmed.

In some instances, however, academics have united and expressed dissatisfaction with a company's approach when that approach appeared to tightly regulate, and, in effect, restrict, the research use of fundamental technologies. In September 1990, US Patent 4,959,317, titled "Site-specific recombination of DNA in eukaryotic cells" was issued, describing a powerful technique for the creation of disease models using an efficient method of DNA editing at a specific site on the mouse genome (the *cre* and *lox* sites). DuPont, the patent holder, had filed the application in 1987 and had spent millions in research, development, and legal costs. As the potential value of the patent became known, DuPont required scientists to sign an agreement that restricted the use of the *cre-lox* system and then severely restricted the transfer of any resulting animal models.

Academicians, accustomed to being able to avoid the vagaries of dealing with commercial entity restrictions, werc furious. A not-for-profit mouse production group, Jackson Laboratories, considered itself unable to conduct business with respect to the new animal models. Jackson negotiated unsuccessfully with DuPont for many years. Finally, with the NIH at the table, DuPont agreed all scientists receiving federal funding were free to share *cre-lox* mice with other nonprofit research laboratories, provided the receiving laboratory signed an agreement that they would not further transfer the models and that DuPont maintained commercial rights. The simplified transfer agreement prohibits "use of any material containing cre DNA and/or lox DNA which is encompassed by the DuPont Patent Rights (Material) for any commercial purpose or for the direct benefit of any for-profit institution." The transfers agreement is similar to the solution DuPont and the NIH reached over DuPont's in-licensed Harvard technology, often referred to as the oncomouse technology.

The research-only license, with its simple requirements of notice of use, notice of transfer, and prohibition of commercial use has gained favor. The issues remaining arise in the areas of supported laboratories, internal commercial entity "research" use, and reach through. Although still a small number, a substantial group of laboratories have accepted financing from commercial entities. Thus a corporate supported academic laboratory may have a research license on a basic technology like *cre-lox*.

Does the laboratory have a right to discuss the results of its research (supported or not) with the corporate benefactor or, alternatively, does the licensor have a right to deny the academic scientist the right to use the technology in any commercially funded research? Currently, two solutions are widely used: (1) silence with respect to notifying the granting party of the funding source for the research and (2) pre-use license negotiations between the commercial entities. The former is problematic, especially when articles acknowledging the research support appear. The latter often creates a discussion around licensing fees for a technology that may not be suitable for the attempted purpose. Given that single technology companies have convinced boards of directors and shareholders to allow research use license to academicians on the promise that big dollars will come from licensees granted to commercial concerns, these early stage commercial negotiations are often difficult and time-consuming. Solutions in this area have arisen when the granting entity reduces its financial expectation in exchange for appropriate data updates (in short, grantor is told whether the technology worked or enabled the experiment to produce the expected results) and payment is postponed until success.

In much the same way, large and small commercial entities conduct ongoing basic research. Some fundamental technologies could be used in these research programs but for financial barriers to use. Again, the solutions have included compromise by both parties. The stakes are very high. If commercial entity A licenses its technology for research purposes to commercial entity B for a set, but low, fee, commercial entity C may request the same deal. This sounds reasonable until reference to A's business plan, wherein A has forecast millions in revenue expected from licenses with other commercial entities, once the technology becomes viable. Entity A must carefully balance the need to develop its technology given its financial and scientific limitations against the possibility of destroying the financial returns on that technology. Solutions normally come from a single early stage partnership with a commercial entity chosen for its expertise, cash, and commitment to the development of entity A's technology. Most commercial entities do not entertain no or low cash partnerships with other commercial entities.

Another outstanding issue is, during development of the technology, certain additional information, such as how the technology works or what makes the technology work better, may come to light. Ownership of these "improvements" is often source of conflict between institutes and companies, an issue that must be resolved in the licensing or material transfer agreement. Entity A wants to secure a financially equitable commercial license while maintaining its absolute ownership over the invention including improvements whereas licensing party B wants ownership over inventions conceived by its scientists. Slowly, partnership discussions have progressed to include assignments of improvements back to entity A with license grants that allow the paying research licensees like B to use the improvement without fee and without the right to sublicense.

Reach through is the final difficult issue with respect to the licensing of new technologies. Disagreement between academicians and commercial entities exists as to when a company's right to demand financial compensation for use of its invention ends. Consider a hypothetical situation in which entity A provides a drug screen to Professor Z. Based on an ongoing relationship between entity C and Dr. Z, the professor runs entity C's library of drugs through A's screen, identifying six viable candidates to treat certain aspects of a particular cancer of interest to C. These new candidates are not new drugs; in fact, only slightly more that one in four drugs in Phase II clinical trials actually become large-market ($100+ million) drugs. However, but for entity A's screen, it is possible that entity C would never have chosen the six compounds from its library to develop. Assume for this purpose that entity A knows of Dr. Z's use of entity C's library. Entity A may receive a certain cash payment for the identification, but the question remains as to whether entity A has a right to royalties if marketed drugs arise from any of the identified compounds. In most instances, entity A would receive a small royalty share of what entity C receives.

Continuing the discussion, if the compounds themselves do not become drugs, but based on the six compounds selected, entity C realizes (or just speculates) that any compound with a particular salt like the six compounds may bind better. Entity C then performs in-house chemistry to existing drugs, adding the salt, has Dr. Z run the screen again, and finds improved efficacy. The question becomes does entity A have a right to payment in this second scenario. Since entity A knows that Dr. Z is using its screen, A would be well served to discuss the issues with entity C. Then a quick response is that entity A will receive payment for anything stipulated in the licensing agreement between A and C. But, the question is raised to discuss the nexus between the use of an invention in the process of developing another invention. This issue will increase in importance as commercial entities continue to increase their financial support of basic scientific research in the United States. As the federal government lowers its spending and commercial entities increase their support, maintaining control over the use of patented property is essential.

Moving one step further, what if Dr. Z publishes the results (with A and C's approval) and entity D's junior chemist, Dr. X, remembers an experiment from a high school chemistry club about this salt. Dr. X experiments with adding a similar salt to D's drug and finds that company D's drug's efficacy is enhanced. Dr. X never used entity A's screen and did not experiment with entity C's compounds; and yet company D will benefit financially based on work supported by those organizations. When does entity A cease to have a right to share in the financial benefits brought about, at least in part, by its patented invention. This is an important issue to consider when licensing, especially when licensing screens, fundamental technologies, and compound libraries.

14.9 Financing the Development

Once a patent application has been filed, the inventor, institution, and/or licensing entity begins the process of financing the development of the described invention. As described above, some inventions are marketable products and their development requires cash for production and sales. However, more often, the time between invention and product revenues is described in terms of years, even decades. To best think through all the issues, an inventor will be well served by doing the preliminary work needed to start a commercial organization whose sole goal is to develop the invention into a saleable product. This method forces parties to consider the elements of development (especially in terms of costs and risks) which then can be avoided through assignment, partnership, marketing agreements or merger.

It is a given that new technologies are being invented, patented, and held out for private investment in good and bad times. Securing investment, even in difficult economical times, requires the same four basic ingredients. The technology must have positive laboratory, preclinical, and/or clinical data; all of the required intellectual property must be exclusively licensed to the company charged with developing it; the company must have pertinent management experience; and the technology must hold financial promise that can be described.

The inventor, along with interested institutional administrators, needs to meet with the commercial community and discuss the potential of the invention, the appropriate members of the management team, and how best to produce a believable business plan, as evidenced by the increasing numbers of university technology transfer offices running incubator space. Scientists turn the invention disclosure over to experts within the university in business, marketing, law, and, if lucky, production and manufacturing for a review of its potential. It is hoped that the team can determine which, if not all, of the patent claims are necessary for product development and what additional technologies, if any, are required.

Individuals and organizations that invest in early stage technology are interested in owning a private organization that has appropriate rights to develop technologies. Early stage investors want experienced scientific and business teams, who are capable of turning business plans into revenues. A business plan is the written description of the basic and applied science, manufacture and production, and legal and business development strategies that the management team intends to implement to make revenue from the new technology. Business plans are organized game plans describing current products available to meet the need that the technology is projected to meet, the financial status of that market place, the current state of and plan for the intellectual property necessary to exploit the technology, the plan of development and introduction of the technology into that market place and the timetable for the plan. Regulatory approvals are estimated based upon current FDA-approval charts. Marketing and sales strategies may be necessary or a licensing strategy may suffice (i.e., once Phase I is completed, the company shall license the candidate to appropriate pharmaceutical or large biotechnology organization).

Once a business plan can be outlined, the inventor and institution would be well served to step back and examine existing companies (pharmaceutical or biotechnology) to determine if any contain the required elements. One challenging discussion to have, once a business plan is written and potential profits described, is whether the development will occur more swiftly and surely in an existing organization. It appears to be difficult to let go of potential millions, even when keen observations can be made that the financing may not occur or may not be sufficient to insure real necessary development. Yet, it is hard to dilute ownership of a freshly studied potential product.

The cost of developing a commercial therapy at a pharmaceutical company is estimated to be between $500 and 800 million and the timeline appears to be 14 years. This enormous figure is based on the estimate that 1 in 10 drugs actually pass Phase I's safety and toxicology and make it to the stage of initial human treatments (from the thousands of potential drugs tested). Furthermore, it appears that the majority of these never create revenues sufficient to recoup their development costs. These staggering costs allow the pharmaceuticals to charge significant prices for their patented drugs, to recoup the otherwise lost developmental costs. Biotechnology companies advertise themselves as able to development, market and sell drugs and therapies at a substantially lower cost in a much shorter time.

Venture capital groups and investment funds (either in private groups or as individuals) have been the traditional basis of the funding for new developing technology. Diversified investment by these funds is made on the belief that 1 in 10 of the new organizations will be successful, but the pay off (taking a privately held organization to the public market, called an initial public offering) will be sufficient to offset individual losses. In the money-rich capital markets of the 1980s and 1990s, venture organizations were able to attract significant funds to invest billions in private, high-risk organizations. The promise of any biotechnology organization is that, with sufficient private investment, the technology, drug and/or device will be on the market in <10 years.

During the 1980s and 1990s, money for basic research came from federal government grants, charities, and other large non-profit institutions, whereas start-up and development financing (through Phase I) came from private venture capital groups. Often, upon filing a first IND, a company would turn to public investments through initial public offerings (IPOs) and finance later-stage FDA trials using secondary public offerings of company stock, sold first in the United States and then internationally, with marketing arrangements made with large pharmaceuticals. As late as 2000, venture capital and other funding groups invested $2.5 billion in biotechnology.

However, after the collapse of the Internet capital market and the continuing post–September 11 economic stagnation, venture funds were low on dollars and lower on new investors (in 2002, an estimated $700 million was invested in biotechnology). At least two venture funds announced plans that would give money back to their investors, as suitable investments were unavailable. In the twenty-first century, new organizations may need to find additional nonpublic market financing so that individual and private groups of investors (angels), state and local government economic development authorities, and charities may finance early development (through successful Phase I); venture groups may enter in later stage development (Phase II and beyond), with the public markets monies available after a product is first marketed. With the capital markets closing, more and more money will need to come from early, middle, and late stage partnerships with pharmaceuticals, large biotechs, and foreign organizations. Given that perhaps just over 25% of approved and marketed drugs make sufficient revenue to cover the current cited development costs ($400–700 million per drug), biotechnology companies may need to create revenues by acquiring smaller market drugs.

14.10 The Future of Patents

The current head of the USPTO has suggested sweeping reorganization, the effect of, which is uncertain. The agency's "21st Century Strategic Plan" (www.uspto.gov), which includes immediate and long-term reforms supported by an action plan that is the road map to guide the transformation of the USPTO over the next 5 years into a quality-driven, highly productive, cost-effective organization, supporting an international market-based intellectual property system. The action plan, which incorporates the values of the president's management agenda (www.whitehouse.gov/omb/budget/fy2002/mgmt.pdf), emphasizes excellence in examiner recruiting, hiring, and training; greater use of electronic initiatives and outside resources to process patents and trademarks; and a menu of products and services tailored to meet the needs of customers, at a cost that is $0.5 billion dollars less than under a business-as-usual scenario.

Another continuing issue seems to be whether the patent system should allow protection of subject matter that is fundamental science discovery rather than the invention leading to a commercial product. Plainly said, many have wondered whether patent law should extend prohibitions against patent protection (e.g., like the prohibition of patenting human functions) to include additional basic scientific discoveries, like the structure of DNA as described by Watson and Crick. Arguments in favor of this restriction assume that basic function and process information is best left in the public domain to be built on by ensuing generations of scientists. However, opponents point out that the end result of the next generation's work could receive patent protection and the potential profits arising therefrom. Currently, the compromises of allowing scientists who work in nonprofit institutes access to novel technologies seems sufficient. Moreover, since patent law is statutory, it seems difficult to expect the legislature to draft restrictions that would promote scientific progress without stifling commercial growth. The best solutions have and probably will come from the compromises worked out by the interested parties, the scientists, institutions and companies developing therapies.

References

1. Eisenberg, E. Patenting research tools and the law. Paper given at the National Academy of Sciences, Feb 15–16, 1996.
2. National Research Council. Intellectual property rights and research tools in molecular biology, summary of a workshop. Paper presented at the National Academy of Sciences, Feb 15–16, 1996b.
3. Silverman, J. O. M. Writing an effective invention disclosure. *JOM* **46,** 70–____ (1994).

Index

Molecular Cancer Therapeutics: Strategies for Drug Discovery and Development, edited by George C. Prendergast.
ISBN 0-471-43202-4

B

E

F

I

K

L

N

Q

R

T

U

V